CONTENTS

KU-024-765

PLANT PHYSIOLOGY

The S203 Course Team

Chairman and General Editor

Michael Stewart

Academic Editors

Caroline M. Pond (Book 1)
Norman Cohen (Book 2)
Michael Stewart (Book 3)
Irene Ridge (Book 4)
Brian Goodwin (Book 5)

Authors

Mary Bell (Book 4)
Sarah Bullock (Book 2)
Norman Cohen (Book 2)
Anna Furth (Book 2)
Brian Goodwin (Book 5)
Tim Halliday (Book 3)
Robin Harding (Book 5)
Stephen Hurry (Book 2)
Judith Metcalfe (Book 5)
Pat Murphy (Book 4)
Phil Parker (Book 4)
Brian Pearce (Book 3)
Caroline M. Pond (Book 1)
Irene Ridge (Book 4)
David Robinson (Book 3)
Michael Stewart (Book 3)
Margaret Swithenby (Book 2)
Frederick Toates (Book 3)
Peggy Varley (Book 3)
Colin Walker (Book 2)

Course Managers

Phil Parker
Colin Walker

Editors

Perry Morley
Carol Russell
Dick Sharp
Margaret Swithenby

Design Group

Diane Mole (Designer)
Pam Owen (Graphic Artist)
Ros Porter (Design Group Coordinator)

External Course Assessor

Professor John Currey (University of York)

General Course Consultant

Peggy Varley

Course Secretaries

Valerie Shadbolt
Christine Randall

PLANT PHYSIOLOGY

Edited by Irene Ridge

BIOLOGY: FORM AND FUNCTION

Hodder & Stoughton   The Open University

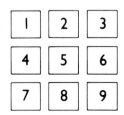

Cover illustrations
1 Dubautia sp., endemic Hawaiian shrub.
2 Ohia (*Metrosideros polymorphia*), yellow-flowered form, from Hawaii.
3 Phlomis italica, from Majorca.
4 Bell heather (*Erica cinerea*) and western gorse (*Ulex gallii*).
5 Hibiscus St Johnii, endemic Hawaiian species.
6 Thrift (*Armeria maritima*) on a sea cliff.
7 Bracket fungus (*Coriolus versicolor*) on a tree stump.
8 Hart's-tongue fern (*Asplenium scolopendrium*).
9 Slime mould.
Back cover: Red campion *(Silene dioica).*

British Library Cataloguing in Publication Data

1. Plants.
I Ridge, Irene II Series
581.1

ISBN 0–340–53186–X

First published 1991.

Designed by the Graphic Design Group of the Open University.

The text forms part of an Open University course. Further information on Open University courses may be obtained from the Admissions Office, The Open University, P.O. Box 48, Walton Hall, Milton Keynes, MK7 6AB.

Typeset by Wearside Tradespools, Fulwell, Sunderland, printed in Great Britain by Thomson Litho Ltd, East Kilbride for the educational division of Hodder and Stoughton Ltd, Mill Road, Dunton Green, Sevenoaks, Kent TN13 2YA, in association with the Open University, Walton Hall, Milton Keynes, MK7 6AB.
1.1

PREFACE

Plant Physiology is the fourth in a series of five volumes that provide a general introduction to biology. It is designed so that it can be read on its own (like any other textbook) or studied as part of *S203, Biology: Form and Function*, a second level course for Open University students. As well as the five books, the course consists of five associated study texts, 30 television programmes, several audiocassettes and a series of home experiments. As is the case with other Open University courses, students of S203 are required both to complete written assignments during the year and to sit an examination at the end of the course.

In this book, each subject is introduced in a way that makes it readily accessible to readers without any specific knowledge of that area. The major learning objectives are listed at the end of each chapter, and there are questions (with answers given at the end of the book) which allow readers to assess how well they have achieved these objectives. Key words are identified in **bold** type both in the text where they are explained and also in the index, for ease of reference. A 'further reading' list is included for those who wish to pursue certain topics beyond the limits of this book.

INTRODUCTION

Plants, the kingdom Plantae, are one of the cornerstones of life on Earth for the simple reason that they can feed themselves: they are autotrophs. Virtually all the major food chains have, as their ultimate source of energy, living or dead plants. In this sense, animal life now depends on plants. Because of their autotrophy and universal position at the base of food chains, plants show less diversity of form and function than do animals. The basic requirements for autotrophic life impose certain constraints on plant lifestyle which apply equally to unicellular algae and giant trees. So we need first to consider what these requirements and constraints are.

BASIC REQUIREMENTS OF PLANTS

There are three of these: light, supplies of essential elements for building tissues, and water. Unlike the majority of animals, plants do not always require a constant, external supply of oxygen for respiration because, in the light, they generate oxygen from water (Chapter 2). However, at certain times or for certain organs, external oxygen is a fourth requirement for most plants.

Light

Plants are *photo-autotrophs*, i.e. they capture light energy from the sun and convert it to chemical energy in the process of photosynthesis. This is why light is a basic requirement of plants but not of animals, most eubacteria or fungi, all of which are heterotrophs, using complex organic molecules ('food') as a source of energy. There are, as usual in biology, certain exceptions to this rule. For example certain angiosperms such as dodders (*Cuscuta* spp) and broomrapes (*Orobanche* spp) are totally parasitic on other plants, having lost the capacity to photosynthesize and having become secondarily hetero-trophic. Similarly there are species of 'green animals', including some corals and giant clams, which use colonies of algae within their tissues as built-in food production systems. These are rare exceptions, however, and the need for light constrains both the habitat and form of plants. They cannot live in permanently dark places, such as the ocean depths or under the soil, and for multicellular plants the amount of green, photosynthetic tissue must be sufficient to support the non-green tissues, such as roots or stems.

There are autotrophs, including photo-autotrophs, which are not plants, however.

◇ From general knowledge or from reading Book 1* of this series, recall the nature of these autotrophs.

*Caroline M. Pond (ed.) (1991) *Diversity of Organisms*, Hodder & Stoughton Ltd in association with the Open University (S203 *Biology: Form and Function*, Book 1).

◆They are all prokaryotes (whereas plants are eukaryotes). Within the archaebacteria there are chemosynthetic bacteria (chemo-autotrophs), which obtain energy through the oxidation of simple inorganic compounds. Within the eubacteria there are chemosynthetic bacteria, photosynthetic bacteria (photo-autotrophs) and also cyanobacteria (blue–greens or blue–green algae), which are all photo-autotrophs.

Cyanobacteria trap light energy by the same process as plants, splitting water molecules and releasing oxygen as they do. In fact they rival plants as the dominant autotrophs in some aquatic habitats: in nutrient-rich lakes, huge populations (blooms) of blue–greens build up in late summer, often releasing toxic compounds that are a hazard to animals. And in the open oceans, minute cyanobacteria (classed as pico- or nanoplankton) may trap as much light as the algae in the phytoplankton. Photosynthetic and chemosynthetic bacteria do not rival plants in this quantitative way because they are even more restricted in where they can live. Photosynthetic bacteria (which are discussed briefly in Chapter 2) require an oxygen-free environment and do not release oxygen during photosynthesis, whilst chemosynthetic bacteria are restricted not by light but by the availability of suitable oxidizable substrates. Photosynthesis is described in detail in Chapter 2.

Essential elements

All the elements which animals need to build new tissues are supplied in their food and, furthermore, most elements are already assembled into basic building blocks — carbon in sugars for polysaccharide synthesis and nitrogen in amino acids for protein synthesis, for example. Plants, on the other hand, obtain all their essential elements as ions or from simple, inorganic molecules in air or water and must then synthesize the basic building blocks for themselves. Carbon (and oxygen) is obtained as CO_2 from the air or, for aquatic plants, dissolved in water; CO_2 is then fixed into sugars during photosynthesis. All the other elements, however, are absorbed only from solution in water, which means that some part of all plants must be in contact with water. Although this poses no problems for algae and other aquatic plants, it places severe constraints on the form and lifestyle of terrestrial plants because the nutrient solution is underground in the soil. This is why the majority of land plants are divided into two functional systems — absorptive roots below ground and shoots which intercept light and absorb CO_2 above ground. Chapter 1 discusses in more detail the structural and functional implications of this split and Chapter 3 describes ion uptake processes, how ions move from roots to shoots and the strategies used by plants in nutrient-poor conditions.

Water

Water is a basic requirement for all life and, because of their two-phase existence on land — roots below and shoots above ground — this has marked effects on the form and function of terrestrial plants. Shoots are essentially waterproof, otherwise they would dry out, but the root surface is mostly water permeable and roots absorb water. However, the waterproof and gas-tight armour of shoots cannot be complete because it must allow entry of CO_2 and there are special pores which open and permit this. The snag is that water vapour inevitably moves *out* through the open pores and, particularly in dry conditions, land plants have to juggle constantly, balancing the conflicting demands for CO_2 uptake (pores open) and water conservation (pores closed).

There are many ingenious adaptations of structure and physiology, especially in desert plants, that act to minimize water loss when pores are open; these are described in Chapters 1 and 5.

From this description it should be clear that the need to obtain basic requirements constrains plants most strongly in terrestrial habitats. This is where the more familiar plants grow — trees, grasses and other flowering plants — and this book is concerned mainly with the physiology of such land plants.

ORGANIZATION OF THE BOOK

In Chapter 1 the basic structure of land plants is described, an essential background to any discussion of how plants function and grow. Because we concentrate on land plants, which are much less diverse than animals and share a common body plan, this can be achieved in one short chapter. The interesting question is how plants with a common body plan and sharing just a few basic cell types can grow in such a wide range of habitats, from arctic tundra to tropical forests.

Chapters 2 to 5 are concerned with how plants obtain essential requirements and function in a variety of habitats. Chapter 2 is about photosynthesis: the trapping of light, its conversion to chemical energy and the use of this to fix CO_2 (carbon fixation). The important biochemical steps are similar in all plants but the chapter describes variations of the basic processes that allow plants to fix carbon in the dim light of a forest floor or the intense sunlight of the tropics and in conditions where CO_2 supply is very low.

Chapter 3 is about mineral nutrition: how plants obtain, metabolize and transport to shoots the nutrient ions taken up from the soil. The transport theme, an important element for the integrated functioning of roots and shoots, is continued in Chapter 4. This deals with transport in phloem, a specialized tissue which carries mainly sugars and some amino acids from sites where they are synthesized (usually leaves) to sites where they are used or stored (e.g. roots or fruits).

Water is the theme of Chapter 5: how plants absorb it from soil, how and why it moves between living cells, and how its loss from shoots is controlled. The chapter also considers the knotty problem of how water from the roots moves to the tops of tall trees. Since newly formed plant cells grow to many times their original size by taking up water, there is a link to Chapter 6, which is concerned with plant growth and, in particular, with the control of cell expansion. The size and shape of a plant are determined firstly by the amount and type of cell division, which is localized to regions such as the shoot and root tips; and secondly by the direction and extent of cell expansion. Cell division and expansion are influenced both by external factors, such as light, water and nutrient supplies, and by internal factors, notably the substances known as plant growth regulators. What these regulators do and how some of them influence cell expansion is a major theme of Chapter 6.

Finally, Chapter 7 looks at the control systems which operate at different stages of the plant life cycle, from birth to death. Among the questions considered are what determines when a seed germinates, when flowering occurs and when whole plants die, leaves fall off or fruits ripen. Most of these questions can be answered only partially; some, for example the nature of the signal that induces flower formation, still have no answer at all. But it is of more than academic interest to seek answers to questions like these and to many more about how plants function and grow. Plants feed, directly or indirectly, most of the organisms on this planet, including human beings, as well as replenishing the oxygen we breathe and dominating the natural landscape. In order to produce enough food from crops and at the same time conserve the richness of plant life and an environment fit to live in, it is necessary to know how plants work and interact with other organisms. This book introduces the subject and we hope that it helps you to begin to understand plants.

◆ CHAPTER 1 ◆ PLANT STRUCTURE AND FUNCTION

1.1 INTRODUCTION

The six chapters which follow this one explore how plant physiological systems work and the way in which growth is initiated and coordinated. In order to follow these properly it is important to understand the anatomical structure of plants and to relate structural differences between plants to their needs in the different environments in which they are found. This is the subject of the present chapter.

This chapter concentrates on land plants. But first we need to look at the main differences between terrestrial and aquatic environments in order to identify the nutritional and structural problems associated with the land habitat. Water is very much denser than air and therefore provides greater support for living organisms. It is true that land organisms can rest on solid ground but raising even small bodies away from the earth requires some kind of supporting skeleton. As you will see, many structural features of land plants are related to this support function.

Another basic difference between land and aquatic environments is that in water, mineral salts and gases diffuse directly from water into plants so absorption usually occurs over the whole plant surface. Things are very different on land because it is a 'two-phase' environment with the nutritional requirements for plants unevenly distributed between the phases, namely air and soil. So, on land, some of the basic requirements for plant growth and survival come from the soil while others come from the air. Air contains oxygen and carbon dioxide (20% and 0.04% by volume respectively) and variable amounts of water vapour. Most mineral nutrients are taken up from the soil, though some such as sulphur (as sulphur dioxide gas) may be available above ground dissolved in rain water. Soil is a semi-solid medium, consisting of particles such as sand, clay and organic debris with air spaces in between. The particles are usually coated with a film of water, and mineral salts may be dissolved in this water film or adsorbed onto the surface of the particles. In contrast to plants, most animals are able to live in a *single* part of the two-phase land environment: either underground, obtaining oxygen from moist air in the soil, or on the surface, taking in their mineral nutrients with their food and water. Plants cannot do this because they need light (for photosynthesis), and light does not penetrate into soil, so *all* land plants must have aerial parts as well as underground parts. (Aquatic plants too are limited to the surface layers of water where light can penetrate.) Furthermore, because water is often patchily distributed in the soil, and concentrations of minerals are usually low, the underground part of a plant (typically a root system) may need to be very extensive if it is to absorb sufficient amounts of water and minerals salts. For plants with large aerial parts (the shoot system) an extensive root system also serves as an anchor. The structure and function of roots are described in Sections 1.6–1.9.

The separation of function into aerial photosynthetic and underground absorptive parts poses a problem for land plants, i.e. transport. There must be some system for transporting water and mineral nutrients from underground to aerial organs and the organic products of photosynthesis in the opposite direction, because underground organs cannot feed themselves. Specialized cells and tissues carry out these transport functions in all vascular plants and these are described chiefly in Section 1.5.

The disposition of resources and the nature of land environments pose yet another problem, which is possibly the most difficult of all to solve — *desiccation*. Living organisms consist typically of about 90% water, and any wet object in the air loses water by evaporation, desiccates and eventually dries out. The rate of drying depends on a number of factors — temperature, wind and atmospheric humidity — but even in cold still air at 99% water saturation (relative humidity) some drying occurs, albeit slowly. The obvious answer is to waterproof the whole surface, or at least any part that is exposed to the air, but this strategy is incompatible with other needs of plants, namely to take up carbon dioxide and release oxygen (in photosynthesis). Compromise is clearly essential for keeping water loss to a minimum while allowing an adequate rate of photosynthesis. The adaptations necessary to maintain this compromise vary according to the environmental conditions in the habitat in which individual plant species are found. Some of these adaptations are described in Sections 1.4 and 1.8.3.

The final difference between land and aquatic environments concerns temperature. Air warms up and cools down much more rapidly than water because of its much smaller specific heat capacity. There are, therefore, greater extremes of heat and cold, both over the year (seasonal) and over the (24-hour) day (diurnal) in terrestrial than in aquatic environments. A diurnal range greater than 20 °C is not unusual in the middle of large continental land masses, but even the surface waters of oceans seldom change by more than 2 °C over this period. Water a few metres below the surface responds even more slowly to diurnal and seasonal changes. Soil resembles the sea with respect to temperature changes because, like water, it is a poor conductor of heat — only surface soil layers show appreciable fluctuations of temperature and the deeper down one goes, the more constant is the temperature. Clearly then temperature characteristics of land environments can pose problems for all organisms but these are more serious for land plants because plants cannot move away to a cooler or warmer place as necessary. Some plant structural features related to temperature are described in Section 1.4.2 and seasonal characteristics are described in 1.4.3 and 1.8.2.

Summary of Section 1.1

1 The main differences between terrestrial and marine environments for plants are:

    (a) the greater density of water compared with air leading to problems of support on land;

    (b) the uneven distribution of essential gases, water, mineral salts and light compared with the sea, leading to (i) separation of function, with well developed transport systems and (ii) problems of desiccation above ground;

    (c) the greater variation in temperature on land compared with the water.

2 Land plants have evolved separate root and shoot systems which have specialist functions in the air and soil phases of the terrestrial environment.

1.2 EXTERNAL FEATURES OF PLANTS

Plates 1a–d show four very different plants: 1a is an alga which grows in thin sheets on the bark of trees; 1b–1d are found in damp shady ditches, in lawns and on bare ground, and in deserts, respectively. Note their relative sizes (each photograph has a different magnification) and the size and extent of the above-ground parts.

◇ Suggest why the colonies of cells in Plate 1a are limited in their distribution to the shady sides of trees which are protected from prevailing winds.

◆ They are in thin layers so they will desiccate quickly in bright sunlight or drying winds.

◇ Describe the above-ground organs of each of the plants shown in Plates 1b–d.

◆ The liverwort, *Pellia*, has only a thin flat green plate of tissue above ground, the dandelion has a rosette of leaves and the cactus is swollen, bulbous, grey-green and spiny.

Below ground the liverwort is anchored by thin rhizoids (root-like structures), the dandelion has a swollen tap root and the cactus a ramifying network of thin, branched roots.

◇ Describe how each of the plants shown in Plates 1b–d is adapted to its particular environment.

◆ *Pellia* is found in damp shady ditches — the thin green sheet of above-ground tissue is relatively unprotected from drying out. Dandelions occur in grassland and on lawns — the flat rosette of leaves with a very short stem helps it to survive mowing. The cactus grows in deserts — it has a small surface area to volume ratio which helps to reduce water loss in the heat.

Answering these simple questions on the above-ground structures shown in Plates 1b–d should have led you to realize that you can learn something about how plants fit their environment, or are limited in their habitats, simply from their external appearance (morphology). Note the thickness of above-ground parts of the cactus (Plate 1d) and their greyish-green colour. These characteristics are external manifestation of anatomical features:

1 The plant is swollen because it contains many cells with large central vacuoles in which water can be stored, hence its bulk.

2 Its surface layer is heavily impregnated with waxes which give it the grey colour. This is a feature which not only improves waterproofing of the surface but also results in light being reflected from the surface, which in turn prevents heating and helps to limit further water loss (you will learn more about this in Section 1.4).

It seems then that morphological features and anatomical characteristics should be studied in parallel. However before considering any details of plant structure the common characteristics of plant cells need to be reviewed.

1.3 PLANT CELL STRUCTURE

Recall the structure and fine structure of plant and animal cells.*

◇ List the structures that are common to cells of both plants and animals and those which are found only in plants.

◆ Plant and animal cells both contain a nucleus, mitochondria, lysosomes, Golgi apparatus, ribosomes and endoplasmic reticulum enclosed within a cell membrane. However, unlike animal cells, plant cells also contain plastids, particularly chloroplasts, and the cell is bounded by a rigid cell wall outside the cell membrane.

Note that in plants the cell membrane is often called the **plasmalemma** and a plant cell minus its cell wall is referred to as a **protoplast**.

Plants, unlike animals, are mainly autotrophs; that is, they trap light energy and synthesize organic material from carbon dioxide and water. In all plants these activities are concentrated in chloroplasts, which are most abundant in leaves. Some of the products of photosynthesis are used in the synthesis of the polysaccharides which make up plant cell walls (e.g. cellulose). To some extent all plant cells, unlike animal cells, have a skeletal function, and it is the cell wall which fulfils part of this need. *Turgor* of cells, that is the pressure the membrane-bound protoplast exerts on the wall, also plays a part. A typical living plant cell has a relatively thin wall. Its skeletal role is consequently small but nevertheless it contributes to the overall shape of the tissue of the plant from which it is derived. Plant cell walls also define the shape of the cells themselves, though where they are thin they can be distorted by the pressure put upon them by surrounding cells.

When plant cells are first formed they are usually cuboid (often referred to as *isodiametric*) with walls about 5–10 μm long. If newly formed cells did not increase in size there would be no tissue growth just an increasing number of smaller and smaller new cells occupying the same volume (Figure 1.1). When plant cells have reached maturity they may be up to 200 times as long as they were when first formed. This is a function of the way cell walls stretch and new cell wall material is laid down.

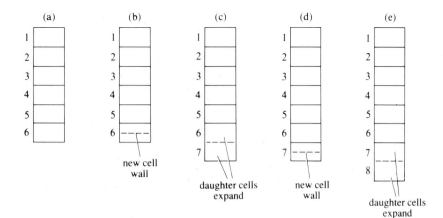

Figure 1.1 How cell division results in tissue growth.

*Eukaryotic cell structure is described in Chapter 2 of Norman Cohen (ed.) (1991) *Cell Structure, Function and Metabolism*, Hodder & Stoughton Ltd in association with The Open University (S203, *Biology: Form and Function*, Book 2).

This basic information about cells, the anatomical structure of tissues and many aspects of the ways in which plants grow and expand have been known for a century or more, but in the last 30 years the study of plant anatomy has been enormously enhanced by the advent of the electron microscope and the development of more sophisticated methods of chemical analysis.

1.3.1 Wall structure in vascular plants

When new cells are first formed as a result of cell division a layer called the **phragmoplast** or *cell plate* is formed between them. Labelling studies with radioactive isotopes of carbon and hydrogen have shown that Golgi bodies move through the cells and bud off vesicles at the new junction, a process called *reverse pinocytosis*. The first materials deposited on the phragmoplast are polysaccharides of various sugars (arabinose, raffinose and xylose) with methyl esters of galacturonic acid; these are referred to generically as **pectins**. Pectins form the **middle lamella** (plural lamellae) which is the first layer of the wall in higher plant cells. Later, long molecules of the glucose polymer *cellulose*, are laid down in bundles called *microfibrils* within a matrix containing more pectins, non-cellulose polysaccharides called *hemicelluloses*, and proteins. These together form the **primary wall**.

Cellulose is the main component of the primary wall. It is a polymer of glucose units linked by β-1,4 glycosidic bonds (Figure 1.2a). It provides strength to the cell wall, due both to the physical properties of the individual cellulose molecules and also the way these are arranged. These long polymer molecules are laid down in groups, head to tail, forming **microfibrils** about 10 nm long Figure 1.2b shows randomly orientated microfibrils and Figure 1.2c shows two separate layers of microfibrils which are clearly at right angles to each other. The tensile strength of cellulose microfibrils is very similar to that of steel. In the longitudinal walls of most young cells the inner microfibrils are arranged transversely around the cell acting like restraining steel hoops and this arrangement explains why stem and root cells expand longitudinally — the inextensible fibrils cannot stretch to allow sideways expansion but can slip past each other and re-orientate to allow longitudinal growth. The essential point to grasp here is that cells with primary walls only are inevitably young *growing* cells and it is the properties of the primary wall that allow cell expansion. There are only two known instances where higher plant cells do not grow evenly over the whole surface; these are the formation of root hairs, which are lateral projections of a single cell, and the germination of pollen grains to form pollen tubes. Both these structures elongate by growing only at their tips. (Cell growth is dealt with in Chapter 6.) Note that cellulose cell walls are by no means impermeable as water can move between the microfibrils. At any time the water content of cell walls is about 80% of the total wall weight.

Both primary cell walls (and the thicker, non-growing *secondary walls* — see below) are not fully continuous and they may also be unevenly thickened. They may have a number of **pits** (thinner areas of wall) which are pierced by plasmodesmata and these pits may be in groups referred to as pit fields. **Plasmodesmata** are membranous strands which maintain continuity between the cytoplasm of adjacent cells and allow large molecules (and possibly viruses) to pass between them without crossing membranes.

The primary wall also provides a mechanical restraint against the hydrostatic pressure exerted by cytoplasm. This in turn creates stiffness in tissues through turgor because of the pressure of the membrane-bound protoplast on the wall. Turgor is therefore one way in which plants obtain structural support

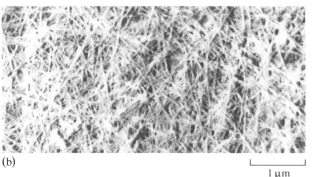

(a)

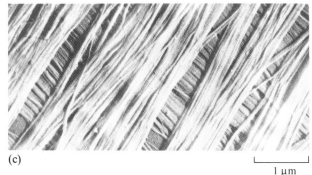

(b)

1 μm

(c)

1 μm

Figure 1.2 (a) Part of a cellulose molecule. (b) Randomly orientated cellulose microfibrils in an algal cell wall (*Valonia* sp.). (c) Cellulose microfibrils at right angles to each other in the cell wall of *Valonia* sp.

and it is especially important in young organs. You can judge its importance by observing the effects of water loss in a young plant: wilting occurs, i.e. the plant collapses.

However, some plant tissues do have a specialized skeletal function (just as in animals). In plants there are specialized supportive cells which have, in addition to and inside the primary wall, a **secondary wall**. The cellulose microfibrils in secondary walls are longer and stronger than in primary walls, containing as many as 14 000 glucose units in each cellulose molecule. When primary cell walls stop growing a secondary cellulose wall is usually laid down and in some cells this may become impregnated with **lignin** which provides additional strength. Lignin is not a simple substance but a complex macromolecule made up of several hundred subunits which are arranged to form a three-dimensional cage-like structure laid down on and between cellulose microfibrils. An important point to remember is that lignified cell walls are waterproof when mature, and cells which are waterproof soon die. More information on lignified cells is given in Section 1.5.

Also laid down in some secondary cell walls are the polymers **suberin** and **cutin**. These are both mixtures of polyesters of hydroxy fatty acids and are much less widespread than lignin. Their function is also waterproofing. Cutin is confined to the outer wall of living cells at the surface of leaves and young stems; whereas suberin is located on certain cells which are described later, but also very importantly on the outermost cell layers of old roots and in the corky layer below the bark of trees (Section 1.8.3). Cells that are fully suberized are dead.

Early plant anatomists used stains to make it easier to identify specific types of cell wall material and hence major cell types and tissues. These are still used in laboratories for diagnostic purposes and some of the more commonly employed stains include the temporary stain *phloroglucinol* (in hydrochloric acid) and the permanent stain *safranin*, both of which colour lignin red (safranin also stains suberin and cutin and nuclei). *Light green* and *haematoxylin* are also permanent stains and they colour cellulose in walls green and purple respectively. As you read later sections of this chapter, you will have

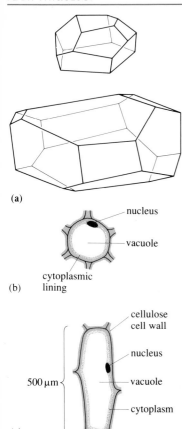

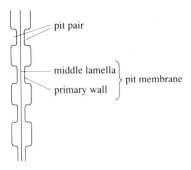

Figure 1.3 Parenchyma cells. (a) Three-dimensional drawings of two different parenchyma cells. (b) Transverse section of a young cell. (c) Longitudinal section of a young cell.

Figure 1.4 Structure of a pit pair: the pit membrane is crossed by plasmodesmata (not shown) linking the two adjoining cells.

the opportunity to learn to recognize the more important cell types which have different staining characteristics visible in both longitudinal and transverse sections of tissues. To do so it is important to remember the diagnostic value of these stains and hence the colours you should look for on the colour plates of plant tissue sections included in this book.

1.3.2 Parenchyma: a relatively unspecialized cell type

It is helpful to look at a relatively simple plant cell in isolation before looking at the different kinds of cells found in tissues. Figure 1.3 shows three-dimensional drawings (a) and transverse and longitudinal sections (b and c, respectively), of an unspecialized type of cell called **parenchyma** (pronounced paren-*ky*-ma or alternatively par*en*-ki-ma). These kinds of cells are often referred to as *packing cells* because they fill space anywhere within a plant and do not to appear to have any specialized features related to function apart from this. Tissue composed of parenchyma cells is often just referred to as parenchyma. You will note that the cells in Figure 1.3a are very roughly cuboid and thin-walled. Because they are thin-walled they can become deformed as a result of pressure from adjoining cells and so sometimes become quite irregular in outline. If stained with a mixture of safranin and light green the walls are light-green coloured, indicating that they do not contain lignin or other waterproofing material. This also tells us that they are living cells. In young parenchyma it is sometimes possible to see the protoplast stained light green and consequently to detect the outline of the cell vacuole towards the centre of the cell.

◇ Look at the transverse and longitudinal sections of a parenchyma cell (Figures 1.3b and c) and describe what you see.

◈ In transverse section the cell is slightly rounded in outline. The wall is very thin and somewhat irregular. In longitudinal section the cell appears considerably longer than it is broad and again it is notable for the thinness of the wall.

Not only is the cell wall irregular in outline but the wall itself is thinner in places (Figure 1.4). These depressions are the pits mentioned in Section 1.3.1. In these regions no thickening is laid down over the primary wall and at this point the wall consists of the middle lamella and the thin layer of primary wall which together form the *pit membrane*. It is quite usual for pits in adjoining cells to coincide forming a *pit pair* when the pit membrane will consist of the middle lamella and some primary wall on each side with plasmodesmata linking the two.

Later you will see parenchyma cells which are much less regular than the examples described above, but the important features to remember are that they always have thin cellulose walls looking *very approximately* square to round in transverse section and that the cells are living. They can vary considerably in size — ranging from 50 to 800 μm in diameter and they can be up to 1 000 μm in length.

Summary of Sections 1.2 and 1.3

1 External characteristics of plant organs can be an indication of the habitat in which the organism is normally found.

2 The external appearance of a plant can often be related to its anatomical features.

3 Plant cells differ from animal cells in containing chloroplasts, the organelles of photosynthesis, and in being bounded by a rigid cell wall, the latter having a skeletal function.

4 Mature plant cells have a three-part wall comprising middle lamella, primary wall and secondary wall. Only the middle lamella and primary wall are present in young growing cells.

5 The major polysaccharide of the primary wall is cellulose, present as short molecules grouped into microfibrils. These are embedded in a matrix of pectins, proteins and non-cellulose polymers.

6 The major components in the secondary wall are longer fibrils of cellulose sometimes impregnated with lignin as a strengthening material.

7 The secondary wall is laid down after cell growth is complete.

8 Adjacent plant cells may be linked by plasmodesmata which cross pit membranes (middle lamella plus primary wall).

9 Stains specific for particular cell wall components are used to identify different plant cell types.

10 Parenchyma is a relatively unspecialized living cell type, thin-walled and without waterproofing. These cells are roughly cuboid when young but they expand and become different shapes and sizes as they mature.

Question 1 (*Objectives 1.1, 1.2, 1.3 and 1.4*) Which of the following statements are true and which are false?

(a) The primary cell wall is deposited on the inside of a middle lamella composed of pectins.

(b) In rapidly elongating cells, cellulose microfibrils are initially laid down parallel to the direction of cell expansion.

(c) Lignin is laid down in secondary cell walls but not in primary walls.

(d) Cells with walls thickened with suberin, cutin and lignin are dead.

(e) Cutin and suberin prevent water entering cells.

(f) Parenchyma cells are found throughout the plant: they are always thin-walled and often rounded in cross-section, and pick up stains which attach to the cellulose in the walls.

I.4 STRUCTURAL ADAPTATIONS FOR LIGHT GATHERING AND WATER CONSERVATION

The aerial shoots of land plants have one main function (in addition to bearing reproductive organs).

◇ Recall what this function is.

◈ They carry out photosynthesis (i.e. they make 'food'); this involves the interception of light and the absorption of carbon dioxide.

A primitive land plant such as *Rhynia* (Figure 1.5), which is an extinct and early relative of modern ferns, lacked the flattened leafy outgrowths that are present in most modern plants; it consisted of an underground stem (called a **rhizome**) bearing simple branched shoots. This shoot structure was probably not as efficient as that of modern angiosperms.

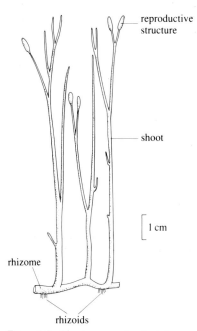

Figure 1.5 Rhynia, an early plant drawn from fossil material from Ordovician rocks.

One measure of plant efficiency is the amount of light reaching the surface of the leaf. For this to be maximized, there should be minimum overlapping and shading of leaves down the stem. Look at Figures 1.6a–c, which show the different patterns in which leaves are commonly arranged. Note that in Figure 1.6a they are arranged in alternate opposite pairs and in Figure 1.6b in a spiral up the stem. Figure 1.6c shows the common pattern amongst grasses and other monocotyledonous plants (see below) where the leaves are often long and relatively thin and are held near vertically. This strategy is particularly effective when plants are low growing and form clumps often grouped together in densely packed swards; less shadow is created. Equally, tiers or spirally arranged expanded flat leaves (Figures 1.6a and b) have a greater absorptive area when the main plant stem continues to grow upward, so carrying the leaves above competing vegetation.

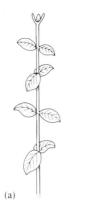  

(a) (b) (c)

Figure 1.6 Arrangement of leaves around a stem axis. (a) Alternate opposite pairs. (b) In a spiral up the stem axis. (c) Linear upright arrangement.

Most leaves are efficient interceptors of light because of their flattened shape and large surface area; but in dry conditions it is quite common to find plants with very small leaves that look rather like those of *Rhynia*; the common broom (*Sarothamnus scoparius*) (Figure 1.7) is an example. Here stems are green (and so photosynthetic) as in cacti and many other *xerophytes* (from the Greek *xeros*, dry and *phyton*, a plant, Section 1.4.2).

◇What is the advantage of small leaf size?

◆Water evaporates from aerial organs so the greater their surface area, the greater the loss of water. A reduction in leaf size thus helps to reduce water loss.

Chloroplasts, the organelles essential for photosynthesis, may occur in most cells near the surface of a shoot, i.e. in stems as well as leaves. However, in large plants such as trees, leaves are the only structures which contain chloroplasts; the stem (or trunk) and branches serve only as a leaf support system (although young tree stems may be green). Leaves are therefore, in general, the chief photosynthetic organs.

1.4.1 Basic leaf structure

Figure 1.7 Flowering shoot of a xerophyte, *Sarothamnus scoparius* (broom).

Though it might seem more logical to start studying leaves by looking at the variety in habit and external morphology, it is easier to understand some of these after examining leaf structure and anatomical variations related to light absorption and water conservation. Leaves are good organs on which to begin anatomical studies as it is easy to separate their internal tissues into the three major layers:

1 **epidermis** (pronounced e-pi-*der*-mis) the outer cell layer.

2 **mesophyll** (pronounced *me*-so-fil, meaning 'middle leaf'), the packing and storage tissues consisting of two kinds of modified parenchyma called *palisade cells* and *spongy mesophyll*.

3 **vascular bundles** or **veins**, which provide the structural skeleton and conducting tissue.

Figures 1.8a and c show the external features of the two major types of angiosperm leaf namely that of **monocotyledons** (or **monocots**) and **dicotyledons** (**dicots**) — so-called because of the presence of one (*mono-*) or two (*di-*) food-storing cotyledons in the seed. Transverse sections of these leaves are shown in Figures 1.8b and d. The monocot leaf (Figure 1.8a) has veins

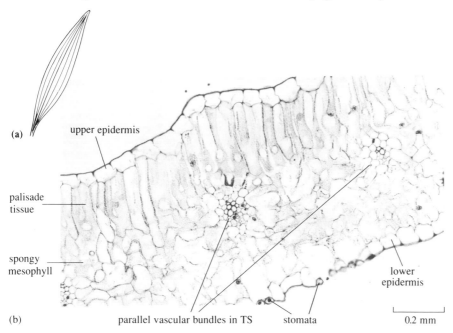

(a)

upper epidermis

palisade
tissue

spongy
mesophyll

lower
epidermis

(b) parallel vascular bundles in TS   stomata   0.2 mm

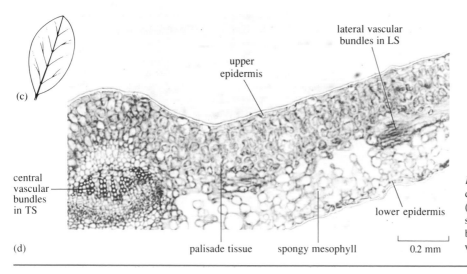

lateral vascular
bundles in LS

upper
epidermis

(c)

central
vascular
bundles
in TS

lower epidermis

(d)   palisade tissue   spongy mesophyll   0.2 mm

Figure 1.8 Leaves of monocots and dicots. (a) Surface view of a lily leaf (*Lilium* sp.), a monocot. (b) Transverse section of a lily leaf. (c) Surface view of a box leaf (*Buxus* sp.), a dicot. (d) Transverse section of a box leaf.

21

(vascular bundles) arranged parallel to each other apart from a few small cross-connections between them. Consequently in Figure 1.8b, the transverse section through this leaf, the vascular bundles are more or less the same size and lie in a row across the middle of the section (for the moment do not worry about the details of cells). In the dicot leaf there is a central main bundle from which arise branches and sub-branches forming a net-like pattern (reticulation or reticulate venation) (Figure 1.8c), so the section shows a large central vein and a number of smaller veins (offten referred to as vascular bundles) of different sizes (Figure 1.8d). In transverse sections of dicot leaves, many of the smaller veins will be cut more or less obliquely, so appearing blurred, while the main vein will be cut transversely and so relatively easy to interpret (as shown in Figure 1.8d); this is a consequence of the net-like arrangement of veins. Some cells within the vascular bundles appear to have thickened walls, indicating that they are lignified. In some dicot species, there may be a ring of unspecialized parenchyma around each vascular bundle (but not so in this case).

Parenchymatous mesophyll tissue fills almost all the remaining space within the leaf. Figure 1.9 shows the mesophyll of a dicot leaf in three dimensions. The cells are, as you would expect, thin-walled and some are unevenly shaped. The **spongy mesophyll** is a loosely packed tissue, with air spaces between cells. Plastids containing chlorophyll (chloroplasts) and starch grains (amyloplasts) are usually found in spongy mesophyll cells where they look like rounded discs. One layer of the mesophyll looks rather different; these uppermost **palisade cells** (called palisade because of their elongated shape) are more regularly and tightly arranged with their long axes at right angles to the leaf surface. Light falls directly on these cells and as they are rich in chloroplasts, they are the main site of photosynthesis.

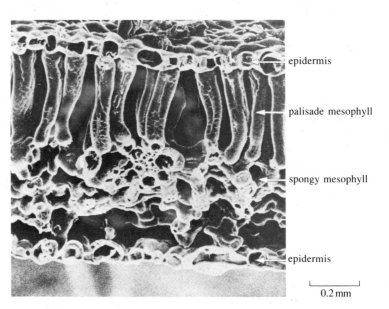

epidermis

palisade mesophyll

spongy mesophyll

epidermis

0.2 mm

Figure 1.9 Scanning electron micrograph of a transverse section of a broad bean (*Vicia faba*) leaf.

The epidermis on both leaf surfaces is made up of small cells that are cuboid in vertical section though they may be rectangular or irregular in surface view. They lack chloroplasts, and form an almost continuous 'skin' around the entire leaf. A layer of **cuticle** made up of cutin (Section 1.3.1) and other material is usually found on the external face of the epidermis. Cuticle keeps the epidermis relatively impermeable to gases and also prevents water loss. Gas exchange between the leaf tissues and the atmosphere is possible because of pores in the surface, called **stomata** (pronounced stow-ma-ta, singular, stoma. They are surrounded by a pair of **guard cells**, which can change shape in such a way that the stomata can open and close. When closed they limit loss of water vapour but also prevent the entry of carbon dioxide (required for photosynthesis) into the leaf. They are more abundant in the lower than the upper epidermis (see Figure 1.8b). (The functions of stomata are described in detail in Chapter 5.)

◇ Can you think why this arrangement of stomata is logical for a typical dicot leaf?

◆ The lower epidermis is relatively protected and so less likely than the upper epidermis to lose water by evaporation (transpiration) from the surface in hot weather or high winds.

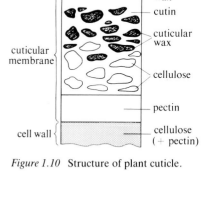

Figure 1.10 Structure of plant cuticle.

The capacity to synthesize a waterproof layer of cuticle must have been a key step in the evolution of land plants, and it is worth digressing briefly to examine this remarkable structure in more detail. Figure 1.10 is a diagram showing the structure of plant cuticle; you can see that it is integrated quite closely with the cell wall. Although the chief waterproofing substance is cutin (a fatty polymer like varnish) there are also variable amounts of **waxes**. When present these are secreted on the outer surface where they form intricate patterns which can help to identify leaf residues to species (Figure 1.11). The deposition of cuticle and wax is often less pronounced in young plants and this partly explains why seedlings wilt more easily than older plants.

0.5 μm

Figure 1.11 Scanning electron micrograph of a leaf surface showing wax deposits.

1.4.2 Modifications of leaf anatomy

Look at Plates 2a and b. They are photographs of sections through a privet leaf that have been stained with safranin and light green. When you think of privet you are probably thinking of the hedging plant (*Ligustrum ovalifolium*) of gardens. However there is a related species, a British native shrub (*Ligustrum vulgare*) which is found in scrubby woodland, especially on chalk and limestone soils which can be fairly dry in summer.

◇ What colours would you expect to see in cells stained with safranin and light green with (a) cellulose in their walls but no lignin and (b) lignified cell walls? (Refer back to Section 1.3.1 if necessary.)

◆Light green will colour cells with cellulose walls green and safranin will stain lignified cells red. (In Plates 2a and b the red colour is rather brownish). Note however that safranin has also stained the nuclei red in Plate 2b.

Now answer the following questions with reference to Plates 2a and b.

◇Describe the shape of the upper and lower epidermal cells. Can you see cuticle in either case?

◆In this transverse section the upper epidermal cells are about twice as wide as they are deep. The cells of the lower epidermis are smaller. The cuticular layer on both the upper and lower epidermis is not well developed; however, because of the thickness of the section, in Plate 2b some of the epidermal cells appear to be standing out from the surface, which makes interpretation difficult.

◇Describe the different mesophyll cells you can see.

◆Both types of mesophyll cell are clearly identifiable: palisade cells make up two layers beneath the upper epidermis while the spongy mesophyll with surrounding air spaces occupies the rest of the space down to the lower epidermis. Note that around the midrib the mesophyll cells are neither palisade-like nor of the spongy mesophyll type.

◇How many vascular bundles can you see in Plate 2a and what do they tell you about the group of plants to which privet belongs?

◆There is one large central bundle within the leaf midrib and about twelve small bundles in the thinner tissue either side of the midrib (the so-called leaf *lamina*). The details of these small bundles are difficult to see, indicating that they are cut obliquely; this suggests that this leaf has reticulate venation and so is from a dicot plant.

The cells immediately above the main bundle do not appear to have thickened walls. However thickened cells are often found in this location and where they are coloured green with the light green stain (i.e. they do not contain lignin) this cell type is **collenchyma** (pronounced collen-*ky*-ma or coll*en*-ki-ma). Figure 1.12a shows collenchyma cells in a ridge of a transverse section of a stem of red dead nettle (*Lamium purpureum*). When grouped together these cells can form an effective small-scale strengthening system, particularly in young growing tissues or in plant organs with limited life expectancy which do not develop other strengthening tissue.

◇Compare Figure 1.12c with Figure 1.3c. In addition to its unevenly thickened cellulose walls what other features of collenchymatous tissue contribute to its strength?

◆The longer, narrow shape with angled end walls means that the cells pack tightly together to give a much stronger tissue than parenchyma.

The lower epidermis of a privet leaf which is shown enlarged in Plate 3 shows several pairs of guard cells of which the central pair is in best focus. Note that they are smaller than the surrounding epidermal cells and have a different shape. The air spaces behind the stomata open into the spongy mesophyll. Notice that the outer walls of the epidermal cells look darker and thicker. This is the cuticular layer. Look again at the upper and lower epidermis of Plate 2b.

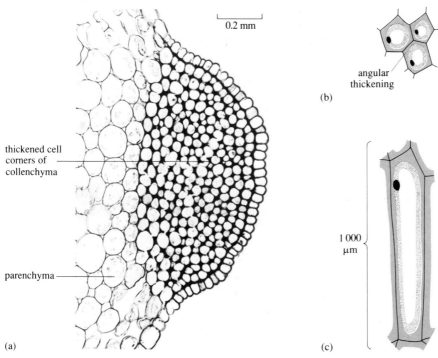

thickened cell
corners of
collenchyma

parenchyma

0.2 mm

angular
thickening

(b)

1 000
μm

(a)

(c)

Figure 1.12 Collenchyma (a) Transverse section showing a clump of cells in the corner of a stem of red dead nettle (*Lamium purpureum*). (b) Diagrammatic transverse section. (c) Diagrammatic longitudinal section of a collenchyma cell.

◇ Compare the frequency of stomata in these two surfaces.

◈ There are no stomata in this part of the upper epidermis. (In fact they are normally relatively rare or absent in this tissue — Section 1.4.1). In the lower epidermis, about three can be seen clearly and four or five are visible but out of focus.

Remembering the large air spaces in the mesophyll of the leaf shown in Figure 1.9, and also in Plates 2 and 3, it is obvious that the stomata will provide a pathway which will allow the photosynthetic mesophyll cells to be bathed in air, and so receive supplies of CO_2 required for photosynthesis. Garden privet is a **mesophyte**, that is a type of plant in which photosynthetic rates are maximized (by keeping the stomata open) at the expense of water conservation. These plants are usually found in habitats which are not exceptionally dry. Plants growing in *very* dry habitats, where the major strategy is to minimize water loss even if photosynthesis is limited as a result, are called **xerophytes**. The obvious way to reduce desiccation is to close stomata before too much water is lost and then, providing losses are small, to sit out the water shortage protected by a thick waterproof cuticle.

◇ Now look at the surface view shown in Figure 1.13. Describe the arrangement of the stomata and the shape of the guard cells.

◈ The stomata seem to be randomly distributed over the leaf surface. The guard cells are more or less kidney-shaped in surface view. Both of these characteristics are typical of dicot leaves.

In this case you can probably see that the stomata are open, i.e. there is a gap between the guard cells (which may be a result of the fixation treatment used for this preparation). In leaves of grasses, the most widespread monocots, stomata are often arranged in rows and individual guard cells are longer and more dumb-bell-shaped. The shape and arrangement of stomata in surface view thus provide one way of deciding whether a plant is a monocot or a dicot.

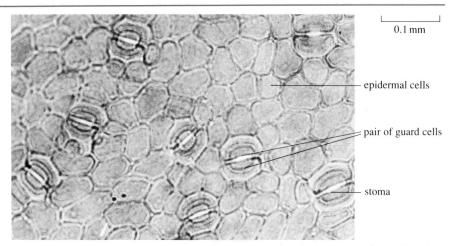

epidermal cells

pair of guard cells

stoma

Figure 1.13 The lower surface of a Box (*Buxus* sp.) leaf showing stomata and guard cells.

Figures 1.14a and b show sections of beech leaves (*Fagus sylvatica*) taken from sunny and shaded parts, respectively, of the same plant.

◇ From what environmental stresses would you expect a sun leaf to need protection?

◈ Heating of the leaf surface could result in excessive evaporation of moisture. Winds could also cause water loss.

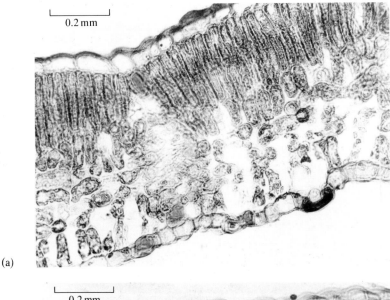

(a)

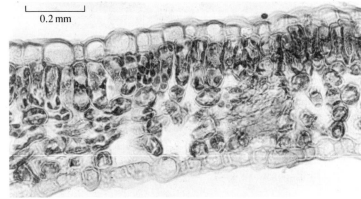

Figure 1.14 Beech (*Fagus sylvatica*) leaves in section. (a) A sun leaf. (b) A shade leaf.

(b)

The epidermis and its cuticle, stomata (especially in the upper epidermis), palisade cells and chloroplasts all show modifications in leaves exposed to bright sunlight.

From Figure 1.14 you may be able to see that:

1 The sun leaf is thicker than the shade leaf.

2 There is cuticle (visible as a dark line) over both the upper and lower epidermis in the sun leaf but in the shade leaf the lower cuticle is not obvious. (In general the upper cuticle of the sun leaf is about twice the thickness of the upper cuticle of the shade leaf.)

3 In the sun leaf, stomata are relatively numerous and confined to the lower epidermis. In contrast, in the shade leaf stomata are less frequent but present in both the upper (very occasionally, so not visible in Figure 1.14b) and lower epidermis.

4 Shade leaf chloroplasts are larger than those in the sun leaf but sun leaf chloroplasts are more numerous.

One feature you cannot see in these photomicrographs is that sun leaf cells are rich in tannins, which protect the leaf against excessive light intensities. These tannins give cells a rich brown cast in sections. They are also a partial cause of the darker colour of whole leaves. In the shade leaf, where light is limited, the larger chloroplasts improve light interception and hence photosynthetic efficiency. So clearly leaf anatomy is different in different circumstances. In these beech leaves, damage due to heating and loss of water from the surface is minimized and absorption of light maximized as a result of leaf characteristics.

Figure 1.15 shows that leaves belonging to the same species may have a larger area in shady conditions; the plant is the bloody cranesbill (*Geranium sanguineum*). You will learn more about this and the physiological and biochemical adaptations that maximize the efficiency of photosynthesis in Chapter 2.

Look at Plates 2 and 3 once more then contrast them with Figure 1.16, which is a section through a drought-tolerant (i.e. xerophytic) leaf, in this case crowberry (*Empetrum nigrum*). First locate the three main tissue types in Figure 1.16 (that is upper and lower epidermis, mesophyll (palisade and spongy cells) and veins.

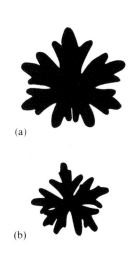

(a)

(b)

(c)

Figure 1.15 Geranium sanguineum leaves from plants growing in three different habitats. (a) Woodland. (b) Semi-wooded. (c) Dry open site.

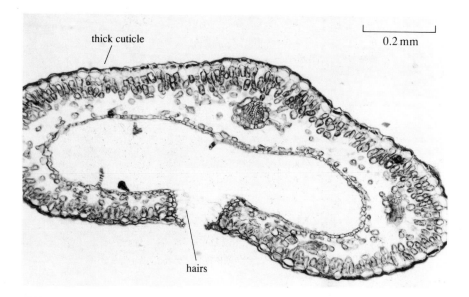

thick cuticle

0.2 mm

hairs

Figure 1.16 Transverse section of a crowberry (*Empetrum nigrum*) leaf showing typical xerophytic adaptations.

◇ What anatomical features do you think limit water loss from the crowberry leaf?

◆ The leaf has a small surface area to volume ratio, an inrolled lower margin and the cavity this forms is partially closed by hairs.

In addition, there are no stomata in the upper epidermis, which has a thick cuticle, but they are present on the thinly cuticularized lower epidermis which lines the cavity. Stomatal guard cells are sunken and the cuticle overhangs the stomata.

There is a second group of plants that are found in dry habitats. These are the **succulents**, which include the cacti (Plate 1d) and the bromeliads (e.g. pineapple). Other examples you may have seen are the so-called 'living stones', house leeks and stonecrops (the genera *Lithops*, *Sempervivum* and

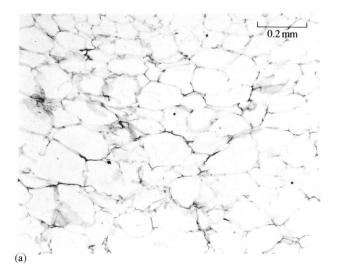

0.2 mm

(a)

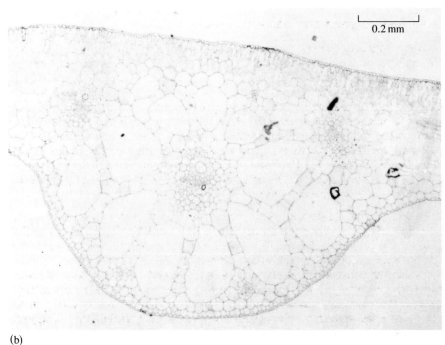

0.2 mm

(b)

Figure 1.17 (a) Section through part of a succulent plant (*Aloe* sp.). (b) Section through a pondweed (*Potamogeton* sp.) leaf.

Sedum, respectively). Succulents are characteristic of deserts and they have specializations that allow them to survive very long droughts. Like xerophytic plants they often have leaves that are very small or even absent. The above-ground parts of succulents are fleshy and swollen, and this also has the effect of reducing the plant's surface area to volume ratio. You have probably also seen cacti covered with dense hairs or spines, which may maintain a moister layer of air around the plant by reducing the flow of air near the surface, thus serving to reduce water loss further. The most obvious anatomical adaptation of succulents is the highly vacuolated water storage tissue which occurs in leaves or stems (Figure 1.17a), which is the reason why the plants look fleshy and swollen. This is coupled with highly efficient water conservation mechanisms. The cuticle is well developed and covered by wax deposits, the stomata are reduced in number and sunk in narrow pits in the epidermis. As external water becomes scarcer, stomata open for shorter and shorter periods, eventually remaining closed altogether until the drought is over. Unlike other plants they never open their stomata during the day. Obviously such water retention mechanisms affect the supply of CO_2 for photosynthesis; succulents have physiological and biochemical adaptations which allow them to photosynthesize even when stomata are closed for long periods during daylight hours and these are described in Chapter 2.

You would not expect a floating leaf of a water lily (e.g. *Nymphaea alba*) or pondweed (*Potamogeton* sp.) (Figure 1.17b) to have a need to conserve water like xerophytes and succulents. These plants are **hydrophytes** (from the Greek *hydro*, water and *phyton*, a plant). Accordingly, the cuticle in leaves of hydrophytes is minimal (Figure 1.17b). Also, stomata occur on the upper epidermis but not on the lower — there would obviously be no point in this! The mesophyll has cavernous air spaces particularly around the vascular bundles and the air in them helps to keep the leaf afloat. The mesophyll tissue that includes these air spaces is often called **aerenchyma**. You may also have noticed that the amount of lignified tissue in the vascular bundles is small in comparison with the total leaf area, so buoyancy, rather than thickening of cell walls, is important in keeping the leaf expanded.

1.4.3 Leaf responses to seasonal variation

In higher latitudes the number of daylight hours decreases in winter. It is also colder. The result is that the amount of photosynthesis in evergreen plants decreases because there is less light energy available, while the rate at which the plant enzymes are working may also be lower at the lower temperatures. In temperate latitudes, many evergreen plants continue to photosynthesize in winter, albeit at slower rates. Evergreen leaves are tough and frost-resistant, whereas the delicate leaves of deciduous trees are more easily damaged by cold and are shed in autumn. (The physiology of the process of leaf fall is described in Chapter 7.) The rate of metabolism in the rest of the plant then falls to a maintenance level until the temperature rises and day length increases in spring, allowing more efficient photosynthesis.

Cold and frozen winter soils also cause a reduction in the amount of water absorbed by roots so the plants may be short of water even if leaves are not killed by cold. Leaf fall is thus a response to shortage of water, low temperature and low light intensity. Obviously this is most critical in alpine and polar environments. In these habitats, survival, even of plants without leaves, is frequently dependent on insulating snow cover. In tropical environments, which are characterized by seasonal uniformity, plants continue to function at the same rate throughout the year. Leaves often live for several years and there is no special time for leaf fall.

Summary of Section 1.4

1 Leaves are composed of three main tissues namely epidermis, mesophyll and vascular bundles (often also referred to as veins); these can be identified in transverse sections.

2 Monocots and dicots can be distinguished by the pattern of vascular bundles seen in transverse leaf sections.

3 The mesophyll is made up of spongy parenchymatous tissue with large air spaces filled with moist air and more closely packed palisade cells with many chloroplasts.

4 In general the epidermis is waterproofed and gas-proofed because it is covered with a cuticular layer of cutin and wax. However the cuticle is pierced by pores called stomata which facilitate gas and water movement in and out of the leaf.

5 Collenchyma, with unevenly thickened cellulose walls, may strengthen leaf tissue particularly in the midrib region.

6 Mesophytes are plants that tend to keep their stomata open so that photosynthetic rates are maximized, and therefore suffer considerable water losses via transpiration. In contrast, xerophytes close their stomata and reduce water loss, thus limiting photosynthesis.

7 Different leaves from a single plant can differ anatomically in response to environmental conditions. Thus shade leaves are relatively thin, have fewer stomata per unit area, and large chloroplasts while sun leaves have well developed cuticle, are thicker, have smaller and more numerous chloroplasts and are rich in protective tannins. Leaves in shady habitats may also be larger.

8 Xerophyte leaves are characterized by thick cuticle, sunken stomata and incurved margins, all of which limit water loss. They may also have low surface area to volume ratios.

9 Succulents have specialized water storage tissues coupled with mechanisms for efficient retention of water, such as thick, waxy cuticles and sunken stomata, which can close for long periods, and which do not open during the day. Like xerophytes they have low surface area to volume ratios.

10 Hydrophytes lack any of the anatomical adaptations usually associated with water retention and have poorly developed vascular bundles; they depend on air within the tissues for buoyancy. This specialized mesophyll parenchyma is called aerenchyma.

11 Plants in temperate and polar habitats may shed their leaves in autumn. This is a response to reduction of the available soil water, reduced amount of daylight and low temperatures. In contrast, plants in tropical environments may have long-lived leaves and leaf fall may occur at any time of the year.

Question 2 (*Objectives 1.1 and 1.4*) Identify the cell types labelled A and B in Plate 1.4 and describe their characteristics and functions.

Question 3 (*Objective 1.5*) Make an outline sketch of the tissues shown in Plate 4. Label these tissues.

Question 4 (*Objectives 1.1, 1.5 and 1.6*) Is the leaf in Plate 4 from a monocot or a dicot? Give reasons for your answer.

Question 5 (*Objectives 1.1 and 1.7*) The saxifrage family has many species which grow in alpine habitats, characterized by high light intensities and high winds, so it is not surprising that members of the family have adaptations related to reducing water loss while maximizing photosynthesis. Identify the ways in which the external characteristics of each of the two plants shown in Plate 5 are adaptations for an alpine habitat.

Question 6 (*Objectives 1.1, 1.5 and 1.7*) List the typical anatomical modifications of (a) leaves of xerophytes, (b) leaves and stems of succulents and (c) floating leaves of hydrophytes.

1.5 MECHANICAL SUPPORT AND TRANSPORT SYSTEMS

Cavities filled with water act as a hydroskeleton in some animals. Plant cells with stiff cellulose walls may act as mini-hydroskeletons when they are turgid i.e. when they contain sufficient water and the protoplast presses against the wall. In fact, cell turgidity is the main support mechanism in lower plants such as bryophytes. However, in young stem tissues of higher plants, support by turgor of individual cells is supplemented by a system of support described as **tissue tension**. The outer layer of cells (the epidermis) is stretched and so held under tension whilst the inner tissues are under compression. This epidermal stretching means that when stems are subject to lateral bending forces by wind the stem is much less likely to buckle. This explains why a longitudinal cut down a stem segment (such as a stick of rhubarb) causes outward bending of the two halves: release of tension allows the epidermis to contract and the inner tissue to expand. This also explains why radishes (which are short, thick stems) form 'roses' when the stem is cut. However, turgidity and tissue tension alone are insufficient to support larger organisms and all vascular plants (ferns and their allies, gymnosperms and angiosperms) have specialized mechanical support systems. These systems consist of groups of cells whose walls are strengthened in various ways. Some of these cells also function as conducting elements, transporting water and minerals from roots to shoots. Figure 1.3 showed parenchyma cells; a type that has no specialist support function. Consider how the rather pliable cellulose walls of young cells of this type could be strengthened once the cells stopped growing, without any change in the wall composition.

◇ What could happen?

◆ The walls could simply get thicker.

A general increase in wall thickness, as a result of laying down more cellulose, does happen in all cells of leaves, roots and shoots which are not actively dividing. However in certain cells, as you saw in Figure 1.12, cellulose is deposited particularly at the *corners* of the cells.

◇ Recall the name for this cell type.

◆ Collenchyma.

In annual and biennial plants like the square-stemmed red dead nettle (Figure 1.12a), there is this type of thickening in the walls of the cells forming the ridges of the stems. Collenchyma also makes up the strands on the outside of celery stalks (*Apium graveolens v. dulce*), (not proper stems but petioles which are leaf stems).

However, there are also several other types of strengthening cell. Figure 1.18 shows four different types. The first two, (a) **sclereids** and (b) **fibres**, are both classified as **sclerenchyma** (scler-*en*-ki-ma or scleren-*ky*-ma). Sclereids are solitary thickened cells which can occur in many expanded leaves and in the stems of hydrophytes as strengthening cells. They can also be found in other organs including fruits for example pears (accounting for their gritty texture). Fibres often occur in groups, for example in the corners of ridged stems in xerophytes (e.g. broom: Figure 1.7). Figure 1.18c shows another type of strengthening cell, a **xylem** (pronounced 'zylem') **vessel element**; like the fibres, these are thickened and give some support to the plant as does the last type, called a **tracheid** (Figure 1.18d). All these cells differ from collenchyma in one major respect: they are thickened with lignin as well as cellulose.

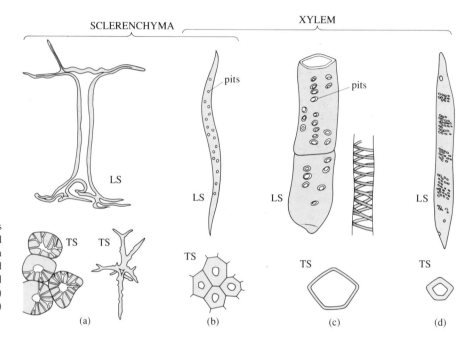

Figure 1.18 Lignified strengthening cells and tissues in transverse section (TS) and longitudinal section (LS). Sclerenchyma is made up of sclereids and fibres and xylem is composed of fibres, xylem vessel elements and tracheids. (a) Sclereids. (b) Fibres. (c) Xylem vessel elements. (d) Tracheids.

◇ Recall the functions of lignin.

◈ Lignin strengthens, hardens and waterproofs.

Remember though that as cell walls become lignified the cell contents disappear so that eventually lignified cells are dead cells.

◇ Look at the cross-section and longitudinal features of the xylem vessel element and the sclerenchyma fibre in Figure 1.18. What characteristics do you think might relate to their function in mechanical support?

◈ Thickness of walls and high ratio of length to width, which also confers strength.

The shape of the end walls of cells can also add to this strength; oblique end walls, like those found in sclerenchyma fibres and tracheids, make stronger tissue than those which are transverse, like the vessel element. This is because oblique end walls pack neatly and tightly together so adding strength to the tissue. Sclerenchyma fibres are long, narrow lignified dead cells whose function therefore is purely supportive. They can be found in ridges, or in packing tissue, or directly outside or even within vascular bundles and they are usually just called fibres. Xylem vessel elements like the one shown in Figure 1.18c are only found in vascular bundles where they form part of a tissue known simply as **xylem**. As well as vessel elements and tracheids, xylem also includes fibres (Figure 1.18b), and parenchyma. Xylem vessel elements are arranged in files, that is columns of cells, but as they mature the horizontal end walls between individual cells disappear and long continuous tubes or **vessels** made from several elements are formed. These vessels have perforated end walls. The main function of vessels is conduction of water and mineral nutrients. Though their walls are lignified the relative length to width does not give them the same strength as fibres and tracheids. Tracheid cells (Figure 1.18d) are narrower than vessels and in this respect more like fibres. The end walls of tracheid cells do *not* break down forming long thin tubes without cross-walls — this is a feature only of vessel elements. Instead tracheids are pointed like fibres (though they are wider than these cells) and are perforated, so that in end view they have a pattern of pores in the pointed transverse end wall. This combination of characteristics — thick walls, high length to width ratio and perforated end walls — makes tracheids most suitable to the dual functions of water transport and support. You will learn later (Chapter 5) that powerful suction forces are involved in the transport process and this may be possible because the conducting cells have very rigid lignified walls that will not collapse inwards.

◇ Think how much suck and blow is needed to move liquid through a narrow tube like a squashed drinking straw in comparison with a wide tube like an open straw. What does this tell you about the suitability of fibres as water conducting cells?

◆ A lot more force is required for a narrow tube than a wide tube. Xylem vessels are wide and tracheids are relatively wide in comparison with fibres; it is the narrow lumen of these last cells that indicates their unsuitability as conducting cells.

It is interesting to note that gymnosperms differ from angiosperms in that they do not have vessels; instead the xylem is made up simply of fibres, tracheids and parenchyma — this is one very obvious way of identifying conifer wood in sections.

The secondary walls of both xylem vessels and tracheids are heavily lignified and this thickening is laid down in set patterns (Figure 1.19) such as narrow rings (**annular** thickening) or **spirals** along the length of the cell. More than one spiral may be laid down in a vessel; and then as the cell ages the spaces in between the layers of spirals become impregnated with more lignin, resulting in the formation of ladder-like (**scalariform**) and **reticulate** or **pitted** thickening. At this stage the wall is fully lignified apart from around the pores and pits which connect the cells laterally. The first-formed xylem elements contain only annular and spiral thickening. As scalariform and reticulate thickening are not laid down in the first-formed xylem (usually called *protoxylem*) in plant stems and roots, this is a useful way of locating this tissue in a vascular bundle. The first-formed vessels and tracheids stretch a long way very rapidly in young growing tissues; this is possible because the rings and spirals stretch out like a piece of string. If scalariform or reticulate thickening was laid down by this stage in development this would not be possible. Later this protoxylem may be crushed as the relatively thin walls do not stand up to pressure from later-formed stronger cells. A tertiary layer of wall may also be laid down as rings, spirals, scalariform or reticulate thickening and it is quite common to observe loose spirals of tertiary wall inside the reticulately thickened secondary wall.

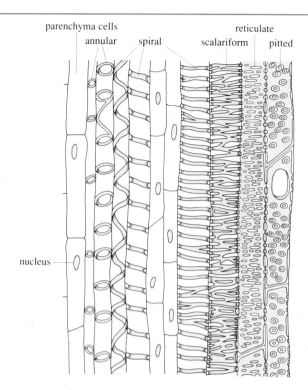

Figure 1.19 Types of lignified thickening in walls of xylem cells.

Recall that when primary walls are laid down, pits pierced by plasmodesmata allow for continuity between adjacent cells (Section 1.3.2).

◇ What do you think could happen to pits when walls become thickened with lignin?

◆ The pits could become sealed and this would mean that lateral movement of water and solutes would not be possible between cells because lignin waterproofs cells.

Clearly this would not be helpful. In angiosperm vessels, thickening is confined to the borders of pits; they are called **bordered pits** (Figure 1.20a). Gymnosperm tracheids are rather different, however. They have more elaborate bordered pits (Figure 1.20b) with the additional refinement of a central plug called a **torus** held in position by fibrils radiating from the raised edge of the thickened wall of the pit (not shown). This is pushed against the pore when water flow up the tracheid is too rapid. The thickened border in angiosperms and the thickened border and torus in gymnosperms also plug the pores when air bubbles (*embolisms*) develop in conducting tissues (Chapter 5).

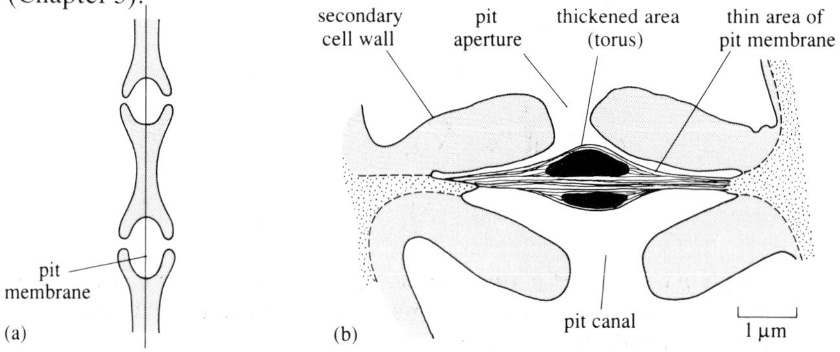

Figure 1.20 Bordered pits. (a) In a dicot vessel. (b) In a conifer tracheid.

The effectiveness of cells as supporting or conducting elements depends not only on their structure but also on their position and grouping in plant organs. As you know, collenchyma cells often form groups to the outside of stems at corners or on ridges (Figure 1.12a), and they are also sometimes found in leaves around the main vascular bundle (i.e. in the midrib: Plate 4); they continue to stretch and so collenchyma is a useful strengthening tissue in growing and expanding organs. Fibres can be found in similar positions, though not usually in very young organs. In leaves they keep the tissue rigid when water is being lost and there is consequently low cell turgor. Remember also that lignified cells are waterproofed, so fibres towards the outside of organs can limit water loss. Not surprisingly then, fibres are common in xerophytic species, particularly where the leaf or stem has a convoluted or ridged surface.

In addition to xylem, vascular bundles also contain a tissue called **phloem**. Unlike xylem, phloem cells have no support function and their main role is to transport organic compounds synthesized in the leaves to other parts of the plant. They are non-lignified and are living. However, there are sometimes groups of fibres within phloem and these do have the usual support function. Some phloem cells are wide and long (Figure 1.21), with perforated end walls called **sieve plates**, and these are called **sieve tubes**. It is the sieve tubes that transport the products of photosynthesis from the leaves to the rest of the plant (Chapter 4). Other phloem cells resemble parenchyma, and are called **companion cells** because they lie close to the sieve tubes and are also involved in the transport of sugars. Companion cells have thin walls like parenchyma but can be as long as the sieve tubes. They are recognizable in electron microscope preparations because they have many mitochondria, and under the light microscope appear to contain dense cytoplasm. Amongst the sieve tubes and companion cells are true *phloem parenchyma* cells.

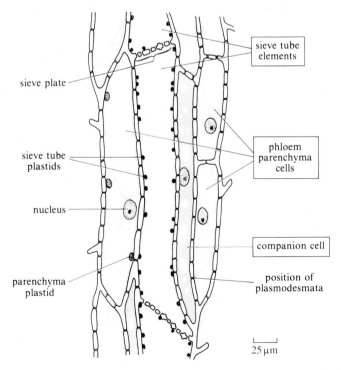

Figure 1.21 Longitudinal sections showing a sieve tube, companion cell and adjacent phloem parenchyma.

Plates 6a and b show part of a longitudinal section of a stem of white bryony (*Bryonia dioica*). First note that there is not much lignified tissue and that it is present at several locations in Plate 6a.

◇ Locate fibres if they are present.

◆ There is a continuous group of fibres (stained red) just inside the epidermis (on the right in Plate 6a).

◇ Are the parenchymatous cells in this section similar in shape and size?

◆ They are similar in shape (rectangular in section) but not in size (varying by a factor of two or more).

◇ Locate the xylem. Can you find vessels and tracheids?

◆ Large red-stained vessels with some scalariform thickening are present at three locations in Plate 6a, but there is one region with obvious tracheids almost at the bottom of the innermost group of xylem (there may be one tracheid in the middle group of xylem but this is not too clear). Tracheids can be distinguished from vessels by their relatively short length and pointed end walls.

◇ How many areas of phloem can you see and can you locate any companion cells? (Use the enlarged section (Plate 6b) to help you to recognize the dark-stained end walls of sieve tubes).

◆ There are three groups of phloem in Plate 6a: one to the outside of the xylem towards the bottom right of the picture, the second to the inside of the central region of xylem and the third to the inside of the third group of xylem cells. Companion cells, lying adjacent to sieve tubes and with obvious cytoplasmic contents, are only visible on the far left of Plate 6b but several cells are overlying each other.

Look at Plate 6b carefully. You should be able to see the sieve plates (end walls) at the ends of the sieve tubes, though these are partly obscured by a densely staining plug of a polysaccharide material called *callose*.

Phloem structure and function are discussed in detail in Chapter 4.

Summary of Section 1.5

1 Turgor pressure and tissue tension are not enough to keep large plants properly supported; this is the function of cellulose and lignin in cell walls of specialized supporting tissues.

2 Young and short-lived stems may depend for rigidity in part on collenchyma with thickened cellulose cell walls.

3 Most stems, roots and leaves depend for rigidity on lignified, and hence dead, tissues such as non-vascular sclerenchyma fibres and sclereids, and the fibres, tracheids and vessels of xylem in vascular bundles.

4 Tracheids and vessels are the main channels for transport of water and mineral nutrients from the root to the stem and leaves.

5 Lignin is laid down in ring, spiral, ladder-like (scalariform) or reticulate patterns on the inside of cell walls of tracheids and vessels.

6 Bordered pits in the walls of vessels and tracheids allow lateral movement of water and solutes within xylem.

7 Phloem sieve tubes with associated companion cells, phloem parenchyma and (sometimes) fibres also occur in vascular bundles, but apart from any phloem fibres which may be present, they have no support function. Instead they carry the products of photosynthesis in solution from the leaves to other parts of the plant.

Question 7 (*Objectives 1.1, 1.4 and 1.5*) Identify the different types of strengthening cell, A, B and C shown in Figures 1.22a and b. Give reasons for your answers.

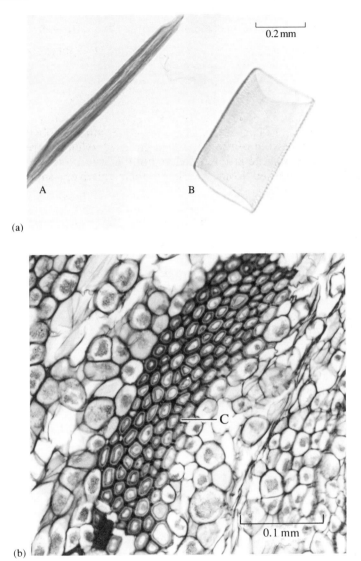

(a)

(b)

Figure 1.22 For Questions 7 and 8.

Question 8 (*Objectives 1.1 and 1.5*) The wall of cell B in Figure 1.22a is thickened in a particular way. What is the name for this kind of thickening?

1.6 STRUCTURE OF YOUNG STEMS AND ROOTS IN DICOTS

Stems and roots are not only subject to different mechanical stresses from the environment but they also have different functions. Stems carry leaves upward, so facilitating light gathering, and transport the products of photo-synthesis manufactured in leaves back down the plant to provide energy for growth (and other energy-consuming activities) and also the precursors of food reserves synthesized in storage organs. Roots channel water and mineral nutrients from the soil to the leaves. Stems (in the atmosphere) and roots (in the soil) are clearly not subjected to the same bending forces and this is reflected in differences in structure. This section describes the different anatomical features of stems and roots which are related to function and also to environmental pressures.

1.6.1 Young stems

Figures 1.23a and b show respectively an outline sketch and a photomicrograph of a transverse section of the stem of a sunflower (*Helianthus annuus*). Figure 1.23a also shows representative cells of the different tissues in longitudinal section. You should note that, like leaves, stems are made up of the three main tissue types, epidermis, packing tissue (parenchyma) and vascular bundles. In sunflower, a dicot, the latter are arranged in a circle.

Note the following points:

1 There is no real functional difference between the tissues labelled pith (in the middle) and cortex (to the outside); they are both parenchymatous packing tissues.

2 Vascular bundles consisting of phloem (with sieve tubes with perforated plates in their end walls, companion cells and phloem parenchyma), and inside this, xylem (including xylem vessels with reticulate and spiral thickening, xylem fibres and tracheids) are arranged in a ring around the stem inside the cortex and outside the pith.

3 Outside the phloem there are groups of fibres.

4 There is an undifferentiated area called **cambium** between the xylem and phloem in the vascular bundles; this tissue contains *meristematic* cells, that is cells which can divide to form new cells. (You will learn more about this tissue in Section 1.8.)

Now look at Plate 7, which is a partial transverse section of the stem of *Anthriscus sylvestris* (cow parsley).

◇ Identify the cells labelled A–C in Plate 7.

◆ A are collenchyma. Note they are thickened but not lignified (as they have stained green). The photograph is not at sufficient magnification to show that thickening is concentrated at the corners of the cells.
B are parenchyma. Note that they are unthickened.
C are xylem. These are lignified cells (stained red) and are either vessels or tracheids (probably the larger ones are vessels and the smaller ones tracheids though it is impossible to be sure from this cross-section).

◇ Identify the tissues in Plate 7 in order (from outside to inside).

◆ Epidermis, collenchyma (confined to ridges), cortex parenchyma, vascular bundle with phloem to the outside and xylem to the inside and pith parenchyma. (The pale green cells between the xylem and phloem are the cambium.)

◇ What do you notice about the centre of this stem?

◆ It is hollow. This is a feature of many plants.

There is very little lignified tissue in this section which would suggest the stem is young and herbaceous.

A term which is often used to describe the central part of a stem or root, i.e. that section containing vascular bundles and pith parenchyma, is the **stele**

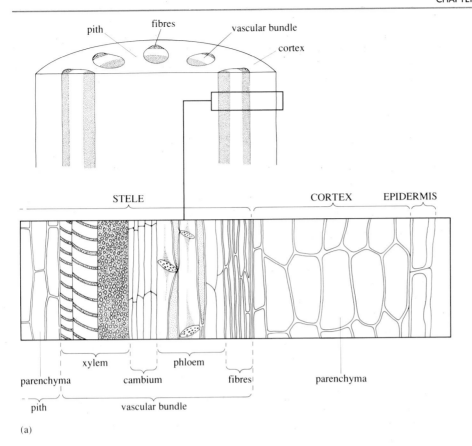

(a)

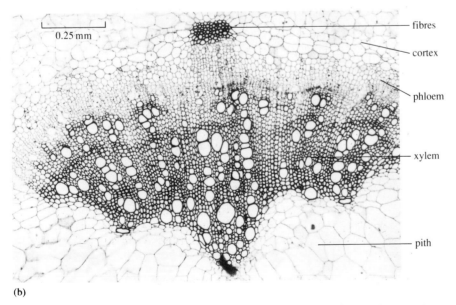

(b)

Figure 1.23 Sunflower (*Helianthus annuus*) stem. (a) Outline diagram of the arrangement of tissues in transverse section and enlarged longitudinal section. (b) Detail of the tissues in transverse section.

(Figure 1.23a). Examination of fossil vascular plants indicates that the most primitive types of stele in young non-woody stems have a solid core of vascular tissue. In more advanced types of stem a continuous cylinder of xylem surrounds a central pith and in the most advanced types the cylinder of vascular tissue is dissected by parenchymatous rays which surround a pith (Figure 1.24). These three kinds of stele are all found among modern fern

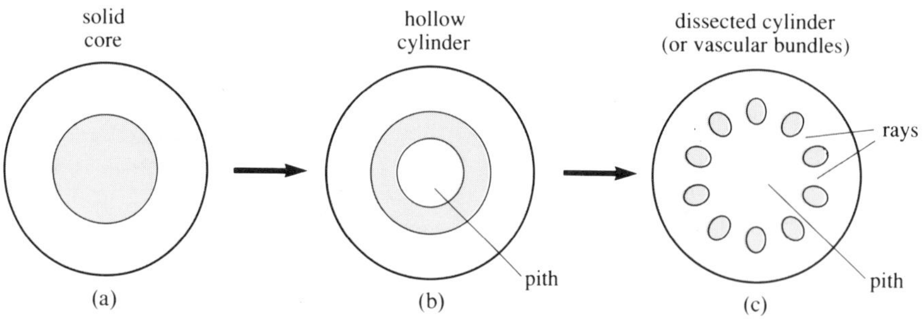

Figure 1.24 Evolutionary trends in the form of steles in plant cells.

solid core

hollow cylinder

dissected cylinder (or vascular bundles)

rays

pith

pith

(a)

(b)

(c)

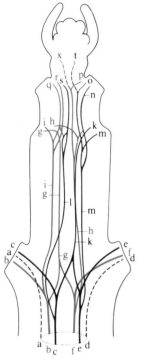

Figure 1.25 Reconstruction to show branching and anastamosis of vascular bundles in a stem.

allies but young angiosperms usually have dissected steles (Figure 1.24c). Individual strands of these dissected cylinders do not form long straight channels; they branch and anastomose throughout the stem (Figure 1.25). Notice how the bundles branch and merge just below the point where the lateral branches emerge.

◇ What kind of stele does *Anthriscus* have (Plate 7)?

◆ A dissected cylinder, that is, one with discrete vascular bundles.

A herbaceous shoot or a young tree seedling has to be pliant and bend in the wind if it is not to be snapped off and, to allow this, a dissected stele has greater resistance to compression and bending, so stems with this structure combine strength with pliability. For similar reasons, engineers building tall concrete structures insert many thin steel rods through the concrete. In larger plants, stronger stems are needed to support the leaves and branches and the arrangement of the strengthening tissues changes to ensure this (as you will see in Section 1.8).

1.6.2 Young roots

Roots are exposed to totally different forces from stems. There is little bending or lateral compression, because they are growing through a supportive medium, but considerable tension (because the root is an anchor) and longitudinal compression as the root pushes through the soil.

◇ Recall why plants need root systems.

◆ Upright plants need firm anchorage. They also need to take in water and mineral nutrients from the soil to supply the needs of the plants.

Plants either have finely divided branched networks or long swollen tap roots, often with thin branch systems, which penetrate into the soil as do carrot roots (Figure 1.26a). **Root hairs** are extensions of epidermal cells of these branched roots and they penetrate into soil spaces from where they absorb water and nutrients. Most succulents have extensive shallow root systems and these absorb water very effectively from the surface soil layers; a light shower or condensation of dew on cold nights can thus be used by the plant. Mesophytes also respond to dry root conditions by developing deeper or more highly branched root systems (Figure 1.26b).

◇ Would you expect water and mineral nutrients to move into the epidermis, across and up through the root cells into the vascular system by diffusion?

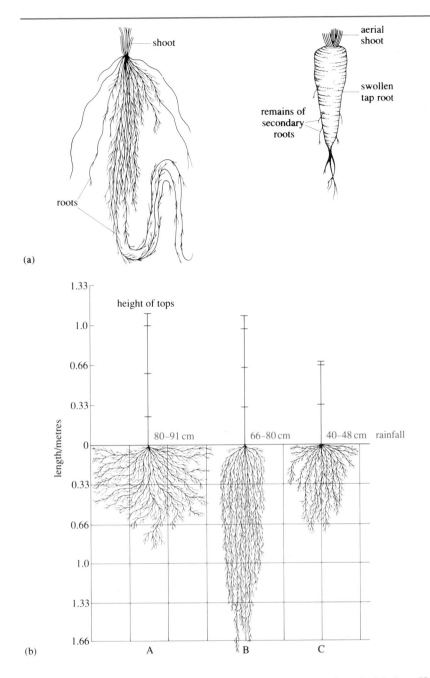

Figure 1.26 Root systems of angiosperms. (a) Fibrous root system of barley (left) and the food storing tap root of carrot (right). (b) Growth of wheat roots in soils with different rainfalls.

◆ They could do so, but the process would be slow and probably insufficient to supply the needs of large plants. In fact, mineral nutrients are actively transported into root cells and water follows by diffusion (Chapters 3 and 5).

Like shoots, roots have a vascular system which transports materials up and down the plant and also has a support role. It is interesting to compare the stele of roots and shoots remembering the difference in stresses affecting these two structures. Plate 8 shows a transverse section of a young root.

◇ Note the cells labelled xylem and phloem. How does the arrangement of these tissues differ from that in the sunflower stem shown in Figure 1.23?

◈ In the root the xylem is at the centre and in two groups ('points') radiating from the centre, whereas in the stem the xylem is the innermost tissue of discrete vascular bundles which are organized as a dissected cylinder around a central pith (Figure 1.23a). In the root the phloem lies in between the xylem points, whereas in the stem it occurs towards the outside and on the same radial axis as each of the groups of xylem.

In the root, the first-formed xylem (protoxylem) is always found at the external point of the xylem. The number of xylem points varies between, and even within the samespecies but numbers between two (as in Plate 8) and six are common. Outside the vascular tissue are one or two layers of thin-walled parenchyma cells called the **pericycle**. This is the outermost layer of the stele. The pericycle is important because this is where lateral roots are initiated (Section 1.8.1). It is relatively unusual to see distinct pith in dicot roots as the xylem fills the centre of the root at a very young age.

◇ Recall how would you recognize early-formed vessels and tracheids.

◈ If you looked at the protoxylem in longitudinal section, lignin would be laid down as rings or spirals and not as ladder-like (scalariform) or reticulate thickening (Section 1.5).

In the root there is a very obvious cortex surrounded by an epidermis.

◇ Comparing Plate 8 and Figure 1.23, can you identify the difference between the cortical tissue of young roots and shoots?

◈ The root cortex is much wider. (Another difference is that root cortical cells often contain starch grains or other storage products, though there are none in the section shown in Plate 8.)

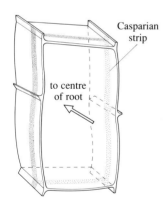

Figure 1.27 An endodermal cell with a thickened Casparian strip.

Roots have an **endodermis** between the pericycle and the cortex (Plate 8). This is a ring of cells with a band of suberized, and therefore waterproofed, thickening round each cell which is called the **Casparian strip** (Figure 1.27). The endodermis is involved in transport (discussed in Chapters 3 and 5). The plasmalemma is firmly attached to the wall adjacent to the Casparian strip and so facilitates water movement into the cytoplasm through the wall on either side. In a few roots (mainly monocots as you will see in Section 1.9), the Casparian strip, which is fully visible when endodermal cells are young, becomes obscured by heavy U-shaped thickening laid down on the lateral and interior walls of the cells. This thickening material consists of lignin and suberin. (Rings of endodermis are sometimes found in stems but they never become thickened.)

Look carefully at the endodermal layer in Plate 8. Sometimes the cells are cut so that you can see the outside of the cell wall, others have the Casparian strip crossing the middle of the cell. However, this is unusual. In Plate 8 the Casparian strip can only be seen as small thickenings in the side walls of the endodermal cells.

Summary of Section 1.6

1 Young dicot stems, like leaves, are made up from parenchymatous packing tissue (cortex and pith) and vascular bundles within an external epidermis.

2 Pith and vascular bundles make up the stele.

3 In primitive stems, steles have a solid cylinder of xylem and phloem. Less primitive stems have a hollow cylinder of vascular tissue around a parenchymatous pith. However a dissected cylinder with a central pith is thought to be the most advanced form of stele, and is characteristic of most young angiosperms.

4 Stems are subject to both bending stress and lateral compression. A dissected stele is more pliant and will bend in the wind so it is less likely to be snapped off by winds.

5 Roots are not subject to lateral and bending stresses like stems and the stele usually has a central core of xylem.

6 In transverse section, roots consist of an epidermis with unicellular root hairs, packing tissues (a broad cortex) and a central stele.

7 The root cortex may act as food storage tissue.

8 In the youngest roots, xylem occurs centrally in the stele, between groups of phloem (which are arranged radially).

9 The outermost layer of the stele is the parenchymatous pericycle.

10 The stele is enclosed within an endodermis in which a thickened Casparian strip runs round each cell. Endodermal cells may become heavily thickened, particularly in monocot roots.

Question 9 (*Objectives 1.1, 1.4, 1.5 and 1.6*) This question relates to Figure 1.28 which shows a transverse section of the root of a French bean (*Phaseolus vulgaris*).

(a) Identify the cell types labelled A–D in Figure 1.28. Explain your reasoning and suggest the functions of these cells.

(b) Identify the tissues labelled E–I and describe their function. Explain your reasoning.

(c) How many clusters of xylem and phloem can you find?

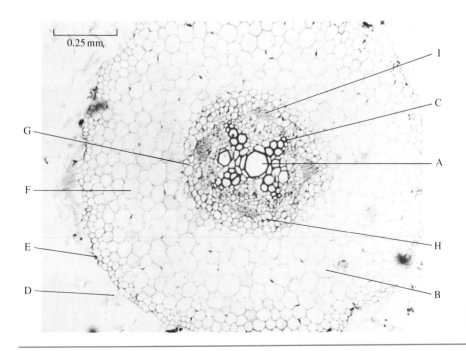

Figure 1.28 Transverse section of a French bean (*Phaseolus vulgaris*) root.

(a)

(b)

Figure 1.29 The positions of growing apices. (a) In a young pea seedling. (b) In a mature bushy plant.

1.7 APICAL MERISTEMS: GROWTH IN LENGTH

As you will recall from Section 1.3, growth involves cell division and expansion. The cells and tissues already described in this chapter are, on the whole, produced by division of meristematic cells localized in root and shoot apices. Subsequent to division these cells elongate and differentiate. The sites of cell division at the apices continue to be the location of apical growth of the main root and shoot axis throughout the life of the plant. So there are two main regions of cell division where young seedlings (such as the pea shown in Figure 1.29a) grow; these are the **apical meristems** at the tip of the root and at the tip of the shoot. Older plants contain many more than two apical meristems, having a growing point at the tip of each branch as shown in the bushy angiosperm plant in Figure 1.29b.

When a meristematic cell divides, *both* daughter cells may retain their meristematic properties and continue to divide. Alternatively, only *one* of the daughter cells may divide again. As the tissue is extended by the addition of more new cells, the non-dividing daughter cell gradually becomes further away from the actively dividing region and may then begin to expand and later *differentiate*, that is become specialized for a particular function.

◇ What would happen if *both* daughter cells started to expand and then differentiate?

◆ There would no longer be a meristem.

Apical meristems are therefore the source of all the cells in the tissues behind them, and as you might expect, isolated apices can re-form whole organs. (You will learn more about this in Chapter 6.) Although many of the cells produced by meristematic activity at the apex rapidly become fully differentiated, it seems as if some retain their potential for meristematic activity. These cells may start to divide again some time after they cease to be part of an apical meristem. Cells of this second type are responsible for a number of processes such as the initiation of lateral (branch) roots, which develop well behind the main root apex, the thickening of roots and shoots (Section 1.8), the expansion of some leaves and the healing of wounds.

If you were given a collection of sections of higher plant tissues stained to show their nuclei and asked to locate regions of cell division, you should have no difficulty in selecting the apical meristems. At any one time, around one in ten of the cells in the apical region are undergoing mitosis (cell division).

◇ How can mitotic cells be recognized under the light microscope?

◆ The individual chromosomes are visible when stained and are arranged in characteristic patterns.

In addition, meristematic cells are very small with dense cytoplasmic contents and large nuclei.

1.7.1 Growth at the root apex

Most root tips (apices) are clearly divided into three zones: the root cap, the dividing (meristematic) region and the region of cell expansion and differentiation, as shown in Figure 1.30a.

At the centre of the meristem, there is a region called the **quiescent centre** (Figure 1.30b) where the cells divide very rarely. This appears to function as a 'reserve bank' of cells, which can undergo mitosis if the normal dividing cells are damaged. On either side of the quiescent centre, however, there are actively dividing cells. Those on one side contribute new cells to the root cap, which is continuously being worn away and replaced as roots push through the soil, while those on the other contribute to growth of the root. The main result of the activity of the meristem is an increase in the length of the root.

◇ After cell division, each of the daughter cells enlarges until it reaches the size of the original parent. Will the direction of growth be mainly parallel or perpendicular to the wall between the daughter cells?

◆ Perpendicular.

Recall that Figure 1.1 showed how an increase in length can be the result of the addition of new cells formed by cell division. Most of the mitotic divisions seen in root tips are in the direction predicted, and vertical lines of cells can be clearly seen. However, the diameter of a young root is greater than that of the zone from which the tissue is derived (compare the lengths of lines a–b and c–d in Figure 1.31). This is partly because the cells are larger, but there are also new cells which are formed by divisions parallel to the length of the root.

This has occurred at many places in Figure 1.31, such as along the planes marked X. Each of the daughter cells continues to divide to give two columns of cells where there was only one before.

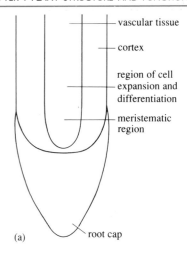

(a)

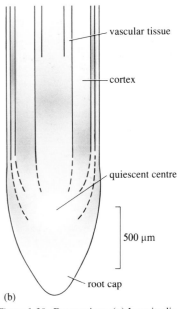

(b)

Figure 1.30 Root apices. (a) Longitudinal section showing the different zones of a root apex. (b) Distribution of cell division in a root apex of an onion (the number of divisions is indicated by the intensity of green tone).

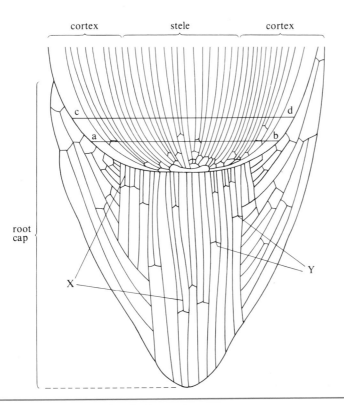

Figure 1.31 Cell lineage map of a root apex constructed from longitudinal sections of maize (*Zea mays*) root.

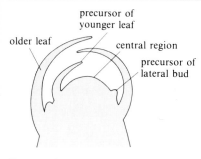

Figure 1.32 Longitudinal section of a shoot apex to show the central dividing region surrounded by precursors of lateral buds and leaves.

1.7.2 Growth at the shoot apex

Figure 1.32 shows the structure of a typical shoot apical meristem with both leaves and buds already formed. These develop early as a result of division of surface cells. In angiosperm shoots, the new cell walls of the surface layer are almost always formed at right angles to the surface; one surface cell therefore gives rise to two surface cells (Figure 1.33). The second layer of cells and sometimes the third divide in the same direction as the outermost layer, but divisions that add new cells to the length of the stem are confined to the central portion of the apex as shown diagrammatically in Figure 1.33. Unlike the root meristem, there is no quiescent centre in the shoot meristem. However, if the apical meristem is removed or damaged, lateral buds may be activated and produce shoots.

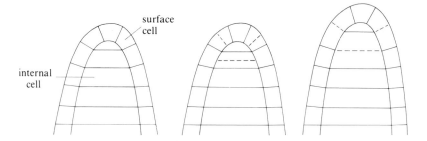

Figure 1.33 Longitudinal section of a generalized shoot apex to show how surface and internal cells divide.

Summary of Section 1.7

1 Growth in length involves production of new cells in apical meristems.

2 The root apex consists of three zones: the root cap, the zone of cell division and the zone of cell expansion and differentiation.

3 The zone of cell division in the root apex surrounds the quiescent centre where the cells divide very rarely.

4 Actively dividing cells contribute to both the root cap and to increase in the number of cells in the expansion and differentiation zone.

5 The shoot apex is divided into surface layers and internal tissues. New cells adding to the length of the shoot are formed from the centre and not in the surface layers.

6 Cell divisions of the surface layer are at right angles to the surface.

7 Unlike the root, the shoot apical meristem has no quiescent centre but inactivation of the apical meristem results in the activation of lateral buds.

Question 10 (*Objectives 1.1 and 1.8*) Identify the quiescent centre, zone of cell division and root cap in Figure 1.34.

Question 11 (*Objective 1.8*) Decide whether the following statements are true or false

(a) Cells of the inner layers of a shoot apex divide at right angles to the surface and add to growth in width and length.

(b) Cells of the dividing zone of the root apex divide both at right angles to the length of the root and also in the direction parallel to it, thereby adding to the number of cells in the root cap.

(c) Cell divisions giving rise to the root cap cells are completed before tissue differentiation begins.

(d) The quiescent centre is never activated.

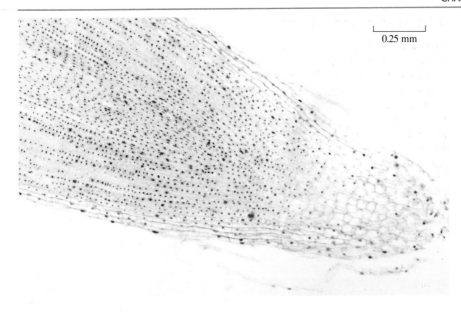

0.25 mm

Figure 1.34 Longitudinal section of French bean (*Phaseolus vulgaris*) root apex.

1.8 SECONDARY THICKENING: GROWTH IN WIDTH

Shoots can grow to considerable size and many trees become massive. Roots therefore have to grow and expand in width, both to anchor massive above-ground structures, and also to carry sufficient water and nutrients to supply the needs of stems and leaves. But size in itself does not satisfy the need for stem tissues to withstand mechanical stresses; a larger proportion of the newly formed tissues (in comparison with young stems) need to be strengthened to achieve this. Secondary vascular tissue is laid down filling up most, or all, of the stem. Tree trunks are massive and solid cylinders of wood which can bear enormous weights and have a limited capacity for bending — their massiveness prevents them from bending. However, if wind speeds are abnormally high, trunks will bend and even break, as occurred in the great storms of 1987 and 1990 in the south and east of England. In 1987 trunks snapped in the wind as the weight of the tree tops was so great (a result of the wet and prolonged summer and consequent late leaf fall); this also caused many very large trees to be uprooted. In the storm of early 1990 the damage was not so great; an obvious explanation for this is that it was winter and so deciduous trees were leafless. Conifers, however, suffered badly.

Wood has probably been of more use to humans than any other natural product.

◇ Think of at least two uses for woody tissues in stems.

◆ Wood is obviously used for construction of buildings and for fuel and paper while natural woody fibres like hemp, sisal and flax are used to make ropes and even spun to make fabrics like linen. (Cotton 'fibre' is not lignified and it also comes from the seed boll of the plant, not from stem fibres.)

As you know, massive strengthened stems are resistant to bending and lateral compression and consequently in trees there is no longer any value in a dissected stele; in fact, this is replaced by a solid cylinder. Unlike stems the pressures on roots are the same throughout life, regardless of size, and solid cylinders which are established at an early stage of growth are retained but of course enlarged.

As discussed in the last section, new cells are added to growing apices by the division of the cells in a limited zone just behind the growing points. On the other hand, increase in girth (called secondary thickening) of differentiated roots and shoots is the result of cell division in a *laterally* placed cambium which is also called a lateral or secondary meristem. The cells thus formed then differentiate to form new tissues. You have already seen cambial cells between the phloem and xylem in the vascular bundle of the sunflower stem (Figure 1.23). In young dicot stems cambium is situated *within* the vascular bundles (as vascular cambium) but may soon develop (from parenchymatous cells) *between* them, so producing cells which differentiate to give extra xylem and phloem tissue to either side (Figure 1.35). These make up *secondary* vascular bundles which may link up to give a complete circle of vascular tissue. Later, most of the centre of the stem becomes filled with xylem tissue making an almost solid cylinder of xylem. This process of **secondary thickening** in stems is summarized in Figure 1.36a. Xylem cylinders are divided by fingers of small parenchymatous cells called **rays**. Five layers of cells were labelled cambium in Figure 1.23a but in reality there is only *one actively dividing layer*; the others are newly formed undifferentiated cells which, as they age, differentiate as xylem cells (to the inside) or phloem cells (to the outside). If the longitudinal section shown in Figure 1.23a were longer you would be able to decide which of the five cell layers was the cambium, as some cells would be halved in width, indicating that they had just divided, leaving two small cells in the space of one parental cell.

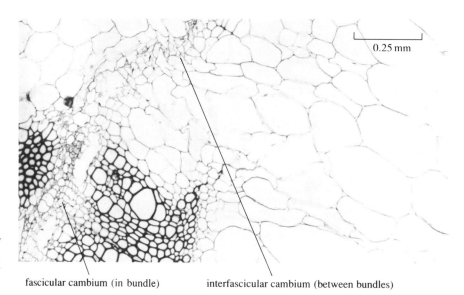

0.25 mm

Figure 1.35 Transverse section of sunflower (*Helianthus annuus*) stem showing newly developing interfascicular (ray) and bundle cambium.

fascicular cambium (in bundle)   interfascicular cambium (between bundles)

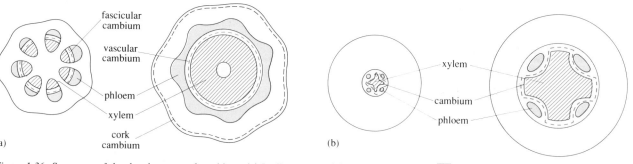

(a)

fascicular cambium
vascular cambium
phloem
xylem
cork cambium

(b)

xylem
cambium
phloem

Figure 1.36 Summary of the development of cambium. (a) In dicot stems. (b) In dicot roots.   ▨ phloem   ▨ xylem

◇ Recall that in young roots the phloem lies in between the points of xylem. Where would you expect to find a lateral meristem in roots these circumstances?

◆ In arcs outside the xylem tips but inside the phloem (Figure 1.36b: left).

Soon these arcs join up and become a continuous, undulating ring. Eventually the undulations are no longer noticeable as the centre is filled with xylem pushing the phloem outwards. Figure 1.36b summarizes the process of secondary thickening in roots.

The shape of cambial cells in both roots and shoots is very significant. Vessel elements, tracheids and xylem parenchyma are all formed from identical spindle-shaped cambial cells (or *initials*), while the parenchymatous rays are formed from wider isodiametric initials (Figure 1.37).

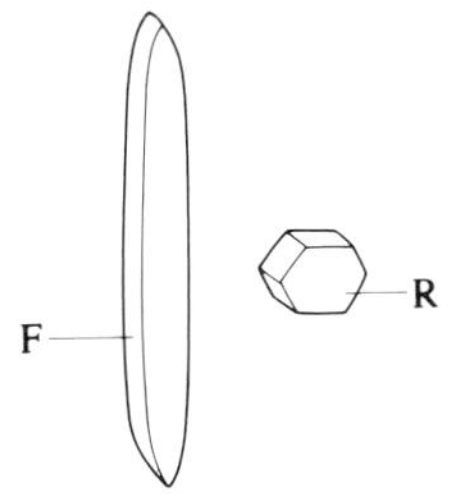

Figure 1.37 Shapes of cambial cells (initials). Spindle-shaped (fusiform) initials (F) form vessel elements, tracheids and xylem parenchyma. Isodiametric initials (R) form ray parenchyma.

1.8.1 Lateral roots

Lateral roots can be found along the parent root anywhere beyond $1–1\frac{1}{2}$ cm behind the root tip.

◇ Would you expect the tissues in this part of the root to be fully differentiated?

◆ From Section 1.6.1 and earlier in this section you could expect this to be so. Xylem and phloem, separated by cambium, will be present.

So how do lateral roots arise? Groups of cells in the pericycle (just outside the phloem and close to the protoxylem vessels; Plate 8 and Figure 1.28) become meristematic and divide to form a protuberance which passes through the cortex and eventually bursts out of the parent root. Eventually the vascular system of the newly differentiating lateral root and the parent root become linked. Figure 1.38 shows a new lateral root still within the cortex and it is clear that the tissue of the cortex has been split by the new root growing through it.

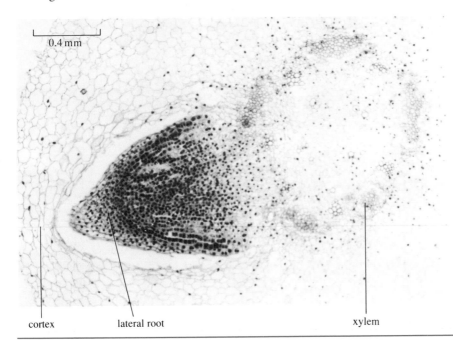

0.4 mm

cortex   lateral root   xylem

Figure 1.38 Transverse section showing a developing lateral root in broad bean (*Vicia faba*).

1.8.2 Old wood

In tree trunks the relative proportions of the different xylem cell types, and the patterns in which they are laid down, varies in different plant species; this is why woods of different species look different when they are made into furniture. The 'grain' of wood is dependent on the relative proportions of vessels and tracheids and patterns of xylem parenchyma and rays which penetrate into the xylem tissue. In the timber trade, close-grained so-called soft woods like pine and dicot species like birch have high proportions of tracheids and sometimes parenchyma, while harder woods have more vessels. In fact conifers (gymnosperms) do not have vessels at all — only tracheids — which is thought to be a primitive feature (Section 1.5).

Many students find it difficult to relate the appearance of slabs of timber in a timber yard (e.g. Figure 1.39, top) to the transverse and longitudinal sections which are commonly studied in anatomical exercises. Figures 1.39 and 1.40a show the relationship between the positions and shapes of rays within wood in tangential, transverse and radial sections.

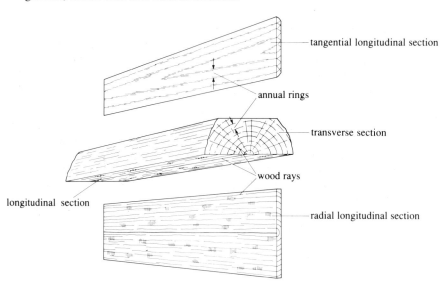

Figure 1.39 Timber cut in various ways, showing the orientation of rays.

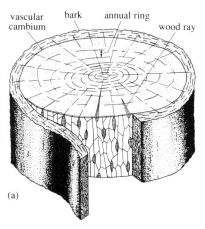

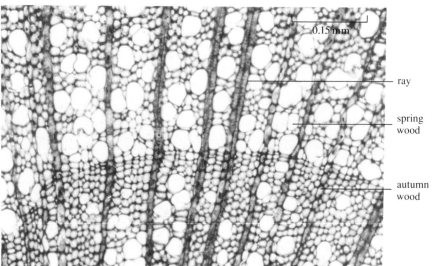

Figure 1.40 Annual rings. (a) In timber. (b) In spring and autumn wood transverse section.

Newly formed vessels and tracheids can vary in size throughout the year; small cells are formed in the late summer in temperate regions when there is less need for rapid transport of water up the stem. Conversely in the spring the newly formed vessels are large and have a wide lumen so allowing maximum rates of water transport. This pattern of growth produces **annual rings** (Figure 1.40a); the small cells and their walls look darker in cross-section because of the greater proportion of wall to cell interior and the smaller proportion of vessels to tracheids, while the zone with larger cells looks lighter in colour because vessels are wider and more numerous than tracheids (Figure 1.40b). It is therefore possible to identify a single year's growth as the distance between two dark-coloured rings and consequently to estimate the age of trees by counting the number of these rings. This science, which is based on removal of narrow cores of wood from trunks and is called **dendrochronology**, is employed by historical ecologists and archaeologists (who can age trees and shrubs), and by archaeologists (who can in some cases date buildings and furniture) from the patterns of annual rings; the width of annual rings and the patterns of these are known to reflect past climatic events.

Only the most recently formed xylem cells in thickened tissues such as tree trunks are actively engaged in movement of water and mineral nutrients. Older tissue is usually blocked. These blockages are called *tyloses*. Young xylem is called **sapwood** because of its function as a channel for water and minerals (i.e. sap). Not surprisingly, the sapwood of timber felled in the spring may bleed. Older wood is often referred to as **heartwood**.

In trees the relative volume of phloem to xylem is small and when the external tissue or bark (see below) is pulled away the thin-walled living phloem and cortical cells tend to come away with the bark. The proportion of active phloem is small because older phloem and cortical cells, having relatively thin walls, are split and crushed by the pressure of other new cells as they are formed. This pressure has two sources: from the xylem internally, and from new external protective tissues which are formed as the original epidermis is split by increased girth of the tissues inside.

◇ If there are to be new external tissues how do you think they originate?

◆ Another cambium must be involved.

1.8.3 Cork and lenticels

The location of the cambium from which new external tissues form can vary from just underneath the epidermis of the young stem to quite deep into the cortex. It is often called cork cambium or **phellogen**.

Cork cambium gives rise to cells destined to make up **secondary cortex** to its inside and cells which become thickened with suberin and lignin (and so waterproofed and strengthened) to the exterior; these form the protective **cork** or **bark** (Figure 1.41a). Bark patterns make it possible to identify trees in winter when there are no leaves. Their scaly patterns are determined by the location and degree of continuity of the cork cambium which can be almost continuous, producing sheets of bark as in London plane (*Platanus occidentalis*), or discontinuous, as in the rough bark of the holm oak (*Quercus ilex*) (Figure 1.41b).

Secondary cortex is also laid down in roots and the external tissues formed from this cork cambium are thickened with suberin.

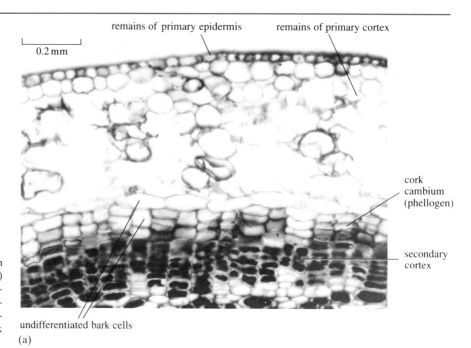

remains of primary epidermis

remains of primary cortex

0.2 mm

cork cambium (phellogen)

secondary cortex

undifferentiated bark cells

(a)

Figure 1.41 Bark. (a) Transverse section of outer stem tissues of apple (*Malus* sp.) showing the position of the cork cambium, secondary cortex and undifferentiated bark cells. (b) Bark of plane (*Platanus occidentalis*) (left) and holm oak (*Quercus ilex*) (right).

(b)

◇ Bark cells are thickened with suberin and hence are gas-proof and waterproof. The living cells of tree trunks need oxygen to survive. How do you think this need is met?

◆ There must be some access for air to internal tissues.

In fact, further division of cells in the phellogen results in formation of masses of loosely arranged cells with numerous intercellular spaces. These specialized cell masses are called **lenticels** (Figure 1.42). The loose packing of the lenticels allows air to pass between the cells and to enter the living externally located tissues of the woody stem. Lenticels are often formed to the inside of stomata. You can see lenticels in young shoots, particularly in elder (*Sambucus nigra*) where they appear as roughened pit-like areas on the bark.

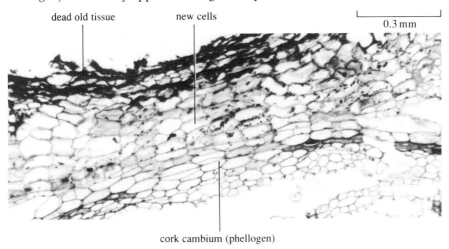

Figure 1.42 Section through a lenticel of elder (*Sambucus nigra*).

1.8.4 Climbing plants

Climbing plants obviously need pliability if they are to twist around supports during growth. They need less strength than non-climbers because they are supported, but as in trees, well developed xylem vessels are necessary to carry water and mineral salts to the leaves and organic compounds to the roots.

◇ Look at Plate 9 which shows a section of a stem of the climbing plant, Dutchman's pipe (*Aristolochia* sp.), and identify the differences between the stele of this stem and the one shown in Figure 1.40b. Then decide if the tissues and cells you see in Plate 9 satisfy the need for strength coupled with pliability.

◆ Though vascular bundles occupy almost the whole width of the stele, they do not form a solid cylinder and yet the stem is three years old — as evidenced by the two dark bands across the vascular bundles. There are broad rays running from the cortex to the pith in the centre. These rays keep the stem more pliable than would be the case if the vascular tissue was arranged in a solid cylinder, in spite of development of large amounts of secondary xylem strengthening the stem. The spring-formed xylem vessels are extremely large. This will also allow considerable amounts of water and nutrients to be moved rapidly up the stem in the early part of the year. (Wine growers are well aware that it is dangerous to prune vines once the sap starts to move up the xylem in the spring. They contain relatively enormous xylem vessels and these are slow to plug when injured so that plants can wilt and die as a result of losing large quantities of sap.)

Many climbing plants, such as bryony and cucumber, have well developed phloem situated on *both* sides of the xylem, that is they have **bicollateral bundles**, Figure 1.43. (Recall the arrangement of phloem relative to xylem in Plate 6a.) This type of bundle arrangement also occurs in some monocot stems and leaves.

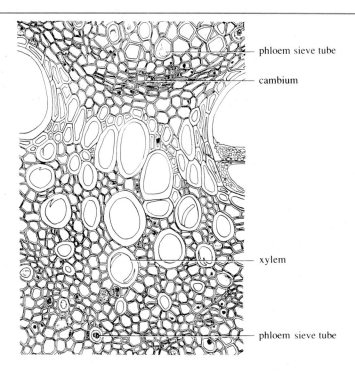

phloem sieve tube

cambium

xylem

phloem sieve tube

Figure 1.43 Bicollateral bundle in *Cucurbita pepo.*

Summary of Section 1.8

1 Trees have massive stems and this protects them in part from wind stresses; the solid stele is appropriate because of its weight-bearing potential.

2 A dissected cylinder — the usual type of stele in young dicot plants — which bends in the wind, often persists if plants are small in stature.

3 Growth in width is the result of a single meristematic layer of cells, the cambium, which produces new cells to either side; these then differentiate to form specialized cells and tissues — phloem to the outside and xylem to the inside.

4 Cambium is found between xylem and phloem tissues in the vascular bundles of most young dicots. Eventually parenchyma between the bundles become meristematic and start to divide so giving a circle of cambium, and hence vascular tissue, around the stem.

5 In roots, the first-formed cambium lies in arcs outside the xylem and inside the phloem but, just as in stems, the actively dividing cells eventually form a complete circle between newly formed xylem and phloem.

6 When the secondary cambium becomes active in roots, xylem soon fills the centre of the root, though the newly formed xylem (protoxylem) continues to lie at the periphery, forming 'points' between the clusters of phloem cells.

7 Lateral roots are formed as a result of meristematic development in the cortex. The young lateral root grows through the tissues of the main root and eventually breaks through the epidermis to emerge at the outside.

8 In woody stems, the diameter and relative abundance of newly formed secondary xylem vessels and tracheids varies throughout the year resulting in the formation of annual growth rings which can be seen in slices of timber.

9 This expansion in width would inevitably split the epidermis. Secondary external layers are formed with a cork cambium (phellogen) cutting off secondary cortex to the inside and an impermeable corky layer (bark) to the outside.

10   The impermeable layer is pierced by lenticels which allow air to enter the stems. Lenticels originate from phellogen.

11   Vines and other climbing plants need to be pliable if they are to twine but they also have to transport water and sugars long distances around the plant. The secondarily thickened stele is broken up by wide parenchymatous rays. Vessels can have a very large cross-sectional area. In some cases they can have bicollateral bundles, with phloem on both sides of the xylem.

Question 12 (*Objectives 1.5, 1.6 and 1.9*) This question relates to Plate 10. The tissue shown has not been stained with safranin and light green. Instead stains have been used which stain unthickened cells red and thickened cells green.

(a) Make an outline sketch of this section and label the tissues.

(b) At how many places in this section is there a cambium? Identify the products of the cambial divisions.

(c) How old was this section when the slide was prepared for photography?

(d) Is it a section of a stem or a root? Give reasons for your answer.

I.9   STRUCTURE OF MONOCOT STEMS AND ROOTS

Grasses and many of the monocot herbaceous plants you are familiar with do not appear to have stems. Until they begin to flower they form an expanding clump of leaves (Figures 1.6c and 1.44). In fact there is a stem but it is very small and buried in the leaf bases at ground level. When the vegetative growth phase is complete this shoot expands to give the flowering stem. Monocots, with their short vegetative shoots, clearly suffer fewer mechanical stresses than plants with tall stems.

However, not *all* monocots are short-stemmed: herbaceous monocots like lilies and monocot trees like date palms are subject to many of the same

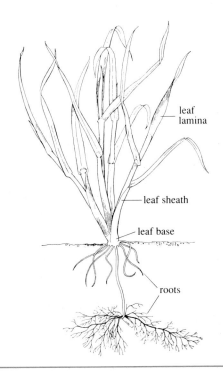

leaf
lamina

leaf sheath

leaf base

roots

Figure 1.44 A grass plant.

stresses as similar dicots. But the arrangement of tissues in all monocot stems is different from that in dicot stems. Figure 1.45a is a transverse section of a typical monocot stem. Note that the many vascular bundles are randomly placed throughout the central parenchymatous pith. Figure 1.45b shows a modification of this arrangement where vascular bundles are roughly arranged in two rings (more like herbaceous dicots). Individual bundles are also differently organized in some species, for instance phloem may occur on both sides of the xylem or may be surrounded by xylem.

◇Recall the name of bundles where phloem is on either side of the xylem.

◆These are called bicollateral bundles (Section 1.8.4).

vascular bundle

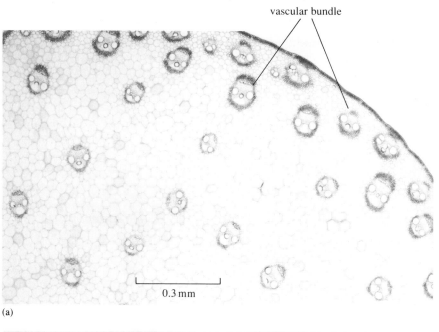

0.3 mm

(a)

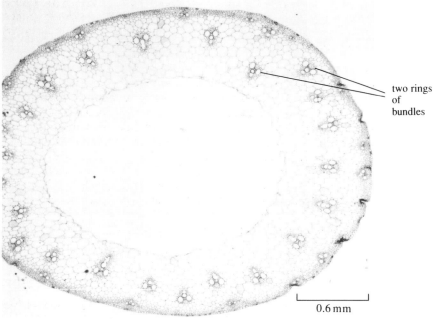

two rings of bundles

Figure 1.45 Transverse sections of monocot stems. (a) Maize (*Zea mays*), with randomly arranged bundles. (b) Wheat (*Triticum aestivum*), with rings of bundles and a hollow stem.

0.6 mm

(b)

◇ Can you identify any other differences between dicot and monocot stems by comparing Figures 1.23 and 1.45?

◈ Figure 1.45b shows a wide ring of sclerenchyma outside the vascular bundles. In Figure 1.45a there is a ring of sclerenchyma around each bundle. In dicots the fibres adjacent to bundles do not enclose them (Figure 1.23b).

The vascular bundles in Figures 1.45a and b are described as **closed bundles** because there is no cambium in the bundle. One consequence of this is that there can be no cambium to extend around the stem so secondary thickening in plants which show this arrangement of bundles is usually impossible. (In contrast, the vascular bundles of dicots are described as **open bundles** because they include cambium which can spread across the rays between bundles).

You are probably aware that individual grass plants continue to grow when the plant is cut to almost ground level whereas dicot plants (except for rosette-type), are less likely to do so.

◇ Suggest an explanation for this.

◈ Upward growth of stems is from apical meristems at the tips of shoots. In grasses, the shoot meristems are at ground level so are not removed by mowing. Rosette-forming weeds resemble grasses, in that they have a very short stem (but with typical dicot anatomy) and so the growing tip is not removed by most mowing operations. Non-rosette-type plants will usually have their apical meristems and lateral buds removed by mowing so will not be able to grow. However, if the cutter bar is set higher, non-rosette weeds may be left with some intact lateral buds, so new growing points can develop from previously dormant side shoots. (This produces bushy plants and is the reason why gardeners tip chrysanthemums and other plants in the seedling stage if they want to encourage them to bush and hence eventually flower profusely.)

However, the flowering shoots of grasses elongate in another way. Small meristematic areas develop in the leaf nodes (the 'joints' where leaves arise from stems). But unlike dicots these so-called *intercalary* meristems do not give rise to branches. Instead they add cells to the stem so allowing the stem to grow in length (Figure 1.46).

Some monocot species are large (e.g. yucca) and some are trees, and hence long-lived (such as the palms mentioned earlier); clearly, random placement of relatively sparse vascular bundles would not give enough strength or conducting capacity to these stem tissues; more supporting tissue is necessary. However, even in the massive palm trunk the vascular system consists of separate bundles. There is simply an increase in the numbers of bundles. These new bundles are formed from a meristem which arises in an arc behind the apical meristem. In many monocot species, parenchyma to the outside of the vascular bundles becomes meristematic and divides to produce cells which differentiate to form a new bundle. This activity is usually short-lived.

◇ Can you suggest what happens to the trunks of palm trees in hurricanes?

◈ They bend in the wind sometimes almost to ground level. Like the stems of herbaceous plants, palm trunks remain pliable and are not snapped off.

Monocot roots do not have secondary thickening though they may develop lateral branches. In long-lived monocots, anchorage and absorptive power is

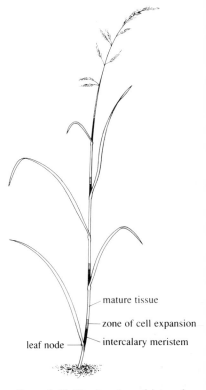

mature tissue

zone of cell expansion

intercalary meristem

leaf node

Figure 1.46 The location of intercalary meristems in a flowering shoot of a grass.

a function of the numbers of roots which are formed at the base of the stem. They can be renewed and increased in number throughout a plant's life.

Look at the transverse section of maize root (*Zea mays*) in Figure 1.47. It shows an epidermis with two rows of thick-walled cortical cells inside. Further in still are thin-walled parenchymatous cortical cells with two lateral roots growing through this layer.

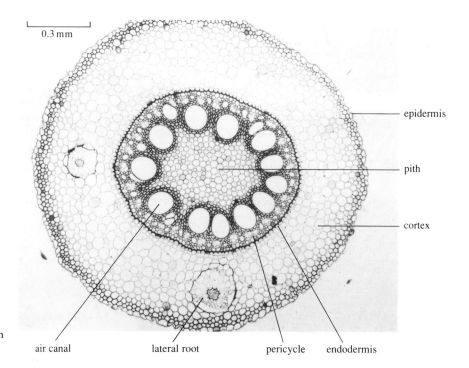

Figure 1.47 Transverse section through a maize (*Zea mays*) root.

0.3 mm

epidermis

pith

cortex

air canal   lateral root   pericycle   endodermis

◇ Describe the endodermis in this root section.

◆ The endodermis has the typical U-shaped thickening on inner and side walls, characteristic of monocots (mentioned in Section 1.6.2).

Within this is the narrow pericycle. There are many vascular bundles arranged in a ring (another monocot feature), as indicated by the arcs of thickened tissue which can be seen adjacent to the pith. There are some large xylem vessels to the inside of this tissue, but the extremely big structures which look like outsize vessels are actually large air canals.

◇ Where would you expect the phloem to lie?

◆ As in dicot roots the phloem alternates with the xylem.

However unlike the typical dicot root structure, the vascular bundles in monocots are arranged radially and the centre of the stem is filled with a parenchymatous pith. *Zea* has an unusually large pith.

Summary of Section 1.9

1 The very short stems that persist during vegetative growth of grasses and some other monocot plants do not suffer the same stresses as most dicotyledonous stems.

2 The arrangement and frequency of vascular bundles is different from that in dicot stems: there are often many discrete vascular bundles randomly located throughout the pith and cortex. The stems of larger monocots owe their strength and conducting capacity to a very high frequency of discrete vascular bundles. Extra bundles are produced as a result of meristematic activity of cells behind the apical meristem or of parenchyma cells outside the existing vascular bundles.

3 Monocots have vascular bundles that are often completely surrounded by fibres and lack a cambium in and between the bundles, which are described as closed. This arrangement precludes the formation of secondary thickening. In contrast dicot vascular bundles, which can become linked by a continuous ring of cambium, are described as open.

4 Unlike dicot roots, monocot roots are never thickened; instead plants produce more roots, from the base of the stem.

5 Monocot roots have a cortex which sometimes contains layers of thickened cells, a thickened endodermis in which the Casparian strip becomes obscured, a narrow pericycle, many vascular bundles arranged radially and a broad pith to the inside of the vascular bundles.

Question 13 (*Objective 1.6*) Decide if the following statements are true or false.

(a) Monocot stems and roots never develop secondary thickening.

(b) Increase in width in some monocot stems occurs when parenchyma cells of the cortex become meristematic.

(c) Only monocot roots have endodermal cells with Casparian strips.

Question 14 (*Objectives 1.1, 1.6 and 1.9*) List the differences between the tissues of monocot and dicot stems and roots.

Question 15 (*Objectives 1.1, 1.4, 1.5 and 1.7*) Figure 1.48 shows a transverse section of a highly modified one year old plant stem. (a) Identify which tissues are specially modified and state how. (b) Is this plant a xerophyte or a mesophyte?

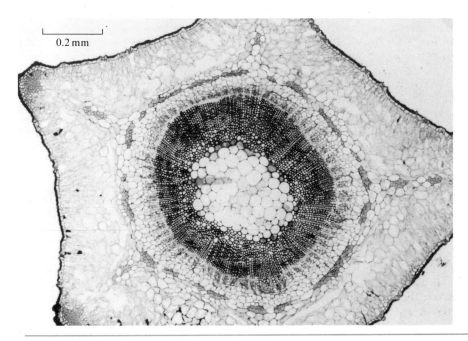

0.2 mm

Figure 1.48 Transverse section of a plant stem (for Question 15).

1.10 CONCLUSION

Plants have evolved by natural selection so higher plants have organs and tissues adapted for effective life on land. Leaves utilize energy in photosynthesis and they are often arranged in such a way that they do not shadow each other and hence can absorb the maximum amount of light. The organelles responsible for trapping light are concentrated in cells or tissues which are in the best position to achieve this, namely the palisade mesophyll tissue of the leaf; they are protected by an epidermis covered by a cuticle and wax, which limits water loss from the tissue. However, the epidermis is not continuous and the openings in it (stomata) are controlled by movements of the surrounding guard cells. The arrangement and frequency of these stomata and the structure of the epidermis and cuticle may allow an observer to make deductions on the habitat in which a plant normally lives. Parenchymatous packing tissues in leaves (and other organs) can be adapted to storage of water or food reserves. In latitudes where light intensity and temperature are low, hence photosynthesis is slow, some plants lose their leaves in autumn. Veins (vascular bundles) and sclerenchyma fibres and sclereids keep leaves rigid. The xylem of the vascular bundles transports water and mineral nutrients from the roots up the stem to the leaves, where photosynthesis takes place, and the phloem (the other conducting tissue of vascular bundles) transports the products of photosynthesis from the leaves to the rest of the plant.

All plant cells have rigid cell walls made up of cellulose, pectins and other materials. In many cells a secondary wall is laid down inside the primary wall and this may be either entirely cellulose, or a mixture of cellulose and lignin. Plant stem tissues function in support of the leaves. They are subject to lateral compression and bending stresses. Short-lived and young stems depend on turgor, tissue tension and sometimes on groups of cellulose-thickened cells called collenchyma, for structural support; these latter cells can still be stretched so they serve a particularly useful function in growing tissues. Support in older stems of small plants is given by separate columns of strengthening tissue arranged in a cylinder and may include lignified cells such as sclerenchyma fibres outside the phloem and vessels, tracheids and fibres within the xylem.

As stems increase in height and the numbers of leaves and branches increase, more strength is required to support them while the flow of water and mineral nutrients up from the soil to supply the leaves also needs to increase. Tall plants therefore require more strengthening and transport tissues. In many dicots almost the entire internal tissue of the stem becomes a solid cylinder of dead lignified cells as the result of secondary meristematic activity. In young stems the meristematic layer (cambium) is located solely within the vascular bundles but later appears between them too. Water transport is limited to the most recently formed and external part of the xylem. External to the cambium is the phloem which conducts the products of photosynthesis around the plant from the leaves and these living cells and any phloem fibres are also increased in quantity and continuously renewed by the cambium. Growth in girth necessarily disrupts the original epidermis, and other secondary meristems produce cells which mature to give secondary epidermal tissues such as bark and specialized areas, called lenticels, in which cells are loosely arranged so allowing gas exchange into and out of the stems.

In monocots, vascular bundles within the stele may be arranged in rings but are often more randomly organized. True secondary thickening does not occur and increase in girth occurs only if a secondary meristem develops behind the shoot apex or alternatively if individual groups of parenchyma cells in the cortex become meristematic.

The main functions of roots are to anchor the plant in the soil and to absorb water and mineral nutrients and make these available to the stems and leaves. The function of anchorage is best served by development of relatively deeply questing thickened roots and these may penetrate so far down that they reach deep water in arid environments. Alternatively, water uptake may require extensive spread of roots at a relatively shallow depth over a considerable distance. (It is the shallow root ball of pine trees that makes them more likely to be blown out of the ground than large oaks.) Increase in height and girth of above-ground plant parts is necessarily accompanied by root growth, branching and, in dicots, by secondary thickening. Again, this last is the result of differentiation of cells from a secondary cambium. The cellular structure of roots allows for maximum absorption of water; in particular, the thin-walled extensions of epidermal cells called root hairs greatly increase the surface area available for absorption. Dicot roots have wide cortical zones which are often sites of food storage. At the endodermis the differential waterproof thickening of the radial walls results in water and mineral nutrients passing from walls to the cell interior from where it is transported up the xylem vessels and tracheids. The stresses on roots do not change as they increase in size so their organization is not changed as the plant matures. Monocot roots are not thickened though they do develop lateral branches. They are sometimes also unusual in having U-shaped thickening in the endodermal cells.

Growth in length of both roots and shoots is by means of apical meristems located in the tips of both roots and shoots. In roots the zone of cell division surrounds a quiescent centre which can become active if the actively dividing area is damaged. In shoots there is no quiescent centre but in the event that the primary meristem is inactivated, lateral buds lower down the axis are activated and produce side shoots.

Plants colonizing similar environments may have similar adaptations, or at least modifications which have similar functions. Careful observation of the internal and external structure of plants can lead to deductions about the external stresses to which plants are exposed and the ways in which plants satisfy requirements for maintenance and growth.

OBJECTIVES FOR CHAPTER I

Now that you have completed this chapter, you should be able to:

1.1 Define and use, or recognize definitions and applications of, each of the terms printed in **bold** in the text. (*Questions 1, 2, 4, 6, 7, 8, 9 and 10*)

1.2 Describe the structure of plant cells and explain how they differ from animal cells. (*Question 1*)

1.3 List the major components of cell walls and the different layers of the wall in which they occur. (*Question 1*)

1.4 Identify and distinguish between the following plant cell types: parenchyma, collenchyma and sclerenchyma (fibres and sclereids), and describe their functions. (*Questions 1, 2, 7, 9 and 15*)

1.5 Identify and describe the function of plant tissues: epidermis (including, guard cells of stomata), spongy mesophyll, palisade mesophyll, bark, phellogen, cortex, endodermis, cambium, phloem (sieve tubes, companion cells, phloem parenchyma and phloem fibres), xylem (fibres, tracheids and vessels) and pith. (*Questions 3, 4, 6, 7, 8, 9, 12 and 15*)

1.6 (a) Distinguish between cells and tissues in microscopic sections of:

(i) leaves of monocots and dicots from morphological and anatomical characteristics;
(ii) stems of young and old monocots and dicots;
(iii) roots of young and old monocots and dicots.

(b) Give explanations for these structural differences. (*Questions 4, 9, 12, 13 and 14*)

1.7 Describe and diagnose those features of leaf anatomy which are associated with mesophytic, xerophytic, succulent and hydrophytic habits. (*Questions 5, 6 and 15*)

1.8 Describe the architecture of root and shoot apices and explain the way in which roots and stems grow in length. (*Questions 10 and 11*)

1.9 Interpret events in the secondary thickening of older roots and stems of dicots. (*Questions 12 and 14*)

▲ 1(a)

▲ 1(b)

▲ 1(c)

1(d) ▼

Plate 1 (a) *Desmococcus*, an alga living on tree bark. (b) *Pellia*, a liverwort. (c) Dandelion (*Taraxacum*). (d) A cactus.

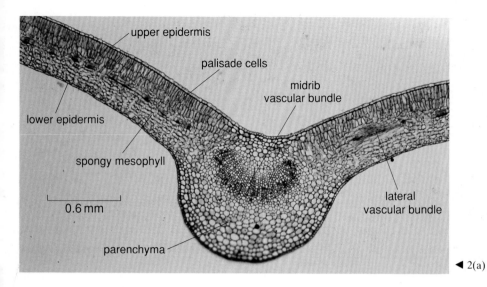

upper epidermis

palisade cells

midrib
vascular bundle

lower epidermis

spongy mesophyll

0.6 mm

parenchyma

lateral
vascular bundle

◀ 2(a)

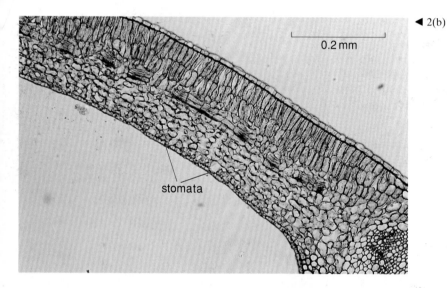

◀ 2(b)

0.2 mm

stomata

3 ▼

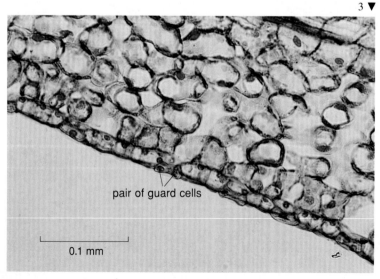

pair of guard cells

0.1 mm

Plate 2 Privet (*Ligustrum vulgare*) leaf. (a) Transverse section. (b) Detail of the lamina.

Plate 3 Transverse section of a privet leaf showing guard cells in the lower epidermis.

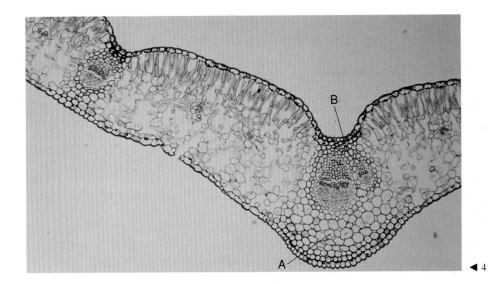

B

A

◄ 4

5(a) ▲

5(b) ▲

Plate 4   Transverse section of a leaf of a dicot.

Plate 5   Saxifrage species. (a) Mossy. (b) Aizoon.

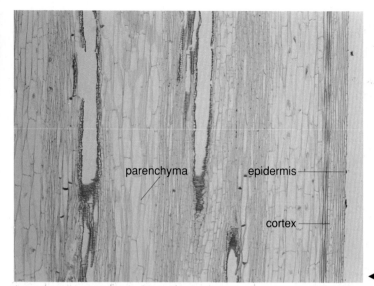

parenchyma

epidermis

cortex

◀ 6(a)

Plate 6 White bryony (*Bryonia* sp.). (a) Longitudinal section of part of stem. (b) Detail of phloem.

Plate 7 Transverse section of the stem of cow parsley (*Anthriscus sylvestris*).

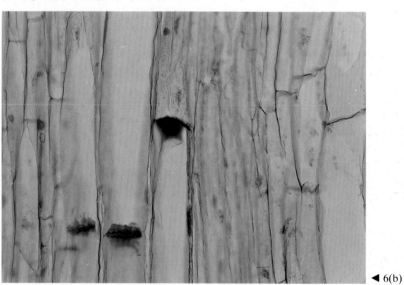

◀ 6(b)

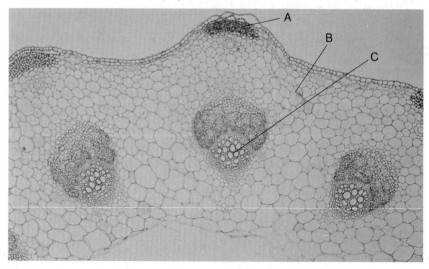

A

B

C

◀ 7

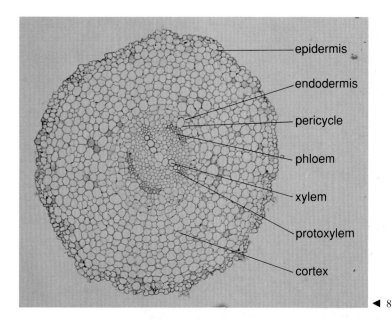

epidermis

endodermis

pericycle

phloem

xylem

protoxylem

cortex

◄ 8

Plate 8 Transverse section of a young sunflower (*Helianthus annuus*) root.

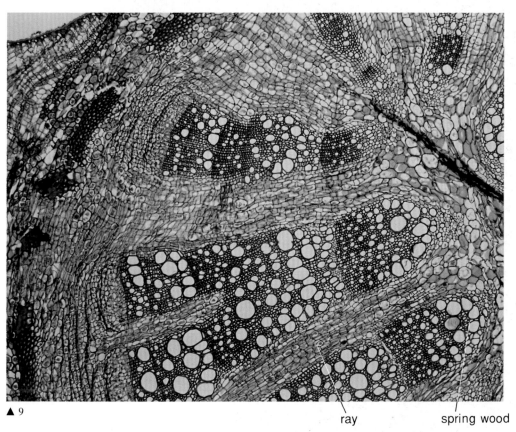

▲ 9

ray   spring wood

Plate 9 Transverse section of part of the stem of a climbing plant, Dutchman's pipe (*Aristolochia* sp.).

▲ 10

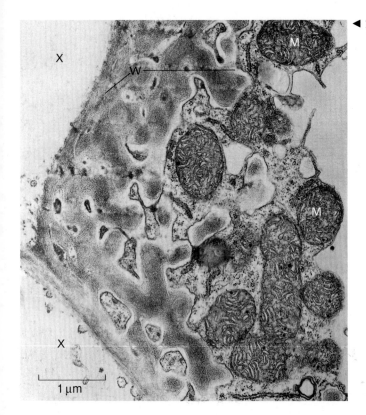

◄ 11

Plate 10 Transverse section of apple (*Malus* sp.).

Plate 11 A transfer cell adjacent to two xylem elements (X). The convoluted cell wall with finger-like projections extending into the cytoplasm of the transfer cell is labelled W. The cytoplasm is dense and contains numerous mitochondria (M). This is a transmission electron micrograph of tissue at the junction between the food-storing 'seed leaves' or cotyledons and the stem of a lettuce seedling.

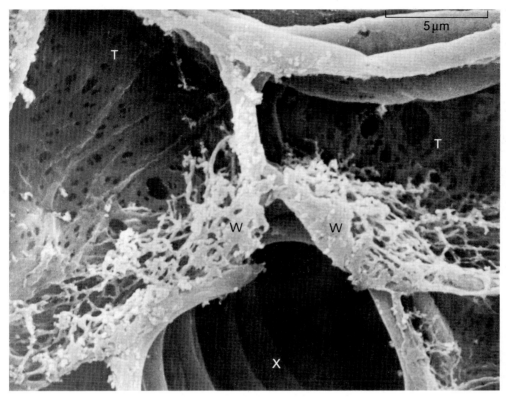

▲ 12

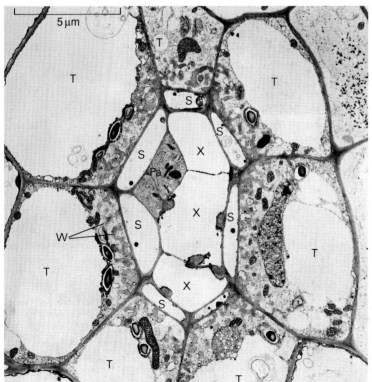

◀ 13

Plate 12 A scanning electron micrograph of part of a vascular bundle at a node (joint) in the stem of *Tradescantia*. Cytoplasm has been digested away so that only cell walls remain. Two transfer cells (T) abut onto a xylem element (X) and the wall ingrowths (W) lie adjacent to this element.

Plate 13 Transmission electron micrograph of a transverse section of a vascular bundle in the leaf of a water fern *Azolla*. Transfer cells (T) with wall ingrowths (e.g. W) are adjacent to sieve elements (S) of the phloem. Cell Pa is a xylem parenchyma cell which acts as a transfer cell and has wall ingrowths adjacent to the xylem elements (X).

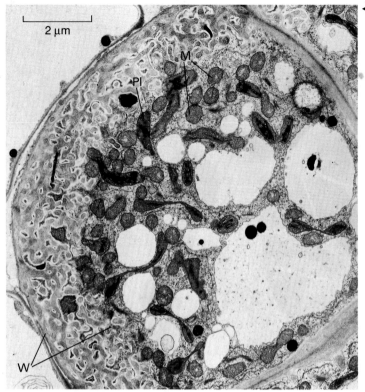

14

Plate 14 Transmission electron micrograph of a transfer cell from the sporophyte of a moss. These cells lie at the junction of sporophyte and gametophyte tissues and facilitate the transfer of nutrients from the gametophyte to the growing sporophyte (capsule and stalk). Cell wall with complex ingrowths (W). Mitochondria (M). Plastid (Pl).

Plate 15 Light micrograph of a transverse section of a vascular bundle in the nitrogen-fixing root nodule of a legume plant. Transfer cells have a fuzz of darkly stained wall ingrowths (e.g. circles). These ingrowths are adjacent to the xylem elements (e.g. X) and the endodermis (where arrows point to the Casparian strip). Sugar passes from the phloem (P) to the cell where nitrogen fixation occurs (not shown). Nitrogeneous components from these cells move to the endodermis, are taken up by the transfer cells and secreted into the cell walls (apoplast), from which they diffuse into the xylem elements.

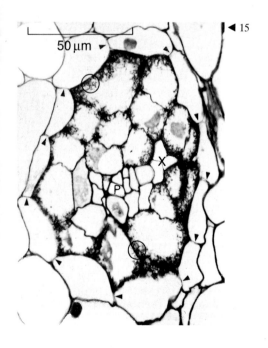

15

PHOTOSYNTHESIS

2.1 INTRODUCTION

The living state requires a constant input of energy, and the most fundamental difference between animals and plants is the way they obtain energy. Animals take in food — organic compounds — and release chemical energy during respiration*; green plants absorb light energy from the Sun, converting it to chemical energy in the process of **photosynthesis**. This is why eukaryotic plants (and some prokaryotes) are described as autotrophs ('self-feeders') whereas animals and most prokaryotes are heterotrophs ('other-feeders').

In many ways aerobic respiration and photosynthesis are mirror images and, indeed, for the biosphere (the sum of organisms on Earth), they balance out. This is illustrated in Figure 2.1.

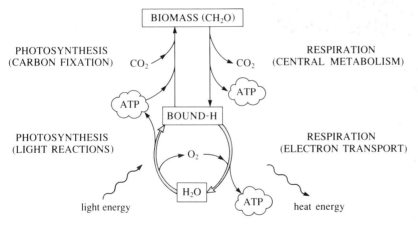

Figure 2.1 Total photosynthesis and total respiration (autotrophs and heterotrophs) are approximately balanced in the biosphere. Carbohydrates in living tissues (biomass) are represented by the general formula CH_2O.

Both processes generate usable chemical energy, stored in the form of ATP, whose synthesis is powered by a transmembrane proton gradient. Aerobic respiration involves oxidation of organic molecules to CO_2 with reduction of oxygen to water and dissipation of energy as heat. Plant photosynthesis involves two linked processes: the light-driven oxidation of water to oxygen (the 'photo' side) during which ATP is produced; and the reduction of CO_2 to organic molecules (the 'synthesis' side), which uses ATP.

◇ Recall the meaning of oxidation and reduction.

◈ Oxidation is the removal of electrons or hydrogen atoms (proton plus electron) or the addition of oxygen; reduction is the addition of electrons or hydrogen atoms or the removal of oxygen.

* The biochemistry of respiration is dealt with fully in Chapter 6 of Norman Cohen (ed.) (1991) *Cell Structure, Function and Metabolism*, Hodder & Stoughton Ltd in association with The Open University (S203, *Biology: Form and Function*, Book 2).

On a global scale, therefore, respiration and photosynthesis are like a gigantic oxidation–reduction couple centred on the splitting and formation of water. For about 2 000 million years, however — half the Earth's history — this was not so. Only prokaryotes existed and the atmosphere contained no O_2, so that all respiration was anaerobic; photosynthesis utilized not water but some other electron donor, such as hydrogen sulphide (H_2S). Photosynthetic bacteria today continue to operate in this way (discussed briefly in Section 2.4.1), which means that photosynthesis can be generally defined only as 'a process in which living organisms transform light to chemical energy and use this energy to fix CO_2 and for other energy-requiring processes'. Note that there is no mention of water and that CO_2 fixation is just one way (although often the major way) in which the trapped energy may be used.

Most of this chapter is concerned with photosynthesis because it is a process unique to plants but plants do, of course, respire.

◇ If photosynthesis generates ATP, why do plants need to respire? Try to think of *two* reasons.

◆ 1 The obvious reason is that photosynthetic ATP is produced only in green cells and only in the light; during the hours of darkness and for non-green cells (e.g. in roots), energy must be supplied through respiration using as a substrate the carbon compounds produced by green cells in the 'synthesis' part of photosynthesis.

2 The other reason why all plant cells, even green ones, respire, is that as well as their role in energy release, the central pathways of respiration — glycolysis and the TCA cycle — also provide many essential precursors for biosynthesis.

In this chapter we shall consider photosynthesis and respiration from two points of view: the physiological one, where the main questions asked are about the internal and external factors influencing the overall rate of the process and how the essential materials (CO_2 and O_2) get to the places where they are needed; and the biochemical one, where the chief concern is the mechanism and control of the process in molecular terms. Particular emphasis is placed on the flexibility of photosynthesis, which allows plants to grow in habitats ranging from the arctic tundra to the Sahara desert and the floor of tropical rainforests. The 'strategies' that have evolved under such different conditions of temperature, water supply and light are among the main themes of this chapter.

2.2 THE BALANCE OF PHOTOSYNTHESIS AND RESPIRATION AND THEIR MEASUREMENT

In Section 2.1 the global balance of photosynthesis and respiration was emphasized (Figure 2.1), but for a *growing* plant there is obviously no such balance.

◇ Explain why not.

◆ Growth requires a net gain of energy, so that inputs (photosynthesis) must exceed losses* (in respiration).

---

* Note that although the biochemical process of respiration releases chemical energy (as ATP), this ATP is then *used* and the energy is ultimately dissipated as heat.

The total energy or CO_2 'fixed' is called **gross photosynthesis (GP)**, respiration (R) being the total released. For growing plants of alfalfa (a forage crop related to clover, see marginal drawing) the ratio of GP to R averages about 7:1 for leaves during daylight, reaches a maximum around 9:1 in full sun at noon and averages 2.5:1 for whole plants over a growing season. The units here are joules or grams CO_2 per unit leaf or ground area covered by plants per unit time, so GP is typically much larger than R but varies with the time over which measurements are made. What most affects the rate at which plants grow or the yield of a crop is the *difference* between GP and R, the net gain or **net photosynthesis (NP)** and

$$NP = GP - R$$

so that

$$GP = NP + R \qquad (2.1)$$

Because of its direct relationship to growth rate and also for practical reasons, data about photosynthesis are usually given in terms of NP rather than GP. The practical reasons relate to the way in which measurements are made and, to understand them, consider first the simple summarizing equations for photosynthesis and aerobic respiration, where carbohydrate (general formula CH_2O) is taken as the end-product of the first and the substrate of the second process:

$$CO_2 + H_2O + energy \longrightarrow (CH_2O) + O_2 \qquad \text{photosynthesis} \qquad (2.2)$$

$$(CH_2O) + O_2 \longrightarrow CO_2 + H_2O + energy \qquad \text{respiration} \qquad (2.3)$$

To a physiologist, the obvious way of measuring either process is to monitor the uptake or release of one of the gases involved, i.e. **gas exchange techniques**, but think what this involves. Photosynthesis can be measured only in the light and, because GP normally exceeds R, you would be measuring CO_2 uptake or O_2 release. Figure 2.2 shows a typical set-up for monitoring CO_2 uptake in the light.

5 cm

Alfalfa or lucerne (*Medicago sativa*)

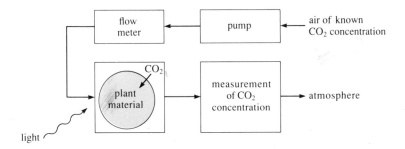

Figure 2.2 A gas exchange system used for measuring photosynthesis for a known weight or area of plant material. CO_2 concentration can be measured with, for example, an infrared gas analyser.

Oxygen output is often measured using an oxygen electrode, which measures $[O_2]$, and oxygen microelectrodes can even be inserted into a leaf to measure photosynthesis of very localized areas.

◇ Suppose that O_2 production by green tissue in the light was $10\,cm^3$ per gram tissue per minute. What does this measurement represent — what does it *mean*?

◆ It means that the *difference* between gross photosynthesis, expressed as total O_2 production, and respiration in that tissue, expressed as total O_2 taken up, is $10\,cm^3$ per gram tissue per minute. In other words it is a measure of net photosynthesis.

Uptake of CO_2 or release of O_2 in the light is, in fact, the operational definition of *NP*. Which leaves us with the problem of how to measure *GP* by gas exchange. To do this you need a measure of respiration in the light, but since this cannot be done directly the usual procedure is to measure *R* in the *dark* when, of course, photosynthesis has stopped.

◇ Suppose the tissues considered in the question above took up O_2 in the dark at a rate of $2\,cm^3$ per gram tissue per minute. What is the apparent rate of *GP* in this tissue?

◆ The sum of net photosynthesis and (dark) respiration, $10 + 2 = 12\,cm^3\,O_2$ per gram tissue per minute.

We use the word 'apparent' in this question because the basic underlying assumption — that the rate of respiration in the dark is identical to that in the light — is not always true. For certain plants under certain conditions (notably high light levels and high temperature), a sequence of interfering reactions, known collectively as **photorespiration**, operates only in the light and results in a considerable uptake of O_2 and release of CO_2. The nature and significance of photorespiration, which occurs only in green tissues, are considered in later sections, but the points to note here are that it bears no biochemical relationship at all to 'normal' or 'dark' respiration 'and that it makes reliable estimates of respiration in the light, and therefore of gross photosynthesis, difficult and sometimes impossible. What is more important from the plant's point of view, photorespiration at temperatures above 30–35 °C may reduce net photosynthesis by 50–60%. Question 3 (at the end of this section) is about this. If *NP* falls to zero, when respiration rates are very high or when *GP* is very low, then clearly $GP = R$ and this is called the **compensation point** — photosynthesis just compensates for respiration.

2.2.1 Other ways of measuring photosynthesis

Gas exchange techniques provide one way of measuring net photosynthesis, dark respiration and (if photorespiration is low) gross photosynthesis. However, there are two other techniques for measuring rates of photosynthesis and we describe these briefly.

$^{14}CO_2$ incorporation

The principle involved here is quite simple. Whole plants or parts of them are enclosed in a chamber into which CO_2 labelled with the isotope ^{14}C is introduced, and after a suitable period in the light, the material is killed, dried and the amount of ^{14}C per unit dry weight or unit area of leaf is measured. There are, however, problems of interpretation because the meaning of results varies depending on how long the material is exposed to $^{14}CO_2$.

◇ Suppose that batches of 10 discs cut from tobacco leaves are incubated under identical conditions with $^{14}CO_2$ for (a) 1 min and (b) 1 h. The results obtained, as counts per minute incorporated per disc per minute, are 70 for (a) and 40 for (b). Why do these results differ? What was being measured in (a) and (b)?

◆ For (a), the short labelling time, gross photosynthesis is probably measured because very little of the CO_2 fixed will have been respired. For (b), however, some of the $^{14}CO_2$ fixed *will* have been respired — although not necessarily all, because some may have been stored as starch — so something close to net photosynthesis is measured. This is why a lower rate of fixation is found with longer labelling times.

The labelling time, the species used and its respiration rate all affect results with this technique so, although it is quick and convenient, data should be interpreted carefully. It is also difficult to obtain *absolute* rates of photosynthesis: you may be able to say, for example, that *NP* in leaf X proceeds twice as fast when the supply of light is doubled, i.e. obtain *relative* or *comparative* estimates of rate, but not what the rates actually are in terms of mg CO_2 fixed per leaf per minute.

Harvest techniques

Since the increase in energy content of growing plants comes from carbon fixation, a simple method for estimating net photosynthesis over a relatively long period (usually weeks or months) is by drying, weighing and combusting in a bomb calorimeter (to measure energy content as heat released) a sample of plants at one time, repeating this at a later date and noting the difference. This is the principle of harvest techniques, which are mainly used by people who are interested in plant growth or production rather than photosynthesis as such. Although simple, there are lots of problems — imagine the difficulty of removing and cleaning up entire root systems, for example — but we mention them so that you have a clearer idea of the full range of methods available.

Summary of Section 2.2

1 Gross photosynthesis (*GP*), the rate at which total energy or CO_2 is fixed, must be distinguished from net photosynthesis (*NP*), the difference between *GP* and respiration, *R*:

$$GP = NP + R$$

2 The only way to measure *R* and the most common way of measuring *GP* and *NP* is by gas exchange techniques — measuring the release or uptake of O_2 or CO_2. *NP* is the release of O_2 or uptake of CO_2 in the light, *R* has to be measured as the uptake of O_2 or release of CO_2 in the dark, and *GP* is their sum.

3 For some plants this method of estimating *GP* is invalid, especially at high light levels or temperatures, because respiration in the dark is much lower than in the light. This is because an additional process, photorespiration, occurs in the light.

4 When $GP = R$, and hence *NP* is zero, a plant is said to be at its compensation point.

5 Measuring the incorporation of $^{14}CO_2$ provides another technique for estimating relative rates of photosynthesis: *NP* is measured for incubations longer than a few minutes and *GP* for shorter incubations. Harvest techniques can be used to estimate *NP* for growing plants over long periods.

Question 1 (*Objectives 2.1 and 2.2*) Classify statements (a)–(d) as generally true, true for some situations or organisms, or generally false.

(a) Photosynthesis can be regarded, overall, as the reverse of aerobic respiration.

(b) Photosynthesis is the only process by which organisms on Earth can tap a source of energy outside the Earth.

(c) In green plant tissues, 'dark' respiration occurs only in the dark and photorespiration only in the light.

(d) A plant at its compensation point will probably not grow.

Question 2 (*Objectives 2.1 and 2.3*) For each of (a)–(e), select appropriate methods of measurement from (i)–(v).

(a) net photosynthesis of a wheat field during one growing season

(b) the effect on net photosynthesis in lettuce leaves of increasing temperature from 20 °C to 30 °C

(c) a comparison under identical conditions of net photosynthesis in young and mature leaves of an oak tree

(d) respiration in roots

(e) gross photosynthesis in green leaves.

(i) gas exchange in the light only

(ii) gas exchange in the dark only

(iii) gas exchange in both light and dark

(iv) $^{14}CO_2$ incorporation

(v) harvest method.

Question 3 (*Objective 2.4*) The data in Table 2.1 were obtained using a gas exchange method with a 10 min incubation period. Measurement (3) was carried out after dipping leaves in a specific inhibitor of photorespiration. For each temperature:

(a) what are the values of net photosynthesis?

(b) what are the values of photorespiration?

(c) what are the values of gross photosynthesis and what would these be if measurement (3) were not available?

Table 2.1 CO_2 uptake and release for tobacco leaves at two temperatures (Data for Question 3)
Values are in μmol CO_2 g^{-1} fresh wt h^{-1}.

	CO$_2$ uptake or release	
	25 °C	35 °C
(1) CO$_2$ release measured in the dark	10	20
(2) CO$_2$ uptake measured in the light	75	65
(3) CO$_2$ uptake measured in the light when photorespiration is inhibited	90	180

2.3 THE PHYSIOLOGY OF PHOTOSYNTHESIS AND RATE LIMITATION

In this section we consider the kinds of environmental factor that affect rates of photosynthesis and respiration, concentrating in particular on those that may *limit* the rate of either process. When you think of the enormous range of environments in which plants grow it is hardly surprising that rate-limiting factors differ within this range. A variety of strategies have evolved that maximize net photosynthesis, reducing the impact of rate-limiting factors in particular environments. What these strategies mean at a *physiological* level is described here and what they mean at a *biochemical* level is considered in Sections 2.4 and 2.5.

2.3.1 Rate limitation of photosynthesis

In a hungry world rates of photosynthesis are a major concern because they relate to crop growth rates and, therefore, food production. Cultivated land produces about 9 billion metric tons (10^9 tonnes) of dry organic matter each year, which is roughly 5% of total world production. What matters in this context is net photosynthesis so the main topics of this section and the following one (Section 2.3.2) are the factors that affect rates of *NP*, the **rate-limiting factors**. It is important to appreciate, however, that these reflect the balance between gross photosynthesis and respiration (Section 2.2): *NP* may increase either because *GP* rises or because *R* falls (or both). Whenever we refer simply to 'rate of photosynthesis', assume that the first holds with rates of *NP* and *GP* varying in parallel — but watch out for situations where this is not true!

Net photosynthesis is most commonly limited by outside conditions (environmental factors) but occasionally conditions inside the cell — levels of certain enzymes, for example — become limiting and this situation will be discussed in later sections. Our immediate concern is to sort out which environmental factors influence most strongly the rate of *NP*.

Environmental factors

From what you know about rate-limiting factors you should be able to work out from Figure 2.3 the three major environmental factors that may limit the rate of photosynthesis (gross or net).

◇ What are these factors?

◆ The level of CO_2, the light level and the temperature.

When a process increases linearly with increase in a particular factor, that factor is rate-limiting — so the initial sloping parts of the graphs show that CO_2 is limiting here. When the graphs flatten, CO_2 supply is saturated and no longer limits photosynthesis but, by comparing curves (1) and (2) (same light level but temperatures differ), it is clear that temperature may be limiting; and by comparing curves (1) and (3) (same temperature but light levels differ), the importance of light can be deduced.

Note that the correct terms to describe level or flux of light are **irradiance**, if using energy units, and photon flux density, if measuring numbers of light quanta. These terms and the units of measurement are described more fully in Box 2.1, which you should read now.

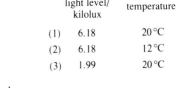

	light level/ kilolux	temperature
(1)	6.18	20 °C
(2)	6.18	12 °C
(3)	1.99	20 °C

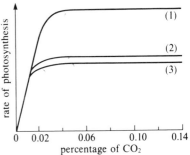

Figure 2.3 The relationship between the rate of photosynthesis and CO_2 concentration in *Hormidium* for different light levels and temperatures.

Box 2.1 Light and its units of measurement

Light in the broadest sense, is electromagnetic radiation with wavelengths in the range 200 nm to about 40 000 nm (1 nm = 10^{-9} m). The wavelengths of visible light (i.e. those perceived by the human eye and commonly referred to as 'light') are in the range 400–740 nm, shorter wavelengths being described as ultraviolet and longer ones as infrared. In the visible range, different wavelengths of light are perceived by the human eye as having different colours (see Figure 2.4). Light absorbed by plants and used in photosynthesis (*photosynthetically active radiation* or PAR) is in the range 400–700 nm.

The dual nature of electromagnetic radiation (wave and particle) means that, as well as having wave characteristics, light also behaves as if it were divided into discrete 'packets' of energy. These are termed quanta (singular quantum) or photons. The amount of energy associated with a photon varies and depends on the wavelength (see Figure 2.4; where the units given are kilojoules (kJ) per mole of quanta: the shorter the wavelength, the greater the energy per quantum.

Light intensity, i.e. the amount of light reaching a surface in unit time, can be measured in three ways: in terms of radiometric units, photon flux density or photometric units.

(a) *Radiometric units* are related to the radiant energy reaching unit area in unit time, and the correct term to describe this is *irradiance*. When concerned with a particular kind of radiation (i.e. of a particular wavelength), the wavelength should be specified, because energy varies with wavelength.

The units here may be J m^{-2} s^{-1} or watts (W) m^{-2} (1 W = 1 J s^{-1}).

(b) *Photon flux density* relates to the number of quanta reaching a surface in unit time and number is expressed as moles of quanta or (an earlier term) einsteins. These units derive from the photochemical principle that absorption of one quantum activates one molecule. Since a mole of any substance contains the same number of molecules (Avogadro's number, 6.023×10^{23}), a mole of quanta is a fixed numer of quanta. Photon flux density, which is sometimes referred to as photon irradiance, is measured in units of moles of quanta (often abbreviated to just moles) m^{-2} s^{-1} or as einsteins m^{-2} s^{-1}. Again the wavelengths of quanta should be specified but, when talking about photosynthesis, they can be assumed to lie between 400 and 700 nm.

(c) *Photometric units* are related to the brightness of light as perceived by the human eye, and the units are quite different and not easily convertible into radiometric units. The standard photometric unit of light intensity or, more correctly, illuminance, is a *lux* (and you may come across an older, non-SI unit, the foot candle). A lux can be defined in terms of a luminous flux (measured in lumens) per unit area:

1 lux = 1 lumen m^{-2}

In broad terms, a lumen is equivalent to a watt, but whereas the latter is a measure of total energy falling on a surface in unit time, the lumen measures energy weighted by a 'visibility' factor that depends on the spectral

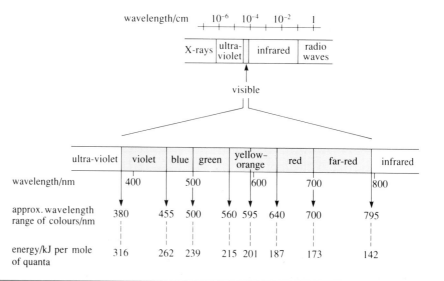

Figure 2.4 Part of the electromagnetic spectrum showing the wavelengths of light. The visible portion is expanded below to show the wavelength bands of colours detected by the human eye and the energy associated with photons (quanta).

sensitivity of the human eye. Figure 2.5 shows that the eye is most sensitive to a wavelength of 555 nm (point A), when the visibility factor is maximum. On the other hand the light intensity (illuminance) for radiation of wavelength B will be zero, because visibility is zero (these longer wavelength are in the far-red/near infrared). However, radiation of wavelength B falling on a surface will have a positive irradiance value.

Because of this visibility weighting, a light source rich in barely visible radiation (such as fluorescent light rich in the blue–violet wavelengths) can have a high irradiance for 400–700 nm but a low illuminance (in lux).

Radiometric or photon flux density units are now standard but some representative measurements of sunlight intensity in all three types of units are given in Table 2.2.

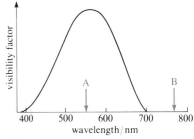

Figure 2.5 The relationship between the visibility of light to the human eye and the wavelength of light.

Table 2.2 Measurements of sunlight in different units

Conditions	Photometric units kilolux	Radiometric units $J\,m^{-2}\,s^{-1}$	Photon flux density (400–700 nm) $\mu mol\,m^{-2}\,s^{-1}$
full sun (noon, cloudless)	100–130	750–1 000 (total radiation) 400–520 (400–700 nm)	1 840–2 400
overcast (noon, cloudy)	14–16	55–65 (400–700 nm)	250–300
deeply shaded (forest floor)	0.8	3 (400–700 nm)	15

An important factor that limits photosynthesis indirectly, through effects on CO_2 supply, is *water*. CO_2 enters and water vapour is lost from leaves through stomatal pores in the epidermis (discussed fully in Chapter 5). These pores are able to close up if water is in short supply, which, of course, cuts off the supply of CO_2, so that plants growing in arid environments such as deserts constantly face the dilemma of desiccation or starvation. One of the interesting strategies that has evolved in one group of desert plants and circumvents this problem is described in Section 2.5.

Finally, there is one rather surprising environmental factor, *oxygen*, which has strong inhibitory effects on net (but not gross) photosynthesis in certain plants and under certain conditions.

◇ Decide from Figure 2.6 what these conditions are.

◆ High temperatures — above 20–25 °C for this plant — and high irradiance.

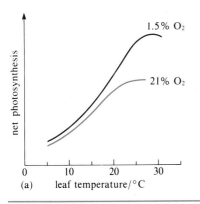

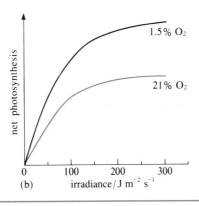

Figure 2.6 Net photosynthesis in a leaf of *Atriplex patula* measured at atmospheric CO_2 concentration, with two concentrations of oxygen. (a) Variation with temperature, measured at light saturation. (b) Variation with irradiance, measured at 27 °C.

These conditions are identical to those described in Section 2.2 as promoting photorespiration and as long ago as 1966 it was shown that this is exactly what oxygen does. Bear in mind that 21% O_2 is the normal level in air, but within the tissues or close to the surface of actively photosynthesizing plants (and with high light and temperature, photosynthetic rates are likely to be high) the concentration can be much higher; so O_2 really is an environmental variable. The full story of photorespiration and oxygen took over 15 years to work out and we shall be telling it later, beginning in Section 2.5.2.

2.3.2 Photosynthesis in different environments

The concentration of CO_2 in the atmosphere is about 350 p.p.m.* (equivalent to 0.035% or a partial pressure of 35 Pa) but conditions of light and temperature in habitats where photosynthesis occurs are much more variable. Leaf temperatures range from below 0 °C in the Arctic to above 50 °C in hot deserts and irradiances from 3 J m^{-2} s^{-1} to 500 J m^{-2} s^{-1} between deeply shaded and open tropical habitats, respectively (see Table 2.2 in Box 2.1). How is this flexibility possible? We examine here some of the physiological characteristics that allow plants to grow in such a wide range of environments.

Shady habitats

Some plants, such as wood sorrel (*Oxalis acetosella*), nearly always grow in shade — on the forest floor, for example — where light is commonly rate-limiting for photosynthesis. These **shade plants** differ in two major respects from **sun plants**, such as daisies (*Bellis perennis*), that grow in open habitats (Figure 2.7).

$\diamond$ From Figure 2.7, how does net photosynthesis differ in sun and shade plants (a) at low irradiances and (b) at high irradiances?

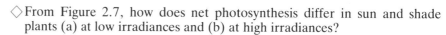 (a) At low irradiances, shade plants have higher rates of net photosynthesis than sun plants and the point at which $NP = 0$ (i.e. GP and R balance: Equation 2.1) is lower. (b) At high irradiances, shade plants have lower rates of net photosynthesis than sun plants.

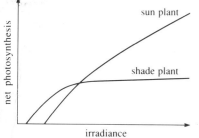

Figure 2.7 Typical photosynthetic responses of sun and shade plants to increasing irradiance.

Answer (a) is the main reason why shade plants can survive and grow in much shadier habitats than sun plants. The light level at which $GP = R$ is called the **light compensation point** and its low value in shade plants results from two features.

First, and this is usually the main reason, the leaves of shade plants have *very low respiration rates*. They have fewer cells per leaf and lower concentrations of proteins than sun leaves, making them, in energetic terms, very cheap to run. However, a second feature is that, at low irradiance, shade plants *absorb the available light with remarkable efficiency* so that with very few leaf cells, few photons are 'wasted' and GP is maximized. Biochemical characteristics (explained in Section 2.4) are involved but some plants of the rainforest floor also have a lens-like arrangement in their upper epidermal cells which focuses light onto the mesophyll cells below: these plants have a beautiful iridescent sheen, which you may have seen in some house plants (e.g. *Hemigraphis* spp.).

* This average concentration of CO_2 is slowly increasing due to human activities and is predicted to exceed 400 p.p.m. during the next century.

On the other hand, shade plants perform very inefficiently at high irradiances compared with sun plants. They rapidly become light-saturated (Figure 2.7) and, furthermore, photosynthesis is actually *inhibited* in full sunlight. This phenomenon of **photoinhibition** occurs in all plants that have grown at low light levels and are suddenly exposed to high levels and it arises because of damage to the photosynthetic apparatus, which in time can often be repaired. Shade plants are not only particularly sensitive to photoinhibition but also less able to repair the damage, and long-term exposure to strong light often causes further and irreversible damage due to **photo-oxidation** of chloroplast pigments. Pigments are destroyed by this process so that leaves become bleached and eventually die.

In obligate shade species, which are never found in sunny habitats, the characteristics just described are genetically fixed and plants never adapt to high irradiances. But many plants that grow well in the open can adapt to shade if they *develop* in shady conditions. This is illustrated for *Atriplex patula* in Figure 2.8: plants raised at low light levels behave subsequently like shade plants, having low rates of net photosynthesis when exposed to high irradiance but relatively high rates at low irradiance. Such environmentally-induced adaptation is described as **acclimation** and even leaves on the same tree may show acclimation to different light levels, with inner or lower shade leaves and outer sun leaves. This difference is reflected in their anatomy (Figure 2.9), sun leaves having a thicker layer of palisade mesophyll (Chapter 1) than shade leaves. Here lies part of the explanation for the higher rates of net photosynthesis at high irradiance in sun leaves: they intercept and absorb a higher proportion of incident light. Equally, however, thick leaves are

Atriplex patula, a plant commonly used in research on photosynthesis.

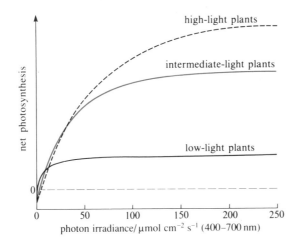

Figure 2.8 Light-saturation curves for *Atriplex patula* raised at irradiances of $200\ \mathrm{J\,m^{-2}\,s^{-1}}$ (high), $63\ \mathrm{J\,m^{-2}\,s^{-1}}$ (intermediate) and $20\ \mathrm{J\,m^{-2}\,s^{-1}}$ (low).

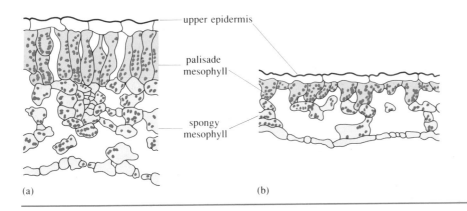

Figure 2.9 Transverse sections of *Impatiens parviflora* leaves. (a) Sun leaf. (b) Shade leaf.

expensive to maintain and their high respiratory costs are a disadvantage at low irradiance. These structural differences occur generally between sun and shade plants.

Open habitats: temperate and tropical regions

Sun plants from open habitats clearly utilize high light levels much better than shade plants (Figure 2.7) but some are more efficient than others. In temperate latitudes the leaves of most plants still show light saturation at around 25% of full sunlight but in the tropical lowlands — where irradiance may be almost twice as great as in the temperate zones — leaves of many herbaceous and shrubby species do not show light saturation up to full sunlight. These tropical species, therefore, have very high rates of net photosynthesis at high irradiances and include some important crop plants such as maize and sugar-cane. They belong to a group known as **C_4 plants**, in contrast to the majority of other species which are known as **C_3 plants**; the names relate to the number of carbon atoms in the first product of CO_2 fixation (discussed in Section 2.5). Responses of typical C_3 and C_4 plants to increasing light are shown in Figure 2.10.

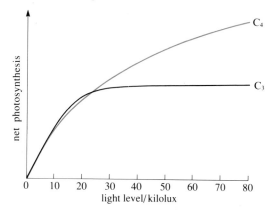

Figure 2.10 Typical effects of increasing light levels on net photosynthesis in C_3 and C_4 plants.

This shows clearly that in C_3 plants some factor *other* than light limits net photosynthesis at high light levels. In fact they are CO_2-limited: there is plenty of light but the rate of supply of CO_2 to chloroplasts is too slow.

Somehow C_4 plants overcome this CO_2 limitation — they use the available CO_2 more efficiently — and consequently have higher rates of net production at high light levels than C_3 plants, although they perform no better (and sometimes slightly worse) at lower light levels. Part of the explanation for these differences can be obtained by comparing the responses of C_3 and C_4 plants to changes in temperature and oxygen level and this is done for two species of *Atriplex* in Figure 2.11.

◇ Study Figure 2.11, compare it with Figure 2.6 and decide what process is missing from *A. rosea* (the C_4 species) that explains its superior net photosynthesis at high irradiance.

◆ Photorespiration — net photosynthesis is *not* inhibited by high $[O_2]$ at high temperatures and high irradiances.

At high irradiance and normal (atmospheric) concentrations of O_2 and CO_2, *A. rosea* (C_4 green line) shows an increasing rate of net photosynthesis up to 30 °C, whereas *A. patula* (C_3 black line) reaches a much lower maximum rate at around 25 °C (Figure 2.11a). This difference largely disappears when the O_2 level is reduced (Figure 2.11b) — the C_3 species now behaves much like

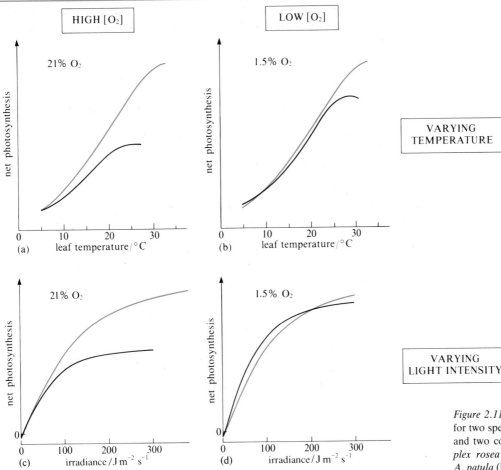

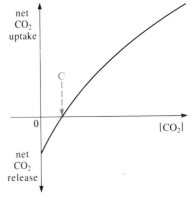

Figure 2.11 Rates of net photosynthesis for two species of *Atriplex* at 0.03% CO_2 and two concentrations of oxygen. *Atriplex rosea* (green) is a C_4 species and *A. patula* (black) a C_3 species. (a) and (b) High irradiance and varying leaf temperature. (c) and (d) Leaf temperature 27 °C and varying irradiance.

the C_4. Earlier (Section 2.3.1, Figure 2.6a) we explained such altered behaviour as being due to the absence of photorespiration at low $[O_2]$, so the C_4 species behaves as though it has no photorespiration even at high $[O_2]$. This is confirmed by comparing Figures 2.11c and d with Figure 2.6b: at low $[O_2]$ (no photorespiration), net photosynthesis in the C_3 species matches that of the C_4 up to high irradiances. In short, the reason why C_4 species perform so efficiently at high temperatures and high irradiances is that carbon losses due to photorespiration do not occur.

Another consequence of this is that a C_4 plant has a very low **CO_2 compensation point**, which is defined as the CO_2 concentration at which gross photosynthesis and respiration balance (Figure 2.12). Table 2.3 shows typical values and those for C_4 plants are close to zero.

Table 2.3 Carbon dioxide compensation points for various plants

Species	Type of plant	CO_2 compensation point (p.p.m.)
sugar-cane (*Saccharum officinarum*)	C_4	≈ 0
maize (*Zea mays*)	C_4	1.3 ± 1.2
Chlorella (a unicellular green alga)	(C_3)	<3
sunflower (*Helianthus annuus*)	C_3	53
barley (*Hordeum vulgare*)	C_3	55–65
temperate tree (*Acer platanoides*)	C_3	55

Figure 2.12 The CO_2 compensation point (C).

What this means is that if a C_4 plant was enclosed in a transparent chamber and allowed to photosynthesize in the light, *NP* would remain positive until virtually all the CO_2 in the chamber had been used up. Now this is surprising even in the absence of photorespiration because dark respiration occurs at similar rates in C_3 and C_4 plants. It suggests that C_4 plants scavenge every scrap of CO_2, including that released through dark respiration. The mechanism by which they do this is described in Section 2.5 but an essential feature is that CO_2 taken up over the whole leaf surface is *concentrated* at the site of carbon fixation in the chloroplasts of comparatively few cells.

Table 2.3 also shows that *Chlorella* (like many other unicellular algae, cyanobacteria and larger aquatic plants too) has a low CO_2 compensation point, *although* classed, biochemically, as a C_3 species. The reason seems to be that these plants are able, as are C_4 plants, to concentrate CO_2 at the site of carbon fixation — in this case by absorbing inorganic carbon from water as the bicarbonate ion (HCO_3^-) and releasing CO_2 inside the cell, assisted by the enzyme carbonic anhydrase:

$$HCO_3^- + H^+ \rightleftharpoons H_2O + CO_2 \tag{2.4}$$

This can raise intracellular levels of CO_2 by as much as 1 000-fold over those of the medium.

◇ Does their low CO_2 compensation point mean that photorespiration does not occur in bicarbonate-using plants?

◆ No: it means that a key symptom of photorespiration — a high CO_2 compensation point — cannot be *detected*. This could be *either* because photorespiration does not occur *or* because any CO_2 released through photorespiration is immediately re-fixed. Without more information you cannot be certain (but actually the second alternative is now thought to be correct).

It appears, therefore, that C_4 plants (and bicarbonate-using C_3 plants) have a strategy that involves concentration of CO_2, which has the effect of abolishing or negating photorespiration; CO_2 is used very efficiently and does not become limiting at high irradiance and temperature. Rates of net photosynthesis and growth are usually far superior to those of C_3 plants in hot, sunny conditions, which is where C_4 plants tend to grow naturally. So why are not all plants in open habitats of the C_4 type? There are basically two reasons:

1 In temperate regions the climate is not hot and sunny enough; maximum irradiance is lower than in the tropics and with long but comparatively cool and cloudy summer days, C_3 plants perform as well as or better than C_4 plants, most of which do particularly badly if temperatures are low (e.g. 13 °C or less) during the growth period.

2 No large trees are of the C_4 type — even in the tropics — probably because photosynthesis for the tree *as a whole* is limited by light and not CO_2. Only the outer leaves are exposed to full sun and the many layers of inner leaves are shaded: the C_4 strategy confers no advantages in this situation.

In terms of their photosynthetic capacity, therefore, herbaceous or shrubby C_3 and C_4 plants are adapted to different environments: a cool and/or low-light climate for C_3 plants and a warm, high-light climate for C_4 plants. This physiological adaptation of plants to their normal environment is an important general principle and we describe below one further example to illustrate it.

Temperature adaptation

Apart from effects on photorespiration and net photosynthesis, temperature affects gross photosynthesis in two distinct ways. First, it influences the rates of chemical reactions — as for any other reaction pathway — and, if temperature is the limiting factor, there is an *optimum temperature* at which the rate of *GP* is maximal. Second, extremely high or low temperatures may damage the photosynthetic machinery (usually by damage to membranes or membrane-bound proteins), so plants are characterized by **threshold temperatures**, above or below which irreversible damage occurs. Typically, there is a correlation between the optimum and threshold temperatures of plants and the thermal regime where they grow naturally. You can see this in Figure 2.13.

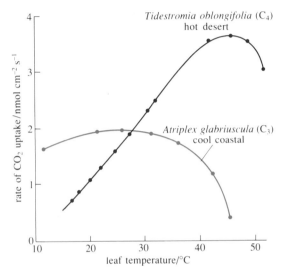

Figure 2.13 Comparison of the temperature dependences of photosynthesis by whole plants of *Tidestromia oblongifolia* during the summer in Death Valley, California, and *Atriplex glabriuscula*, grown under a temperature regime simulating that of its native coastal habitat.

◇ What are the optimum temperatures and the approximate threshold temperatures for the two species in Figure 2.13?

◆ For *T. oblongifolia* (a C_4 species which grows in summer in the baking heat of Death Valley, California), the temperature optimum is about 46 °C; the upper threshold is above 52 °C and the lower is well below 20 °C (thought to be around 10 °C). For *A. glabriuscula* (a C_3 species native to cool, coastal environments), the optimum is about 23 °C and the upper and lower threshold temperatures are around 46 °C and well below 10 °C, respectively.

Rather flat temperature–response curves, similar to that for *A. glabriuscula* (Figure 2.13), are typical for plants from cool or cold environments, with optima in the range 15–25 °C. However, just as plants show acclimation to different light environments, depending on the conditions under which they are grown (Figure 2.8), so also do they show temperature acclimation. Ability to do this varies widely between species but is best developed when there are large changes in temperature over the growing season. At Barrow, Alaska, for example, the temperature increases by 14 °C from early spring to the warmest time and the optimum temperature for photosynthesis of three mosses increased from 12–13 °C to 19–21 °C over this period. Evergreen desert shrubs that are active in both cool winters and hot summers show a similar shift, illustrated for *Atriplex lentiformis* (a C_4 species) and *Nerium oleander* (a C_3 species) grown at different temperatures in Figure 2.14a and b. By contrast, species such as *A. glabriuscula* and *T. oblongifolia* (Figure 2.13) which have a limited growing season with relatively constant temperature,

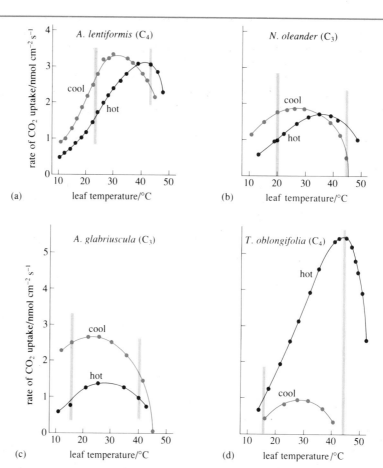

Figure 2.14 Effect of growing under a 'cool' and 'hot' temperature regime on the rate and temperature dependence of light-saturated photosynthesis. For the 'cool' regime (green curve), the pale green bar shows the daytime temperature experienced; for the 'hot' regime (black curve), the grey bar shows the daytime temperature. (a) and (b) Curves for two evergreen desert shrubs active all the year round. (c) and (d) Curves for two species active only in summer.

show little ability to acclimate (Figure 2.14c and d): when grown at a high temperature, the cool-climate *A. glabriuscula* does not have an improved photosynthetic performance at high temperatures and *T. oblongifolia* performs miserably at all temperatures when grown under cool conditions.

The important conclusions from Section 2.3.2 are that (a) environmental factors may interact in complex ways to influence the rate of photosynthesis, both gross and net; and (b) different environmental factors may limit the rate of photosynthesis in different environments or for different plants in the same environment — depending on their physiology, biochemistry and life history. In general, whatever the major stress or dominant rate-limiting factor for photosynthesis in a particular environment, native plants tend to have adaptations that minimize its effects. This does *not* mean that all plants grow in optimal conditions for photosynthesis. Bluebells (*Hyacinthoides non-scriptus*), for example, occur typically in woodland but actually attain higher rates of photosynthesis and growth in open habitats. They are usually excluded from open habitats by competition from other species which cannot, however, tolerate as much shade as bluebells.

Summary of Section 2.3

1 The rate of photosynthesis (gross or net) is determined mainly by environmental factors, chiefly light (irradiance), the availability of CO_2, and temperature. Water supply has indirect effects, by influencing the availability of CO_2, and oxygen affects net photosynthesis by stimulating photorespiration, especially at high temperatures.

2 Plants show adaptations to different environments with respect to photo-synthesis. In shady habitats, where light is rate-limiting, shade plants have low rates of respiration and harvest light very efficiently at low irradiances. Consequently they have lower light compensation points and, in shade, grow better than sun plants. However, shade plants have a poor photosynthetic capacity in the open. Light saturation occurs at a relatively low irradiance and they show rapid photoinhibition with high susceptibility to irreversible photo-oxidation. Some plants or leaves on a plant may adapt to shade (light acclimation) if grown under low irradiance. Their physiology and leaf anatomy then resemble those of obligate shade plants.

3 When leaves are exposed to full sun, the most common limiting factor for photosynthesis is CO_2 concentration within cells ($[CO_2]$). Two broad types can be distinguished among herbaceous or shrubby species: C_3 plants tend to become light-saturated and limited by $[CO_2]$ at relatively low irradiances, photorespiration (and inhibition by O_2) severely reduces NP at high irradiances, and CO_2 compensation points are high. C_4 plants do not become light-saturated nor limited by $[CO_2]$ even in full sunlight; they do not show photorespiration nor inhibition by O_2, and CO_2 compensation points are very low. The main factor responsible for these characteristics of C_4 plants is their ability to concentrate CO_2 at the site of carbon fixation. Certain aquatic plants that use bicarbonate ions as a carbon source have the same ability to concentrate CO_2 within cells and, although they are biochemically designated as C_3, possess physiological characteristics typical of C_4 plants. In hot, high-light conditions (e.g. open, tropical habitats), the C_4 strategy gives superior photosynthetic performance, except for large trees, where many of the lower leaves are heavily shaded. In cool moderate-light conditions (e.g. temperate regions), the C_3 strategy gives as good or better photosynthetic performance.

4 Both the optimum and threshold temperatures for gross photosynthesis vary between species. They are lower for plants growing in cool climates compared with plants in hot climates. Temperature acclimation is common, especially for plants that experience wide variations in temperature during the growth season.

Question 4 (*Objective 2.1*) Are statements (a)–(d) true or false?

(a) A factor is rate-limiting for a process when an increase in that factor causes no increase in the rate of the process.

(b) Shade plants have high rates of net photosynthesis in the shade and sun plants have similar high rates in full sun.

(c) The CO_2 compensation point of a C_3 plant will decrease as the $[O_2]$ decreases, all other conditions remaining constant.

(d) C_4 species are sun plants and C_3 species are shade plants.

(e) Plants grown under particular environmental conditions always show acclimation to those conditions.

Question 5 (*Objectives 2.1 and 2.5*) What is likely to be rate-limiting for:

(a) gross photosynthesis in a tomato (C_3 plant) growing in full sunlight in July in an English garden at 28 °C?

(b) net photosynthesis in a subtropical variety of maize, *Zea mays*, (C_4 plant) growing under the conditions described for (a)?

(c) net photosynthesis in a plant from the floor of a tropical rainforest immediately after transfer to full tropical sun?

(d) gross photosynthesis for snowdrops (*Galanthus nivalis*) growing in a deciduous woodland on a sunny day in January (temperature 4 °C)?

Question 6 (*Objectives 2.5 and 2.18*) Table 2.4 shows the distribution of C_3 and C_4 grasses and sedges at different altitudes (but the same latitude) in open habitats in Java. Suggest an explanation for this distribution pattern. What factor(s) probably limits net photosynthesis of the plants at 3 000 m?

Table 2.4 Distribution of C_3 and C_4 grasses and sedges at four open habitat sites in Java*

Site (all open)	Height above sea level /m	No. of species in 500 m² areas	
		C_4	C_3
seaside	1	15	0
riverside	200	15	1
plateau	2 000	3	6
mountain top	3 000	0	7

*Data of Hofstra, J.J. *et al.* (1972) *Annales Bogoriensis*, **5**, pp. 143–57

Question 7 (*Objective 2.6*) Two species, one C_3 and the other C_4, grow in the same hot desert environment in North America. (a) What similarities and differences would you expect to see in the photosynthetic responses to temperature of these species? (b) Under what environmental conditions would you expect the differences to disappear?

2.4   THE BIOCHEMISTRY OF PHOTOSYNTHESIS: THE LIGHT REACTIONS

Section 2.3 described the physiology of photosynthesis, at the level of whole plants or leaves but, for a deeper understanding, you have to go to the molecular level. In eukaryotes, the central reactions of photosynthesis take place in the organelles called **chloroplasts** and there are two distinct stages. First, light of certain wavelengths is trapped and converted to chemical energy by a series of steps called the **light reactions**; these take place on the internal membranes of the chloroplasts. Second, CO_2 is 'fixed' and reduced to organic compounds, typically sugars, by a series of steps called the dark reactions or **CO_2 fixation**; these occur in the fluid matrix or **stroma** of chloroplasts and we describe them in Section 2.5. Figure 2.15a summarizes the light reactions for green plants; note from this figure that chemical energy is trapped in two forms — as ATP and as 'reducing power', the reduced pyridine nucleotide NADPH (NADP$^+$ is a phosphorylated form of the coenzyme NAD$^+$). The energy is then used to power the fixation of CO_2 (Figure 2.15b).

A point worth remembering, although we do not discuss it in detail here, is that ATP and NADPH generated by the light reactions can be used in several other energy-requiring processes besides CO_2 fixation. Nitrate and sulphate reduction in green leaves, ion uptake in aquatic algae and nitrogen fixation by cyanobacteria are examples of such processes. CO_2 fixation is much the commonest and most important process coupled to the light reactions, but it is not the only one.

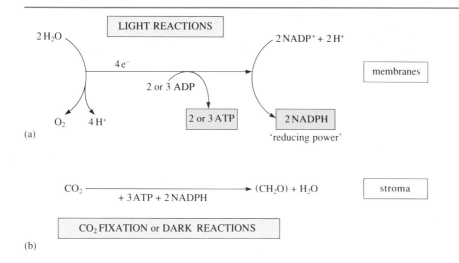

Figure 2.15 Summary of photosynthesis. (a) The light reactions. (b) CO_2 fixation or the dark reactions.

In this section we consider first (2.4.1) the general features of light reactions, which apply not only to green plants but also to photosynthetic bacteria. Then (2.4.2) the light reactions of green plants are described in more detail and, finally, (2.4.3) these are related to adaptations in sun and shade plants.

2.4.1 Basic features of the light reactions

Photosynthetic light reactions do one crucial thing: convert (or transduce) light energy to chemical energy — so how is this done? Figure 2.16 sets out the four steps involved, and you will notice that steps 3 and 4 are essentially identical to steps in aerobic respiration.

STEP =	REACTION by	MEMBRANE-BOUND AGENTS
1 Light harvesting	Absorb photons, excite (i.e. energize) electrons, and transfer *energy* between molecules	Photosynthetic pigment–protein complexes
2 Photochemical reaction	Separate electrical charge and transfer energized *electron* from pigment to non-pigment acceptor	Reaction centre complex
3 Electron transport coupled to proton pumping	Transfer electrons along a chain of carriers, using the energy released to set up a proton-motive force (H^+ gradient)	Reaction centre complex plus other carriers
4 Drive energy-requiring reactions using the proton-motive force	ATP synthesis	ATP synthetase (coupling factor)

Figure 2.16 Four steps common to all photosynthetic light reactions.

◇ What names are given to these steps in respiration?

◆ Step 3 is called *respiratory electron transport* and step 4 *oxidative phosphorylation*: both these processes are essentially the same in animals and plants.

Simplified general diagrams of respiratory electron transport and one type of photosynthetic electron transport (Figure 2.17) make their relationship clearer. The only real difference between photosynthesis and respiration is the way in which high energy electrons are produced in the first place. In respiration it is by the production of reduced pyridine nucleotides (NADH or $FADH_2$) during glycolysis and the TCA cycle; in photosynthesis, it is by steps 1 and 2 in Figure 2.16. First, therefore, we examine the nature of these steps.

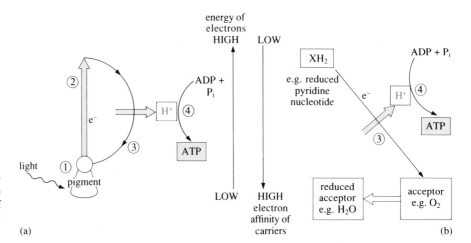

Figure 2.17 Simplified diagrams of electron transport. Numbers relate to the steps in Figure 2.16. (a) One type of photosynthetic electron transport. (b) Respiratory electron transport.

Light harvesting

Plants are green because the **photosynthetic pigments** which harvest the light used in photosynthesis absorb essentially all the visible wavelengths of light except those in the green part of the spectrum. The most important of these pigments in higher plants is **chlorophyll *a*** (replaced by bacteriochlorophyll *a* in photosynthetic bacteria) and there are also a number of **accessory pigments**, including chlorophyll *b* and various yellow–orange carotenoids. All are associated with proteins and embedded in the membranes within chloroplasts.

Light comes in discrete packets of energy (photons), the amount of energy per photon being related to the wavelength of the light (Box 2.1). A pigment molecule can absorb only one photon at a time and, in theory, every photon absorbed could initiate a photochemical reaction (step 2, Figure 2.16). But if this were so then the complex molecular machinery which backs up each photochemical reaction would remain idle for much of the time: in moderate shade, for example, a pigment molecule absorbs only about one photon per second although the back-up machinery can cope with the absorption of hundreds of photons per second. This inefficiency is circumvented by having a collecting system or **antenna**, whereby the energy absorbed by anything from 50 to 1000 chlorophyll molecules plus accessory pigments is channelled to one place, the *reaction centre*, where step 2 then proceeds more or less continuously. To understand how this 'energy funnel' works, we have to consider what happens when a photon is absorbed.

Electrons in atoms or molecules can exist in a number of discrete energy states: the 'normal' level, with the lowest energy, is called the ground state. Absorption of a photon by a chlorophyll molecule 'excites' an electron, raising it from the ground state to an excited state and what happens then is shown in Figure 2.18.

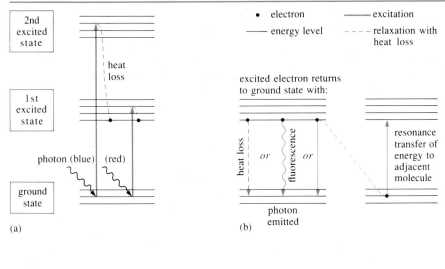

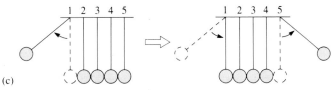

Figure 2.18 Light harvesting. (a) Absorption of a photon and excitation of an electron in a photosynthetic pigment. (b) Excited electron returns to ground state and energy is lost as heat or light (fluorescence) or transferred to an adjacent molecule (resonance transfer). (c) A desk toy, the mechanically coupled pendulum, which illustrates the concept of resonance transfer.

The excited electron first relaxes to the lowest of the excited states, emitting infrared radiation, i.e. heat (Figure 2.18a); then it returns to the ground state with either loss of heat, emission of a photon (**fluorescence**), or — the important one — transfer of energy to a neighbouring membrane-bound molecule, where another electron is excited (Figure 2.18b). A model illustrating this 'energy hopping' or **resonance transfer** is shown in Figure 2.18c and it is the mechanism by which the energy harvested by dozens or hundreds of pigment molecules is funnelled to one site — step 1 in Figure 2.16.

◇ Having studied Figure 2.18 explain: (a) why a photon of blue light is no more 'useful' in photosynthesis than one of red light (which has a longer wavelength and less energy); (b) why a solution of chlorophyll extracted from a leaf looks deep red when held up to the light.

◆ (a) Even though blue light excites electrons to a higher level than red light, all excited electrons relax to about the same level before resonance transfer occurs (Figure 2.18a). (b) It is because of fluorescence; the wavelengths of light emitted is, on average, longer (i.e. redder than that of the exciting photons.

The photochemical reaction

The sites or 'sinks' to which energy is funnelled by antennae are the **reaction centres**, where a true photochemical reaction occurs i.e. an excited *electron* is transferred from a pigment molecule to an adjacent, non-pigment acceptor molecule, resulting in a separation of electrical charge. If we represent the pigment involved (a type of chlorophyll *a* in green plants and of bacteriochlorophyll *a* in photosynthetic bacteria) as Chl, Chl* in the excited state, and the acceptor as A, then the photochemical reaction can be expressed as:

$$\text{Chl A} \xrightarrow{\text{light energy}} \text{Chl* A} \longrightarrow \text{Chl}^+ \text{A}^-.$$

This is the essence of step 2 in Figure 2.16. What follows (step 3) is the transfer of the high energy electron from A^- to a series of other electron transport molecules and donation of an electron to Chl^+. Reaction centres able to carry out photochemical reactions have been isolated from several types of bacteria (although not, so far, from green plants) and they comprise large complexes of protein, pigment and acceptor molecules. With a better understanding of how they work it may one day be possible to make synthetic analogues that could power light-driven fuel cells.

Electron transport and ATP synthesis

The principles of photosynthetic electron transport are basically the same as for respiration, and were illustrated in Figure 2.17b. Electrons are transferred from substrates of *low* electron affinity (e.g. NADH, which readily donates electrons) to substrates of *high* electron affinity (e.g. O_2, which readily accepts electrons). Another way of describing this process is to say that electrons of high energy are transferred in a 'downhill' direction, losing energy along the way.

In photosynthesis, high energy electrons are ejected by reaction centres after absorption of light and Figure 2.17a shows one pathway of electron transport that may occur — via a series of carriers back to the oxidized Chl^+. For obvious reasons this is described as **cyclic electron transport**. It occurs in photosynthetic bacteria, where it generates a transmembrane proton-motive force (PMF) which, in turn, can be coupled to ATP synthesis — the process of *cyclic photophosphorylation*. The enzyme or coupling factor, ATP synthetase, which catalyses ATP synthesis is very similar to that which operates in respiratory electron transport. Figure 2.19 illustrates in a simplified way the sequence of steps from light absorption to cyclic photophosphorylation in such bacteria.

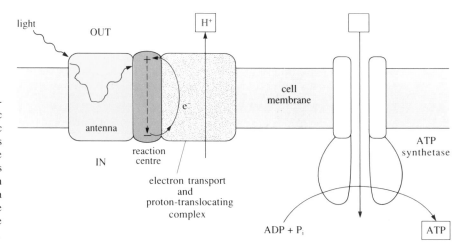

Figure 2.19 A simplified diagram of cyclic electron transfer coupled to cyclic photophosphorylation in photosynthetic bacteria. The antenna absorbs photons and transfers excitation energy to the reaction centre. Charge separation occurs and the reaction centre ejects a high energy electron which cycles back to it via a proton-translocating complex. Reverse flow of protons via the ATP synthetase (coupling factor) leads to ATP synthesis.

In effect, ATP is the only 'product' of cyclic electron transport — but CO_2 fixation requires both ATP *and* a powerful reducing agent, whose synthesis requires electrons of high energy level. Only a system of **non-cyclic electron transport** can provide the reducing power and Figure 2.20 illustrates one system, found in a group of photosynthetic bacteria called the green sulphur bacteria; the scale on the left gives a measure of the energy of electrons — the more negative the value, the higher the energy. Note in Figure 2.20 that electrons are supplied to the reaction centre by an external electron donor, sulphide (S^{2-}). This usually derives from ferrous sulphide (FeS) or hydrogen sulphide (H_2S) and is oxidized to sulphur, S. Many of the Earth's sulphur deposits originated in this way. There is no photophosphorylation associated with this non-cyclic pathway but ATP for carbon fixation can be generated from the cyclic pathway (the green pathway in Figure 2.20).

A problem with these primitive bacterial systems is that electron donors such as sulphides are available only in certain habitats (e.g. anaerobic, organic-rich mud); this greatly restricts where the bacteria can grow. Green plants (and cyanobacteria or blue-greens) have an elegant solution to the problem, which is illustrated in Figure 2.21. Compare this with Figure 2.19, noting that the same three basic components — membrane-bound antenna pigments, reaction centres and proton-translocating electron-transfer chains — are involved.

◇ How is the problem described above solved in the green plant system?

◆ By using *water* — one of the commonest substances on Earth — as the external electron donor for this non-cyclic system.

However, water does not readily donate electrons; it is difficult to oxidize and compared with the electrons of sulphide, which have an electrical potential of about zero volts (Figure 2.20), the electrons of water are of much lower energy (around +0.8 volts). To oxidize water, therefore, and at the same time generate reducing power, green plants link together *two* light reactions, each with its own antenna and reaction centre (called photosystems I and II: Figure 2.21). Very low energy electrons can now be transferred from water to $NADP^+$ using two 'booster' steps, each equivalent to the thick grey line in Figure 2.20, and sufficient energy is available to generate a PMF for ATP synthesis. Furthermore, when the water molecule is split (oxidized), oxygen is released, a reaction from which derives nearly all O_2 in the atmosphere. These light reactions of green plants are, therefore, supremely important for maintaining life on this planet and we consider them in more detail next.

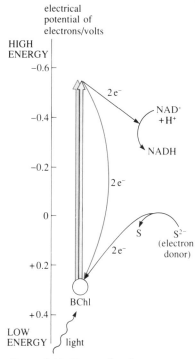

Figure 2.20 Non-cyclic electron transport and the generation of reducing power (NADH) in green sulphur bacteria, where an external electron donor such as sulphide (in the form of H_2S or FeS) is utilized. BChl is bacteriochlorophyll; the thick vertical grey arrow represents the photochemical reaction. The green pathway represents cyclic electron transport.

Figure 2.21 A simplified diagram of non-cyclic electron transport in the chloroplasts of green plants. There are two reaction centres, photosystems (PS) I and II. Electrons are extracted from water by PSII, transferred via a proton-translocating electron-transfer chain to PSI and ultimately to $NADP^+$.

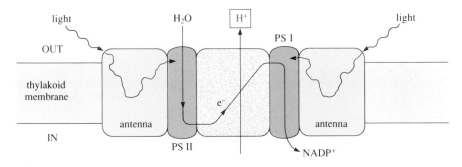

2.4.2 Light reactions in green plants

Location

We mentioned earlier that all components of the light reactions are membrane-bound. In photosynthetic bacteria the outer (cell) membrane is involved and proton pumping occurs from inside the cell to the exterior. In green plants an array of flattened membranous vesicles called **thylakoids** (Greek *thylakos*, sac or pouch) are involved and these lie within the matrix or stroma of chloroplasts, which is bounded by a double membrane envelope. When proton pumping occurs in chloroplasts it is from the stroma to the space inside thylakoids. Figure 2.22 illustrates this arrangement together with the

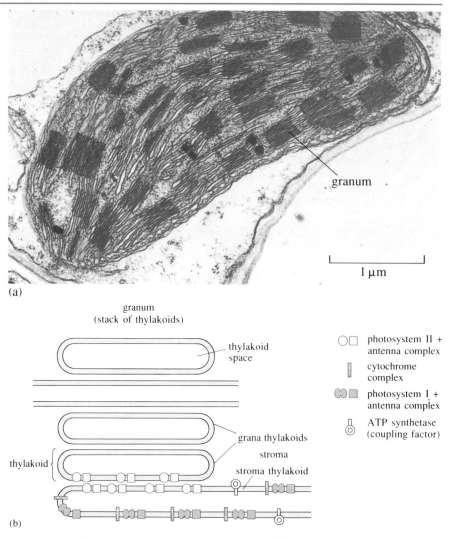

(a)

granum

1 μm

granum
(stack of thylakoids)

thylakoid
space

○□ photosystem II +
antenna complex

cytochrome
complex

photosystem I +
antenna complex

ATP synthetase
(coupling factor)

grana thylakoids

stroma

stroma thylakoid

thylakoid

Figure 2.22 (a) Electron micrograph of a chloroplast. (b) Diagram of a granum with the positions of various components of the light reactions shown in the lower thylakoids.

(b)

grouping of thylakoids to form stacks or **grana** (singular granum), rather like piles of hollow pancakes. Notice in Figure 2.22b that different pigment and protein complexes are associated with grana thylakoids and the longer stroma thylakoids that link up grana. We shall return to this point later.

Arrangement of components

Figure 2.23 shows diagrammatically how the electron carriers of the light reactions are arranged in the thylakoid membrane. There is no need to memorize all the components shown but we describe below the steps involved.

Starting off with photosystem II, absorption of light by the associated antenna pigments excites P_{680} (a special form of chlorophyll that absorbs light of wavelength 680 nm) and leads to a photochemical reaction — ejection of high energy electron. The now oxidized P_{680} is an extremely powerful oxidizing agent — strong enough to extract electrons from water. These are transferred from water to oxidized P_{680} via a manganese–protein complex lying next to the thylakoid space and, for each water molecule split, two H^+ ions are released into this space (contributing to the proton gradient) and two electrons released.

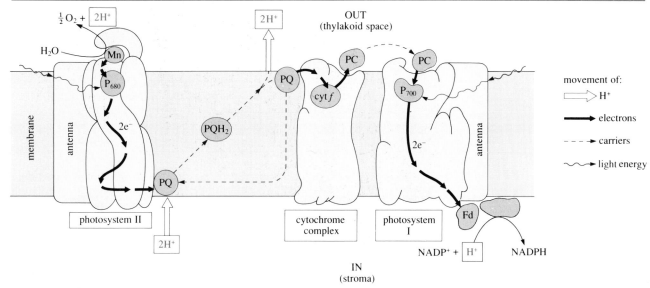

Figure 2.23 Arrangement of electron carriers in the thylakoid membrane. Mn, manganese; P_{680} and P_{700}, chlorophyll complexes where the photochemical reactions occur; PQ, plastoquinone; PC, plastocyanin; Fd, ferredoxin; $NADP^+$/NADPH, oxidized/reduced nicotinamide adenine dinucleotide phosphate.

◇ How many quanta of light must be absorbed before one molecule of O_2 is released?

◆ Four: two water molecules must be split, yielding four electrons and one electron is energized for each quantum absorbed.

Electrons from activated P_{680} are transferred via other carriers (not shown) to **plastoquinone (PQ)**, which is similar to ubiquinone (coenzyme Q) used in respiratory electron transport. PQ is a hydrogen carrier and, having accepted two electrons, picks up two protons from the *stroma* side of the membrane. Like ubiquinone it is very mobile within the membrane and PQH_2 shuttles off — sometimes quite long distances — before donating electrons one at a time to the cytochrome complex and losing two H^+ ions to the *thylakoid space*.

Electrons from one cytochrome, cytochrome *f*, are then transferred to plastocyanin (PC), a copper-containing protein which can move a short distance along the surface of the membrane to photosystem I (PSI). The primary electron donor in PSI, P_{700}, accepts electrons from PC once it has been energized by light absorbed by the associated antenna. P_{700} transfers electrons to ferredoxin (Fd, an iron-containing protein) at the stroma side, which in turn reduces $NADP^+$ via a reductase enzyme.

This non-cyclic electron pathway, therefore, produces NADPH and achieves a net transfer of protons into the thylakoid space, which becomes acidified (pH 5) relative to the stroma (pH 8).

◇ At which points in the pathway are protons released into the thylakoid space?

◆ During the oxidation of water and during electron transfer from reduced plastoquinone (PQH_2) to cytochrome *f*, as shown in Figure 2.23.

The proton gradient is harnessed to ATP synthesis via a coupling factor in exactly the same way as in bacterial photosynthesis (Section 2.4.1) and respiratory electron transport.

The Z scheme and cyclic electron transfer

Another way of representing photosynthetic electron transport is in relation to the electron affinity of the carriers and the energy of electrons (as in Figure 2.15). This is done in Figure 2.24 and shows clearly how electrons are moved 'uphill' by the two photochemical reactions and transferred 'downhill' between electron carriers. Because the shape of the pathway is roughly like a letter Z on its side, it is referred to as the **Z scheme**. This representation makes it easier to grasp why coupling two photoreactions together makes possible the reduction of $NADP^+$ by water: PSII copes with the oxidation of water and PSI with the reduction of $NADP^+$.

Two groups of powerful herbicides act by interfering with this pathway. One group, urea derivatives and triazines such as simazine, blocks electron transfer to PQ. The other group, which includes diquat and paraquat, intercepts electrons going to ferredoxin and transfers them to O_2, to form toxic free radicals* which damage membranes.

Also shown in green in Figure 2.24 is a cyclic pathway which involves only PSI. Like the cyclic pathways in bacteria (Figures 2.19 and 2.20), there are no external electron donors but a proton gradient that may be coupled to ATP synthesis is generated. This is termed **cyclic photophosphorylation** as distinct from the **non-cyclic photophosphorylation** linked to the Z scheme. Its importance is not clear at present but it would allow some flexibility in balancing the demand for ATP with that for reducing power.

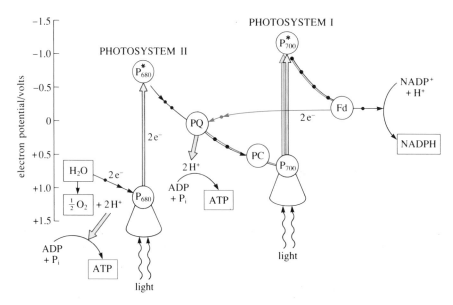

Figure 2.24 The Z scheme of non-cyclic photosynthetic electron transport in green plants. The green line shows the pathway of cyclic electron transport. Black blobs represent electron carriers and the abbreviations are as in Figure 2.23. The scale gives a measure of the relative electron affinity of carriers: the more negative the value, the higher the affinity.

2.4.3 Regulation and adaptation

When light is rate-limiting for photosynthesis, as it often is for plants growing in shade or for the lower leaves in a canopy, efficient use of the available light is important. On the other hand, the biochemical machinery of the light reactions is rather easily damaged by exposure to high irradiance. Unable to walk into or out of the sun, plants adapt as best they can to a constantly fluctuating light environment and this section is about how they do it.

*Free radicals are atoms or molecules with one or two unpaired electrons, e.g. chlorine atoms, $Cl^\bullet$, or superoxide (negatively charged oxygen molecules), O_2^-. They are highly reactive.

Optimizing the use of light

When light is rate-limiting for photosynthesis, the two main requirements are that:

(a) as much as possible of the 'usable' light (i.e. **photosynthetically active radiation**, PAR: see also Box 2.1) reaching chloroplasts should be absorbed and the supply of electrons to the photosynthetic electron transport chain should be as close as possible to the maximum;

(b) the two photochemical reactions at PSI and PSII should be balanced, proceeding at the same rate and with neither receiving substantially more light than the other.

(a) involves mainly long-term shifts in the ratios of pigments and electron carriers and relates to differences between sun and shade plants (Section 2.3.2). (b), however, can also be achieved by rapid, short-term changes in thylakoid organization.

Long-term changes — sun and shade plants

The most striking adaptation of chloroplasts to low irradiance is an increase in the ratio of light-harvesting pigments — the antenna complexes — to other components of the light reactions. This is often accompanied by increased synthesis of thylakoid membranes and, in deep shade plants, chloroplasts are packed with thylakoids. An additional effect is seen when the shade is created by plant leaves, because they preferentially absorb certain wavelengths and the remaining light is enriched in the longer (far-red) wavelengths. Now, on average, PSI absorbs slightly longer wavelengths than PSII, so this 'red enrichment' in shade favours PSI excitation.

◇ Bearing in mind requirement (b) above, what change would you expect in the ratio between PSI and PSII in red-enriched shade?

◆ The ratio of PSII + antenna to PSI + antenna should *increase*.

If you look back at Figure 2.22b, it shows that PSII is mainly associated with the inner (appressed) membranes of the grana thylakoids whereas PSI is mainly located on the stroma thylakoids. Not surprisingly, therefore, in shade plants with a higher PSII/PSI ratio the grana are bigger, both wider and with more stacks.

Short-term changes: fine control

Balancing the excitation rates of PSI and PSII can be achieved by rapid, reversible changes in grana stacking and by a very interesting mechanism. This is illustrated in Figure 2.25.

The mechanism hinges on the relative amounts of oxidized and reduced plastoquinone (PQ/PQH_2; see Figure 2.23). If PSII is over-excited, PQH_2 levels will be high and this leads to the sequence of changes shown on the left of Figure 2.25: activation of a membrane-bound protein kinase which phosphorylates certain antenna proteins in the PSII–antenna complex. This increases the electrical (negative) charge on the appressed grana membranes causing mutual repulsion and partial unstacking of grana. In addition, the phosphorylated antenna proteins and the residual PSII–antenna complex are thought to separate and both migrate laterally into non-appressed membrane regions, where energy may be transferred directly to PSI. The whole process goes into reverse when PSI is over-excited and PQ levels are high: a membrane-bound phosphatase causes dephosphorylation of the antenna II proteins, reversing the reaction catalysed by the protein kinase (right of Figure 2.25). The essential features of this elegant control system can be

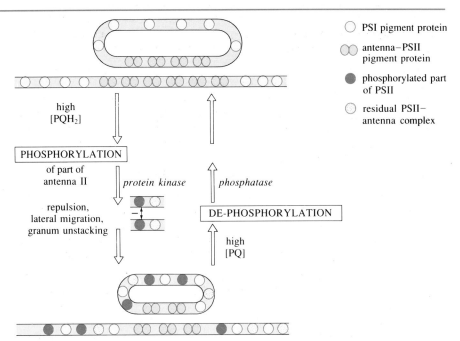

Figure 2.25 A model of how phosphory-lation/dephosphorylation of antenna proteins in the antenna-photosystem II complex may regulate the distribution of this complex, energy transfer to photo-system I and the degree of membrane stacking in grana.

summarized as: enzyme specificity, membrane fluidity (allowing lateral migration) and electrical repulsion/attraction. The ratio of $PQ : PQH_2$ 'senses' the excitation state of the two photosystems and the control system adjusts the activity of PSII appropriately.

Too much light

We mentioned in Section 2.3.2 that strong light may inhibit photosynthesis and that this involves two processes.

◇ Recall the processes.

◆ Photoinhibition, a rapid, reversible process and photo-oxidation, a longer term irreversible process.

Photoinhibition involves damage to the reaction centre complexes, especially PSII, when they are grossly over-excited. What seems to happen in PSII is that a protein with a very rapid turnover that is involved in electron transfer between P_{680} and plastoquinone is lost but can be replaced later. This protein is rather like a weak link in the electron transport chain, which breaks when the going becomes very rough.

◇ Why do you think that shade plants experience photoinhibition at much lower light levels than do sun plants (Section 2.3.2)?

◆ Because they have very large antenna complexes which, in strong light, funnel a great deal of light energy to the reaction centre.

Photo-oxidation is an irreversible and, therefore, more damaging process and involves the light-harvesting pigments directly. When these absorb too much light, relatively long-lived excited states are formed which then interact with O_2 producing free radicals such as superoxide (O_2^-), which can destroy pigment molecules. There are some biochemical defences, for example the enzyme **superoxide dismutase (SOD)** can destroy free radicals, but these defences are swamped if exposure to strong light is prolonged.

There are also physiological changes which reduce very effectively the risk of damage in high light. Chloroplasts may move to the sides of cells, for example, and the orientation of leaves may change, so that they are aligned parallel to the incident rays and thus absorb less light. When the deep shade species *Oxalis oregano* is exposed to flickering light patches (sunflecks) that penetrate through a tree canopy, it shows rapid leaf movements and no photoinhibition; but when its leaves are mechanically constrained, making movement impossible, strong photoinhibition occurs in sunflecks.

Plants that grow in high-light environments often have structural and chemical characteristics that reduce the amount of light reaching chloroplasts. Leaves may have shiny, reflective surfaces or dense coatings of silvery hairs, for example, and epidermal cells may contain red anthocyanin pigments that absorb the shorter (and potentially most damaging) wavelengths.

Overall, the light reactions are a very efficient piece of biochemical machinery. They may convert up to 20% of absorbed light into ATP, which is a high rate for such a complex system, and when combined with some morphological adaptations, can accommodate a wide range of light environments. Their main function, of course, is to supply ATP and NADPH for carbon fixation and we consider next what this involves.

Summary of Section 2.4

1 The photosynthetic light reactions occur on the internal membranes of chloroplasts and convert light to chemical energy. In green plants, these membranes form hollow, flattened vesicles (thylakoids).

2 These reactions involve four steps: light harvesting, a photochemical reaction, electron transport and, in parallel with the last step, proton pumping. The proton-motive force (PMF) that develops can be used for ATP synthesis (photophosphorylation).

3 Light harvesting is carried out by photosynthetic pigments, which are organized in groups (antennae). Light absorbed within one antenna is transferred between pigment molecules by resonance transfer and funnelled to a single reaction centre.

4 The photochemical reaction occurs in the reaction centre, a pigment–protein complex containing a modified form of chlorophyll *a* or bacteriochlorophyll *a*. This pigment transfers excited electrons to the photosynthetic electron transport chain.

5 Two types of chain occur: the cyclic electron transport chain, which transfers electrons back to the oxidized reaction centre, and the non-cyclic electron transport chain, which transfers electrons to NAD^+ (in photosynthetic bacteria) or $NADP^+$ (in green plants and cyanobacteria), producing NADH and NADPH respectively.

6 In non-cyclic chains, the oxidized reaction centre receives electrons from an external donor. This may be sulphides in photosynthetic bacteria but in green plants it is always water, which is oxidized to O_2.

7 Because the electrons in water are at a very low energy level, *two* photochemical reactions are required to transfer electrons to $NADP^+$.

8 Both cyclic and two-step (water$\rightarrow NADP^+$) non-cyclic electron transport generate a proton-motive force. Protons are transferred from the bacterial

cell or chloroplast stroma (= inside) to the bacterial medium or space between thylakoid membranes (= outside). As in respiration, the PMF drives ATP synthesis via membrane-bound coupling factors (ATP synthetases).

9 The two light reactions that occur in green plants are associated with two photosystems: PSI and PSII. Oxidation of water is linked to PSII, which transfers electrons to plastoquinone. Electrons are then transferred in an energetically 'downhill' fashion to the oxidized reaction centre of PSI. PSI transfers electrons to ferredoxin and thence to $NADP^+$. This pathway for non-cyclic electron transport is described as the Z scheme. PSI is also the centre for cyclic electron transport.

10 Optimal use of the available light is achieved:

(a) in continuous shade, by increasing the size of antennae and the number of thylakoids. If light shows 'red enrichment', an increase in the ratio of PSII to PSI occurs;

(b) with fluctuations in the level or quality of light, a balance between the amount of light absorbed by PSI and PSII can be maintained by varying the stacking of grana thylakoids.

11 Strong light may damage the photosystems in two ways:

(a) by inactivation of a labile, 'weak-link' protein in PSII. This photoinhibition is reversible;

(b) by causing the formation of free radicals that destroy pigment molecules irreversibly.

12 Protection against the effects of strong light may be biochemical (e.g. superoxide dismutase, epidermal anthocyanin pigments), physiological (chloroplast and leaf movements) or structural (reflective leaf surfaces).

Question 8 (*Objectives 2.1 and 2.7*) Complete statements (a)–(d) by choosing one or more items from (i)–(viii).

(a) Antenna pigments form ___ which channel light energy to ____.

(b) The energy of absorbed photons is transferred between antenna pigments by the process of ____ and may also be lost as ___ or ___.

(c) Photochemical reactions in the light reactions occur at ___.

(d) Charge separation occurs during a photochemical reaction because an excited ____ is transferred from a __ molecule to an __.

(i)   reaction centres
(ii) resonance transfer
(iii) electron
(iv) fluorescence

(v)   pigment
(vi) light-harvesting systems
(vii) heat
(viii) acceptor molecule

Question 9 (*Objectives 2.1 and 2.8*) In green plants, which of (a)–(d) applies to (i) respiratory electron transport, (ii) the light reactions, (iii) both and (iv) neither.

(a) Electron transport is coupled to the generation of a proton-motive force (PMF).

(b) Water acts as a primary electron donor.

(c) Oxygen acts as a terminal electron acceptor.

(d) The PMF is used to synthesize ATP when electrons flow through membrane-bound ATP synthetases or coupling factors.

(e) Reducing power, usually in the form of reduced pyridine nucleotides, is generated.

Question 10 (*Objectives 2.1, 2.9 and 2.11*) Are statements (a)–(e) about the light reactions of photosynthesis true, partly true or false?

(a) Photosynthetic bacteria have only one light reaction whereas green plants have two.

(b) Photophosphorylation may be either cyclic or non-cyclic in photosynthetic organisms.

(c) In green plants, every stage of the light reactions, from light-harvesting to the synthesis of ATP and NADPH, involves highly ordered membrane-bound components.

(d) Shade plants harvest light more efficiently than sun plants because their chloroplasts contain more photosynthetic pigments.

(e) The ratio of oxidized to reduced plastoquinone (PQ/PQH_2) acts as a fine control in balancing the activity of photosystems I and II: a high ratio favours PSII and membrane stacking in grana and a low ratio favours PSI and membrane unstacking.

Question 11 (*Objective 2.10*) (a) Draw a simple version of the Z scheme, including arrows to indicate the direction of increasing energy of electrons or electron affinity of carriers. Which component is the strongest oxidizing agent (has the highest electron affinity) and which is the strongest reducing agent?

(b) Which components or stages in the Z scheme are involved in steps 1 to 4 shown in Figure 2.16?

2.5   THE BIOCHEMISTRY OF PHOTOSYNTHESIS: CARBON FIXATION

Until the early 1950s little was known about the reactions by which CO_2 is reduced or 'fixed' in photosynthesis. Today the central pathway is fully worked out (Section 2.5.1), its relationship to photorespiration is reasonably understood (2.5.2), additional pathways that occur in certain environments are well known (2.5.3 and 2.5.4) and the overall control and coordination of all these pathways plus the light reactions are just beginning to be explored (2.5.5). The story is complex but a vital part of understanding plant function.

2.5.1 The Calvin cycle and product synthesis

Carboxylation reactions where CO_2 is used as a substrate are common in most organisms but the CO_2 is always lost again in a later reaction. There is only *one* pathway which achieves a *net* fixation of CO_2, the **Calvin cycle**, which is universal among green plants and occurs in many autotrophic bacteria. The pathway is named after an American, Melvin Calvin, leader of a group of researchers who first described it in 1953. Calvin was awarded a Nobel prize for this work in 1961. An alternative name to the Calvin cycle is the C_3 pathway, because the product of the first reaction contains three carbon atoms.

The reactions of the Calvin cycle

Each turn of the cycle fixes one molecule of CO_2 and requires an input of *three* ATP and *two* NADPH molecules. The product which is tapped off from the cycle and used to synthesize all the end products of photosynthesis —

starch, sucrose etc. — is a 3-carbon sugar, triose phosphate*. So the net production of one molecule of triose phosphate requires *three* turns of the cycle. The steps involved are summarized in Figure 2.26 and fall into three phases:

1 carboxylation (i.e. addition of CO_2) of an acceptor

2 reduction of the product

3 regeneration of the acceptor.

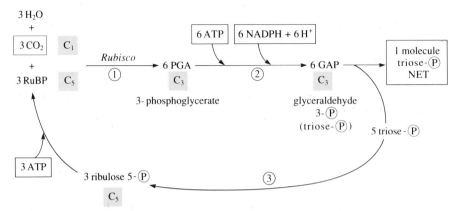

Figure 2.26 A simplified version of the Calvin cycle. RuBP, ribulose bisphosphate; Rubisco, ribulose bisphosphate carboxylase/oxygenase. 1 is the carboxylation reaction, 2 the reduction step, and 3 the regeneration step. *Note*: the symbol ℗ denotes a phosphate group.

Carboxylation

In this first reaction (1 in Figure 2.26) CO_2 combines with a 5-carbon (pentose) acceptor, **ribulose 1,5-bisphosphate (RuBP)**, to give two molecules of a 3-carbon acid, phosphoglyceric acid (PGA). It is because the first product of CO_2 fixation has three carbon atoms that the Calvin cycle is also referred to as the **C_3 pathway**. The reaction is catalysed by an enzyme which has been described as the most abundant protein on Earth — it may comprise up to half the total soluble leaf protein — and is commonly called **Rubisco**. Rubisco's full name is ribulose bisphosphate carboxylase/oxygenase because not only does it interact with CO_2 (the carboxylase function) but also with O_2 (the oxygenase function), the significance of which will become clear when we consider photorespiration. The full carboxylase reaction is shown in Figure 2.27.

Figure 2.27 The carboxylation reaction in the Calvin cycle.

$$
\begin{array}{c}
CH_2O\,℗ \\
| \\
CO \\
| \\
HCOH \\
| \\
HCOH \\
| \\
CH_2O\,℗
\end{array}
\quad + \; CO_2 + H_2O \quad \xrightarrow{\textit{Rubisco}} \quad
2\begin{array}{c}
CH_2O\,℗ \\
| \\
CHOH \\
| \\
COO^-
\end{array}
\quad + \; 2H^+
$$

ribulose 1,5-bisphosphate   3-phosphoglycerate

Reduction

This phase (2 in Figure 2.26) achieves the reduction of the carboxylate ($—COO^-$) group on PGA to an aldehyde ($—CHO$) group, producing glyceraldehyde 3-phosphate (GAP), a triose phosphate. A large input of energy is required (Figure 2.26) and this is, in effect, the reverse of the first energy-releasing step of glycolysis. The two reactions involved are shown in

*In case you find the sugar terminology confusing: *hex*oses have six carbon atoms (C_6), *tri*oses three (C_3), *pent*oses five (C_5) etc.

Figure 2.28. One-sixth of the triose phosphate produced can be used to synthesize more complex end-products but five-sixths is required to regenerate the primary CO_2 acceptor, RuBP.

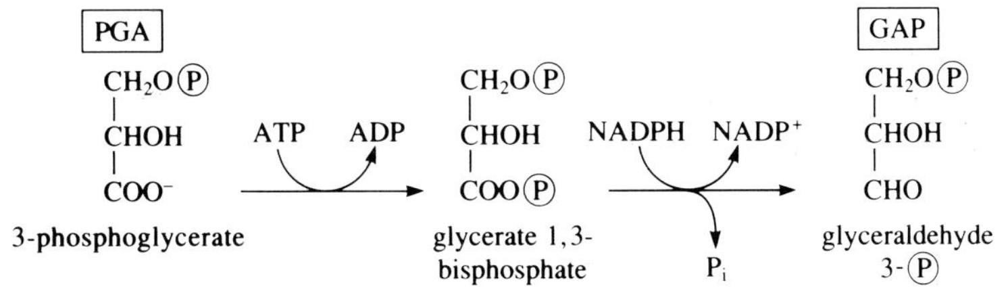

Figure 2.28 The reduction step in the Calvin cycle.

Regeneration

The reactions (3 in Figure 2.26) by which five molecules of triose phosphate generate three molecules of pentose phosphate are complex and we shall not describe them in detail: sugar phosphates with 3, 4, 5, 6 and 7 carbon atoms are involved.

◇ From Figure 2.26, is triose phosphate the only input to this regeneration phase?

◆ No: ATP is necessary for the last step, the conversion of ribulose 5-phosphate to ribulose 1,5-bisphosphate. One third of the total ATP required for carbon fixation is used in this phosphorylation step.

◇ Now write a summarizing equation, showing all the inputs, for the formation of one molecule of triose phosphate from three molecules of CO_2. (RuBP can be omitted because it appears on both sides of the equation.)

◆ $3CO_2 + 3H_2O + 9ATP + 6NADPH + 6H^+ \longrightarrow GAP + 9ADP + 6NADP^+$

Product synthesis

The triose phosphate produced by the Calvin cycle is the starting point for a wide range of biosynthetic pathways. Strictly speaking product synthesis is not part of the Calvin cycle but it is closely linked and so important for the overall control of photosynthesis that we consider it here. It is also important because we consume large amounts of the two major end-products formed in green leaves — *starch* and *sucrose* (table sugar).

Sucrose is a disaccharide made up of glucose and fructose units and is the main form in which carbohydrates are exported from leaves and transported to other parts of the plant. It is synthesized *outside* chloroplasts and the main steps are shown in Figure 2.29.

Notice that the step first is the export of triose phosphate from the chloroplast stroma via a carrier, the **phosphate translocator**. This can work in either direction but always moves triose phosphate one way and inorganic phosphate (P_i) the other, so it does not change the total amount of phosphate in either compartment. Formation of fructose 6-phosphate and glucose 1-phosphate is by a series of reactions which is effectively the reverse of the early steps in glycolysis. But these sugar phosphates cannot form sucrose until the glucose 1-phosphate has been activated by interaction with a close relative

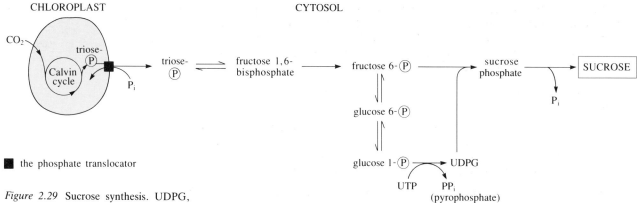

Figure 2.29 Sucrose synthesis. UDPG, uridine diphosphate glucose ('activated glucose').

of ATP, uridine triphosphate (UTP), to give **UDP-glucose**. This type of 'activated sugar' is widely used in the synthesis of polysaccharides.

By contrast with sucrose, **starch** is synthesized in the chloroplast stroma where it commonly builds up during the day and is used as an energy source at night. It is also produced and stored in many non-photosynthetic cells within the colourless plastids known as amyloplasts. The molecule is simply a glucose polymer but, as with sucrose, activated glucose is needed to make it. In this case ADP-glucose, formed from ATP and glucose 1-phosphate (derived from triose phosphate), is the activated form and glucose units are simply added one at a time to the polymer chain (which can be represented as n-glucose units, n representing a number:

$$\text{ADP-glucose} + (n \text{ glucose units}) \longrightarrow (n + 1 \text{ glucose units}) + \text{ADP}$$

From this brief account you can see that some energy (ATP) is required to synthesize end-products from triose phosphate but the amount is small compared to that required for GAP synthesis in the Calvin cycle.

2.5.2 Photorespiration

We discussed the physiology of this process earlier (Sections 2.2 and 2.3.2): it involves uptake of O_2 and release of CO_2 by green leaves in the light and is most pronounced in conditions of high light and temperature.

◇ In what types of plants is photorespiration (a) clearly apparent and (b) undetectable?

◆ (a) In C_3 species, which include nearly all trees and most plants from temperate regions; (b) in bicarbonate-using species and C_4 species, which include many herbaceous plants from sunny, tropical regions (Section 2.3.2).

Since the late 1970s, the full biochemical explanation for this rather puzzling phenomenon has been worked out and it centres on Rubisco, the first enzyme of the Calvin cycle. Recall that this can function as an oxygenase and, in fact, CO_2 and O_2 compete at the active site in the reaction with RuBP. The oxygenase reaction is:

$$\text{RuBP} + O_2 \longrightarrow \text{PGA} + \text{phosphoglycolate} \qquad (2.5)$$
$$(C_5) \qquad\qquad (C_3) \qquad\quad (C_2)$$

PGA can enter the Calvin cycle but phosphoglycolate, the anion of phospho-glycolic acid, cannot and represents a loss of carbon from the cycle. Phosphoglycolate is rapidly converted to **glycolate** in the chloroplast and photorespiration is simply a complex series of reactions which metabolize glycolate, which can be summarized as:

COO^-
$CH_2O\circledP$
phosphoglycolate

COO^-
CH_2OH
glycolate

$$2\,\text{glycolate} + O_2 + ATP + Fd_{red} \longrightarrow \text{glycerate} + CO_2 + ADP + P_i + Fd_{ox}$$
(Fd$_{red}$ and Fd$_{ox}$ are reduced and oxidized ferredoxin, respectively; see Figure 2.24)

In terms of carbon atoms: $2 \times C_2 \rightarrow C_3 + C_1$, and the glycerate ($C_3$) is converted to PGA by ATP and re-enters the Calvin cycle. So although this pathway *uses* energy (the complete opposite of 'dark' respiration), it salvages three-quarters of the carbon from glycolate and returns it to the Calvin cycle.

Converting glycolate to glycerate is not a simple process and you can see from Figure 2.30 that three organelles are involved. In **peroxisomes** (organelles containing oxidative enzymes), glycolate from chloroplasts is first oxidized, producing toxic hydrogen peroxide. The H_2O_2 is broken down to water and O_2 by catalase, an enzyme confined to peroxisomes, and oxidized glycolate is converted to the amino acid, glycine, by transfer of an amino (—NH_2) group (transamination). Glycine is further metabolized in mitochondria, where two glycine molecules yield one serine, with loss of ammonia and CO_2 (this is where all the photorespiratory CO_2 is lost). Serine moves back to peroxisomes, passes on its amino group (to oxidized glycolate) and is converted to glycerate; while NH_3 passes to chloroplasts and is re-incorporated into amino acids, which is where the major energy input is required. Nitrogen cycling between the chloroplasts, peroxisomes and mitochondria ties up large amounts of amino acids — which cannot be tapped off and used for other biosynthetic reactions — and one can only assume that there is no other way of metabolizing the glycolate.

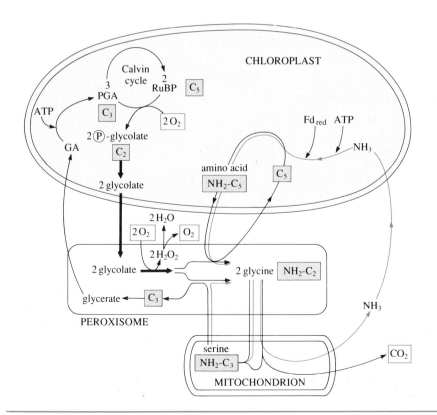

Figure 2.30 A simplified diagram of photorespiration. Black arrows indicate carbon flow, green arrows nitrogen flow and pairs of green and black lines show where carbon and nitrogen move together (e.g. as amino acids).

The real puzzle is why the glycolate is formed in the first place, i.e. why does Rubisco have oxygenase activity? One suggestion is that it is a historical accident. Rubisco evolved first in photosynthetic bacteria which operate in the complete absence of oxygen; but the enzyme isolated from these bacteria shows oxygenase activity with a rather low discrimination in favour of CO_2. Discrimination (CO_2/O_2 substrate specificity) is higher in the oxygen-producing cyanobacteria and even higher in green algae and higher plants, but it is never 100%. Plants appear to be stuck with the oxygenase reaction and photorespiration is the consequence.

Explaining the physiological effects

Knowing the nature and cause of photorespiration, we can now begin to make sense of the physiological effects described in Section 2.3.2. The temperature effect, for example (Figure 2.11), is because the substrate specificity of Rubisco shifts markedly in favour of O_2 as temperature increases. Similarly, the stimulation of net photosynthesis in C_3 species either by lowering $[O_2]$ or raising $[CO_2]$ is readily understood in terms of competitive interactions at the active site of Rubisco.

◇ Why do you think that O_2 inhibition of net photosynthesis in C_3 species is often more severe at high irradiances (Figure 2.11)?

◈ Because CO_2 is likely to be rapidly fixed and become rate-limiting under these conditions; thus $[CO_2]$ within the leaf will be low and competition from O_2 severe.

The only really satisfactory way of minimizing the Rubisco oxygenase reaction and photorespiration is that which occurs in bicarbonate-using and C_4 plants.

◇ Recall from Section 2.3.2 what this way is.

◈ Concentrating CO_2 at the site of Rubisco action and so overcoming competition by O_2.

The explanation of how C_4 plants do this, pieced together during the late 1960s and 1970s, turned out to involve an elegant combination of biochemical and structural features.

2.5.3 C_4 plants and the C_4 pathway

C_4 species comprise less than 0.5% of the known species of flowering plants but they include some of the most abundant tropical grasses and some important crop plants. Maize and sugar-cane, for example, both noted for their high productivity, are C_4 species (although rice is not). The two distinctive features of C_4 plants are: (a) a special kind of leaf anatomy and (b) an extra biochemical pathway, which fixes CO_2 and then re-releases it for incorporation into the Calvin cycle. We show here how these characteristics achieve concentration of CO_2 and explain the physiological properties of C_4 plants that were described in Section 2.3.2.

Leaf anatomy in C_4 plants

If you compare leaf sections of typical C_3 and C_4 species (Figure 2.31), the most obvious difference is the presence in the C_4 leaf of a distinct layer of cells around the vein (or vascular bundle). These **bundle sheath** cells contain more

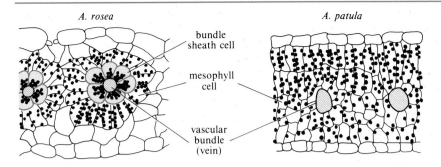

Figure 2.31 Sections of leaves of *Atriplex rosea* (a C_4 plant) and *A. patula* (a C_3 plant).

and larger chloroplasts than the other (mesophyll) cells and starch accumulates almost entirely in bundle sheath chloroplasts.

A shorthand way of referring to the cell arrangement in C_4 plants is as **Kranz anatomy** (after the German for wreath, which is what the bundle sheath looks like) and you can often detect it by holding a leaf to the light and looking for darker green bands round the veins. All C_4 plants show Kranz anatomy and it is genetically, not environmentally, determined. The interesting thing is that the division of chloroplast-containing cells into two types in C_4 plants is matched by a division of biochemical activities.

The C_4 pathway

Bearing in mind that C_4 plants concentrate CO_2 at the site of the Calvin cycle — which it must be emphasized is the only pathway that achieves a *net* fixation of CO_2 — their strategy becomes easy to understand. They fix CO_2 initially in the numerous mesophyll cells by means of the **C_4 pathway** and re-release it in the chloroplasts of a relatively few cells (the bundle sheath) where Calvin cycle enzymes are located.

◇ From information in the previous section, how might you have guessed that the Calvin cycle is located only in bundle sheath cells?

◈ Because only here does starch, an end-product of the Calvin cycle, accumulate.

The C_4 pathway is thus an addition to the Calvin cycle and its name derives from the first product of fixation, a 4-carbon acid (in contrast to the 3-carbon acid of the Calvin cycle). The pathway is shown in outline in Figure 2.32.

Initial fixation of CO_2 is by carboxylation of phosphoenolpyruvate (PEP), a C_3 compound which is also an intermediate in glycolysis. The reaction is catalysed by a cytosolic enzyme, *PEP carboxylase*, with a very high affinity for CO_2, and the 4-carbon product, oxaloacetate (OAA), is then rapidly converted into the anions of other C_4 acids, either malate or aspartate (depending on the species). These acid anions act simply as carriers of CO_2 to the bundle sheath cells, where decarboxylation and release of CO_2 occurs.

◇ From Figure 2.32 describe what happens in the C_4 pathway after malate decarboxylation in the bundle sheath cells.

◈ The 3-carbon product, pyruvate, is transported back to the mesophyll cells where it is converted to PEP, the primary CO_2 acceptor of the pathway.

The conversion of pyruvate to PEP requires an input of energy (ATP) so there is clearly a metabolic 'cost' attached to the C_4 pathway. This is an

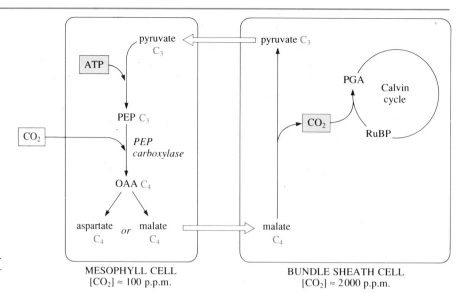

Figure 2.32 Outline of the C_4 pathway. PEP, phosphoenolpyruvate; OAA, oxaloacetate.

MESOPHYLL CELL
$[CO_2] \approx 100$ p.p.m.

BUNDLE SHEATH CELL
$[CO_2] \approx 2000$ p.p.m.

important point to appreciate and helps to explain why C_4 plants perform rather poorly when light rather than CO_2 is rate-limiting for photosynthesis. Under conditions of high light and temperature, however, the 'cost' is far outweighed by the advantages, as you saw in Section 2.3.2. The CO_2-concentrating action in C_4 plants means not only that Rubisco in the bundle sheath cells is supplied with optimum levels of CO_2 but also that glycolate formation and photorespiration are minimized.

Strangely, all the enzymes necessary for glycolate metabolism are present in C_4 leaves and it seems likely that some photorespiration occurs but cannot be *detected* as CO_2 release in the light. The explanation for this is that PEP carboxylase has such a high affinity for CO_2 that any CO_2 released in photorespiration is immediately re-fixed.

◇ What other characteristic of C_4 plants described in Section 2.3.2 does this help to explain?

◆ Their low CO_2 compensation point (the $[CO_2]$ at which photosynthetic uptake of CO_2 balances respiratory output).

Low CO_2 availability is also often correlated with shortage of water because the pores (stomata) through which CO_2 enters leaves are also the route by which water is lost through evaporation (the process of transpiration). The aperture of these pores can be varied (described in Chapter 5) so that, in dry conditions, pores can be partially or entirely closed and water is conserved. This, of course, reduces the availability of CO_2 in leaves, precisely the conditions under which C_4 plants perform well, which helps to explain why they are especially common in seasonally dry areas such as the African savannah. By comparing grams water lost per gram dry weight increase (the **transpiration ratio**), it is possible to assess quantitatively the relative efficiency of C_3 and C_4 plants at conserving water during photosynthesis: the ratio is 450–950:1 for C_3 species and 250–350:1 for C_4 species — quite a difference. However, for plants growing in *extremely* dry habitats, a slightly different photosynthetic strategy has evolved for which the transpiration ratio is only 18–125:1. This strategy, which allows photosynthesis to occur with the absolute minimum loss of water, is discussed next.

2.5.4 CAM — Crassulacean acid metabolism

C_3 and C_4 plants generally open their stomata during the day (allowing CO_2 entry) and close them at night. But this is a poor strategy when reducing water loss is of paramount importance because transpiration is greatest when air is hot and dry (day) and least when air is cool and humid (night). A third group of plants, nearly all of them succulents with fleshy leaves or stems (e.g. cacti), have evolved an ingenious way round this problem: they open their stomata mainly at night and close them during the day — and yet still achieve net fixation of CO_2. The mechanism which allows this is termed **Crassulacean acid metabolism** or **CAM** for short, after the family of succulents (Crassulaceae) in which it was first discovered.

CAM plants fix atmospheric CO_2 mainly at night when stomatal pores are open (Figure 2.33) but they cannot then use the Calvin cycle because this operates only in the light.

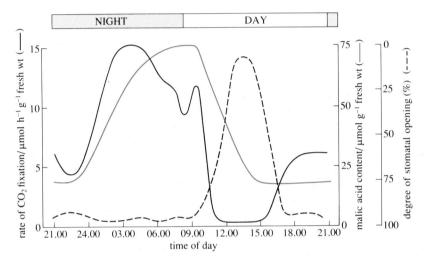

Figure 2.33 Changes in a CAM plant during night and day of rate of CO_2 fixation from the atmosphere, malic acid content and degree of stomatal opening.

◇ From Figure 2.33: (a) what happens to the CO_2 fixed at night and (b) what do you think happens subsequently to this dark-fixed CO_2 during the day?

◈ (a) It is *stored* as malic acid — the acid and not the anion this time, whose levels increase roughly in parallel with CO_2 fixation during the night. (b) Bearing in mind that malate is one of the primary fixation products of C_4 plants, the decrease in malic acid during the day in CAM plants can be interpreted as a decarboxylation: the dark-fixed CO_2 is re-released and can then be fixed by the Calvin cycle in the light.

In CAM plants, therefore, initial CO_2 fixation and the Calvin cycle operate at *different times* but in the same cells, in contrast to C_4 plants where they operate at the same time but in different cells. Decarboxylation of malate is similar in both types of plants and in CAM, as in C_4 metabolism, high intracellular levels of CO_2 may be reached during the middle part of the day, effectively preventing photorespiration. In the early morning and late afternoon, however, when CO_2 release is starting up and slowing down respectively, and stomata are partially open (see Figure 2.33), CAM plants have some fixation by Rubisco directly into the Calvin cycle and photorespiration may then occur. Other similarities and differences between CAM and C_4 metabolism can be worked out by comparing Figures 2.34 and 2.32).

Both, for example, use PEP carboxylase for initial CO_2 fixation.

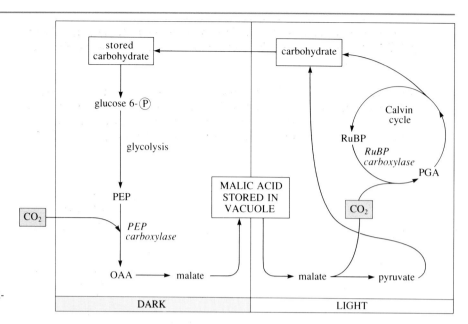

Figure 2.34 Crassulacean acid metabolism (CAM) in the light and dark.

◇ What two important differences are there?

◈ (i) Malic acid is stored in cell vacuoles in CAM plants compared with rapid decarboxylation of cytosolic malate in C_4 plants; (ii) pyruvate formed by decarboxylation of malic acid is metabolized (by a reverse-glycolysis pathway) to carbohydrates in CAM plants but is used to regenerate PEP in C_4 plants.

Thus diagnostic features of CAM are diurnal (day–night) fluctuations of malic acid and vacuolar pH with reciprocal diurnal fluctuations of storage carbohydrate, such as starch. Accumulation of malic acid in the vacuole requires energy and is yet another example of a system linked to proton pumping: two H^+ are pumped into the vacuole (requiring ATP) and one anion, $malate^{2-}$, enters passively down the electrochemical gradient: the high cytosolic pH causes dissociation of malic acid to malate and this is why malate is the form transported. The additional 'cost' of malate transport in CAM means that it is energetically even more expensive than the C_4 pathway — this is the price paid for conserving water. Furthermore, there is a limit to the amount of malic acid which can be stored and this usually determines the overall photosynthetic capacity of CAM plants. CAM, however, is a much more flexible photosynthetic strategy than the C_4 pathway. Some plants, for example pineapple, a bromeliad, behave as normal C_3 species when well watered but shift to CAM if droughted so, unlike the C_4 pathway, CAM is *inducible*. During severe drought, many CAM plants show a phenomenon called *CAM-idling*: stomata remain closed all the time but there is diurnal fluctuation of malic acid linked to fixation and release of respiratory CO_2. There is no *net* fixation of CO_2 and these CAM-idlers do not grow; rather, their metabolic machinery ticks over, poised for action when water is once more available. Even stranger is *CAM-cycling* which has been discovered in 43 species, including the British succulent, *Sedum acre* (wall-pepper) common on rocks, walls and dry ground. Here the stomata show day opening/night closure, with C_3 gas exchange patterns, but accompanied by CAM-like fluctuations in malic acid.

◇ What must be causing these acid fluctuations?

◆ Fixation of respiratory CO_2 into malate at night when stomata are closed, using PEP carboxylase, and decarboxylation of malate during the day, re-releasing CO_2.

But *why* CAM-cyclers behave in this way is still unclear. A recent suggestion is that CAM-cycling may save from 5–44% of daytime water loss because the same amount of CO_2 can be fixed with stomata open less widely or for a shorter time in the day. Figure 2.35 summarizes these various forms of CAM.

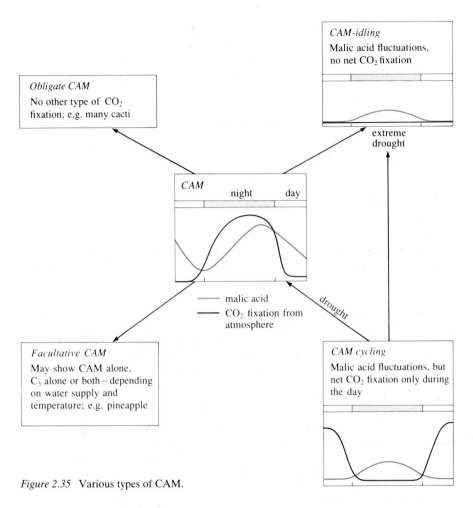

Figure 2.35 Various types of CAM.

About 20 000 species are thought to show CAM or one of the variants just described — roughly 8% of all angiosperms. These include many orchids and the bromeliads that grow as epiphytes (attached to other plants) in the canopy of tropical forests. The majority of these have fleshy, succulent tissues which store considerable amounts of water and possession of CAM can be regarded as a water-saving strategy. However, some outstanding exceptions have been found: certain plants, for example species of the fern ally, *Isoetes*, which grow *submerged* in shallow acid lakes also show CAM. It appears that in these lakes the availability of CO_2 becomes very low during the day but is high at night and possession of CAM is, therefore, related to the fluctuations of inorganic carbon sources. Clearly the adaptive significance of CAM may vary for different species.

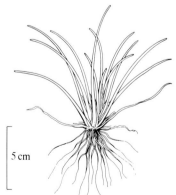

5 cm

Isoetes sp. (a Quill-wort), a CAM plant that grows submerged.

2.5.5 Regulation of carbon fixation

In C_3 plants, carbon fixation involves coordination of at least four metabolic sequences — the light reactions, the Calvin cycle, photorespiration and product synthesis — whilst in C_4 plants the C_4 pathway and in CAM plants the decarboxylation of malate must also be coordinated with everything else. Not surprisingly the control systems are remarkably complex and still not fully understood, so we shall not attempt to describe them in detail here. What follows is a brief description of what control systems achieve in particular circumstances without saying how they do it.

Light–dark transitions

When it becomes dark carbon fixation does not gradually slow down, depleting all the substrate pools, rather it stops quickly — freezes almost — and remains poised to start up again as soon as it becomes light. This is achieved by **light control of enzymes**, which are inactivated in the dark and activated in the light. Several enzymes in the Calvin cycle are involved, most notably Rubisco, and, in C_4 plants, PEP carboxylase (the primary fixation enzyme) is also light-regulated.

Light does not activate these enzymes directly but by influencing the levels of regulatory substances; the mechanism varies for different enzymes. In the case of Rubisco, for example, either or both of the two systems may operate: (i) in some species a specific inhibitor is present in the dark and breaks down in the light; (ii) a Rubisco-activase may operate in the light. The Rubisco inhibitor, discovered in 1986, is a sugar alcohol, carboxyarabinitol-1-phosphate, whose breakdown seems to be linked to the turnover of photosystem II — hence its disappearance in the light. Operation of the Rubisco activase depends on the ATP level and requires high levels of Mg^{2+} ions and high pH — both conditions which apply in the chloroplast stroma only in the light.

◇ From Section 2.4, why would you expect a high stromal pH in the light?

◆ Because light-induced proton pumping into the thylakoid spaces *raises* pH (lowers H^+ concentration) in the stroma.

Flux control in the light

Once photosynthesis has got going in the light the prime requirements are that the available light and CO_2 be used efficiently and a steady, continuous supply of sucrose be maintained for export to developing tissues. In low light, the supply of ATP and NADPH determines the rate of carbon fixation and turnover of the Calvin cycle is tuned in to match but never limit the rate of the light reactions. At the other extreme of light saturation, Rubisco is fully activated with the Calvin cycle going full blast and it is either the supply of CO_2 or the actual level of Rubisco that becomes limiting. Fine tuning of the light reactions and ATP synthetase activity now occur, damping down the supply of ATP and NADPH to match the rates of carbon fixation and photorespiration — although if severe overshoot happens, then photoinhibition starts to operate (Section 2.4.3).

A third situation arises when the bottleneck is either the synthesis or export of end-products from the leaf and an important regulator here is the level of inorganic phosphate (P_i). If, for example, sucrose accumulates in the cytosol, then the first thing to happen is a series of inhibitory feedback reactions which result in a rise of cytosol P_i. This affects the phosphate translocator (Section

2.5.1), which exports less triose phosphate from the chloroplasts and, with the important proviso that *total* phosphate (P_i + esterified phosphate) in the stroma is constant, stromal P_i then rises. Starch synthesis in the chloroplast is strongly promoted by P_i, so there is an immediate switch from sucrose to starch synthesis, which is often sufficient to relieve the original bottleneck. However, if this does not solve the problem and stromal P_i continues to increase, then P_i-mediated inhibition of both the Calvin cycle and the light reactions comes into play.

In C_4 plants, it is important to link the rate of primary CO_2 fixation by PEP carboxylase with turnover of the Calvin cycle. This involves not only light regulation of PEP carboxylase but also feedback control by various Calvin cycle intermediates. This is a striking contrast to CAM plants where PEP carboxylase is active in *darkness* and is inhibited by low pH and high levels of cytoplasmic malate in the light.

Despite great advances in the last five years, the control of photosynthesis is still not well understood. It is an area of very active research because breeding or genetically engineering more productive crops should be easier if the limits and controls on photosynthesis were known. Rising levels of atmospheric CO_2 for example, could shift the control from CO_2 supply to Rubisco levels in C_3 plants under high light and it might be possible to breed plants containing more Rubisco. The conversion efficiency of crops (percentage conversion of photosynthetically active radiation, PAR, to plant tissues) could then be increased: currently the best crop growth rates, averaged over the entire life cycle, are equivalent to only about 6% conversion efficiency of PAR — so there appears to be room for improvement. Bear in mind, however, that in many natural habitats it is not photosynthesis but factors such as mineral nutrients or water supply that limit plant growth: the mechanisms by which plants obtain, transport and conserve these valuable resources are the subjects of the next two chapters.

Summary of Section 2.5

1 The Calvin cycle or C_3 pathway is the only sequence of reactions that gives a *net* fixation of CO_2. Each turn of the cycle fixes one molecule of CO_2 and requires three molecules of ATP and two NADPH. One molecule of the end product, a triose phosphate, is produced after three turns of the cycle.

2 The three phases of the Calvin cycle are: (i) carboxylation, (combination of CO_2 and a 5-carbon acceptor, ribulose bisphosphate or RuBP), (ii) reduction (of the first product, the 3-carbon acid, phosphoglycerate, PGA), and (iii) regeneration (of the acceptor RuBP).

3 The triose phosphate produced by the Calvin cycle may be converted to two major end-products: sucrose (for export from the leaf) and starch (for storage in plastids). Their synthesis involves conversion of triose phosphate to activated glucose: UDP-glucose (from uridine triphosphate) for sucrose and ADP-glucose (from ATP) for starch synthesis.

4 Photorespiration occurs because Rubisco, the first (carboxylating) enzyme of the Calvin cycle also acts as an oxygenase. CO_2 and O_2 compete at the active site but when O_2 combines with RuBP, a 2-carbon acid (phosphoglycolate) is formed that takes no further part in the Calvin cycle. Phosphoglycolate is rapidly converted to glycolate, which is the substrate for photorespiration.

5 Glycolate metabolism involves three organelles: peroxisomes, mitochondria and chloroplasts. An energy input is required (ATP and reduced

ferredoxin), O_2 is consumed, and cycling of a large pool of amino acids occurs. One-quarter of the glycolate carbon is lost as CO_2 but three-quarters are returned to the Calvin cycle as PGA.

6 High rates of photorespiration occur at high temperature and high irradiance because: (a) the substrate specificity of Rubisco shifts in favour of O_2 as temperature increases and (b) $[CO_2]$ in the leaf is rate-limiting and very low at high irradiance because photosynthesis proceeds rapidly.

7 C_4 plants minimize the Rubisco oxygenase reaction and photorespiration by concentrating CO_2 at the site of the Calvin cycle. They fix CO_2 initially into a 4-carbon acid (malate or aspartate) in the mesophyll cells, whose chloroplasts do not contain the enzymes of the Calvin cycle. The 4-carbon acid moves to bundle sheath cells where CO_2 is released and fixed into the Calvin cycle.

8 Kranz anatomy and the C_4 pathway characterize C_4 plants. The pathway utilizes PEP carboxylase for the initial combination of CO_2 and a 3-carbon acceptor, phosphoenolpyruvate (PEP). ATP is required for the regeneration of PEP so there is an extra 'cost' when the C_4 pathway operates alongside the Calvin cycle. Nevertheless, the strategy appears to be advantageous when CO_2 is rate-limiting and the rate of photorespiration is high. C_4 plants occur mainly in hot, sunny habitats.

9 Crassulacean acid metabolism (CAM) has evolved in habitats where shortage of water is the dominant factor affecting survival. In this photosynthetic strategy stomata open at night when transpiration losses are minimal and CO_2 is fixed into the 4-carbon acid, malate, as in C_4 plants. Malate is then stored in cell vacuoles as malic acid, which is released into the cytosol and decarboxylated during the day, when stomata are closed. The CO_2 released is fixed into the Calvin cycle.

10 CAM plants are all succulents (e.g. cacti) and are characterized by diurnal fluctuations of malic acid and of starch (which is used at night to synthesize the CO_2 acceptor, PEP). The CAM pathway may be obligate or facultative during drought. It may operate when stomata are closed all the time, when respiratory CO_2 is re-fixed at night (CAM-idling). CAM cycling is another variant.

11 The light and dark reactions of photosynthesis are regulated such that, as far as possible, the turnover of the Calvin cycle matches the supply of ATP and NADPH. Both the Calvin cycle and the C_4 pathway stop abruptly when it becomes dark because several enzymes, including Rubisco and PEP carboxylase, are controlled by light. Accumulation of sucrose in the cytosol initiates feedback control via changes in phosphate levels: first, the synthesis of starch is promoted but, if chloroplast P_i levels remain high, then inhibition of the Calvin cycle and light reactions occurs.

Question 12 (*Objectives 2.1, 2.12, 2.13, 2.14 and 2.17*) Are the following statements true, partly true or false?

(a) The only energy-requiring step in the Calvin cycle is the reduction of PGA to triose phosphate.

(b) Rubisco is a universal enzyme for CO_2 fixation but catalyses the formation of a 3-carbon acid in C_3 plants and 4-carbon acids in C_4 and CAM plants.

(c) The C_4 pathway can be regarded as an ATP-driven pump which concentrates CO_2 at the site of the Calvin cycle.

(d) Sucrose and starch are major carbohydrate products of the Calvin cycle and both require activated glucose for their synthesis in the cytoplasm.

(e) Photorespiration releases energy from oxidized ribulose bisphosphate, formed when Rubisco acts as an oxygenase.

(f) Carbon fixation is most commonly limited by the level or activity of Rubisco.

Question 13 (*Objective 2.15*) How would you determine whether a plant growing in its natural habitat was a C_3, C_4 or CAM type if your only equipment was a microscope (plus slides, razor blades, etc.) and pH indicator paper?

Question 14 (*Objectives 2.16 and 2.18*) Decide from Figure 2.36 which of species X and Y is a C_3 type and which is a C_4 and explain how you reach your answer.

Question 15 (*Objectives 2.15 and 2.18*) The data in Figure 2.37 relate to a tropical succulent. What type of carbon fixation is occurring? Would you expect this plant to be growing in very dry conditions or with an adequate supply of water?

Question 16 (*Objective 2.16*) Which of (a)–(g) occur (i) only in C_4 plants, (ii) only in CAM plants and (iii) in both types.

(a) Rubisco

(b) light-activated PEP carboxylase

(c) large diurnal fluctuations in malic acid levels

(d) primary fixation of CO_2 into 4-carbon acids

(e) carbon fixation into the Calvin cycle during the day

(f) regeneration of PEP from stored carbohydrate

(g) a low CO_2 compensation point.

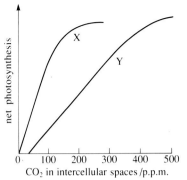

Figure 2.36 The effect of increasing CO_2 concentration on net photosynthesis in two species. Measurements were made at 27 °C and a high (saturating) irradiance.

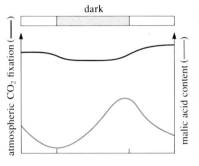

Figure 2.37 Patterns of net CO_2 fixation and malic acid levels in a tropical succulent.

OBJECTIVES FOR CHAPTER 2

Now that you have completed this chapter, you should be able to:

2.1 Define and use, or recognize definitions and applications of, each of the terms printed in **bold** in the text. (*Questions 1, 2, 4, 5, 8, 9, 10 and 12*)

2.2 Give operational definitions (i.e. ones related to the method of measurement) of gross photosynthesis, net photosynthesis, (dark) respiration, photorespiration and compensation point. (*Question 1*)

2.3 Describe the principles of three methods used to measure *GP*, *NP* and *R*, recognize limitations and possible sources of error in the methods and situations where their use is appropriate. (*Question 2*)

2.4 Calculate the rates of *GP*, *NP*, *R* and photorespiration from appropriate data. (*Question 3*)

2.5 List the environmental factors that may limit rates of net or gross photosynthesis, deduce, from appropriate data, which factor is rate-limiting, and predict the conditions when each factor may operate. (*Questions 5 and 6*)

2.6 Give examples to illustrate how plants are adapted to different environments in ways that maximize net photosynthesis and minimize the effects of rate-limiting factors with reference to: C_3 and C_4 plants, sun and shade species, and plants from hot and cold climates. (*Question 7*)

2.7 Describe the principles of light harvesting and the photochemical reactions in photosynthesis. (*Question 8*)

2.8 Recognize the main similarities and differences between respiratory electron transport and the photosynthetic light reactions in green plants. (*Question 9*)

2.9 Compare and contrast the light reactions in photosynthetic bacteria and green plants. (*Question 10*)

2.10 Describe in words or using simple diagrams the Z scheme and cyclic electron transfer in green plants. (*Question 11*)

2.11 Describe the ways in which plants adapt to low and high irradiances such that the use of light in photosynthesis is maximized and damage caused by high light is minimized. (*Question 10*)

2.12 Outline in words or by drawing a simple diagram the main steps in the Calvin cycle and indicate which reactions require an input of energy. (*Question 12*)

2.13 Describe how and where sucrose and starch are synthesized in photosynthesis. (*Question 12*)

2.14 Show how the properties of Rubisco explain: (a) the occurrence of photorespiration and (b) some of the responses of C_3 plants to environmental factors. (*Question 12*)

2.15 Describe the characteristic structural and biochemical features of C_4 and CAM plants and relate these to their physiology and habitats. (*Questions 13 and 15*)

2.16 List the main similarities and differences between C_4 and CAM plants. (*Questions 14 and 16*)

2.17 Outline the main ways in which carbon fixation is regulated. (*Question 12*)

2.18 Interpret and draw relevant conclusions from experimental data relating to the material in this chapter. (*Questions 6, 14 and 15*)

MINERAL NUTRIENTS

3.1 INTRODUCTION

In the Introduction to this book we said that green plants need supplies of three major items for growth and maintenance: (a) energy (as light) which is used to synthesize organic compounds from carbon dioxide, (b) mineral nutrients and (c) water. (a) was the subject of Chapter 2 and in this chapter we look at (b) — how plants obtain, distribute and, sometimes, dispose of inorganic ions. In addition to carbon, hydrogen and oxygen, all vascular plants require at least 13 essential elements in order to build tissues and function properly. These elements can all be obtained from the soil as inorganic ions or molecules, and they are referred to as *mineral* nutrients (in contrast to *organic* nutrients such as sugars). They are listed in Table 3.1.

The first six elements in Table 3.1 are present in relatively large amounts in plant tissues and you may recognize N, P and K as the three components of 'general' fertilizers; they are referred to as **macro-** or **major nutrients** and you should be able to remember these. Nitrogen, phosphorus and sulphur are major components of macromolecules such as proteins and nucleic acids and are present also in a wide range of smaller molecules. Potassium, however, is important chiefly as the free K^+ ion in cell vacuoles, where it contributes substantially to osmotic pressure and is, therefore, an important osmotic

Table 3.1 Essential elements that are obtained by plants mainly from the soil.

Element	Chemical symbol	Form absorbed by plants*	Average relative concentration in dry tissues
Macronutrients			
nitrogen	N	NO_3^- (NH_4^+)	1 000
potassium	K	K^+	250
calcium	Ca	Ca^{2+}	125
magnesium	Mg	Mg^{2+}	80
phosphorus	P	$H_2PO_4^-$ (HPO_4^{2-})	60
sulphur	S	SO_4^{2-}	30
Micronutrients			
chlorine	Cl	Cl^-	3.0
boron	B	H_3BO_3, boric acid	2.0
iron	Fe	Fe^{2+} (Fe^{3+})	2.0
manganese	Mn	Mn^{2+}	1.0
zinc	Zn	Zn^{2+}	0.30
copper	Cu	Cu^+ (Cu^{2+})	0.10
molybdenum	Mo	MoO_4^{2-}, molybdate	0.001

Micronutrients essential to some but not all plants are:
  sodium (Na), silicon (Si) and nickel (Ni).

* Forms in brackets are those less commonly available to plants.

solute. The remaining elements in Table 3.1, often called **micronutrients**, are needed in much smaller amounts. They function in various ways — as enzyme cofactors or components of electron transport proteins, for example. Shortages of any of these elements lead to reductions in growth which, for crops, means a fall in yield. Experienced plant-growers may use **deficiency symptoms**, specific for each element, to diagnose which element is in short supply: for example, phosphorus deficiency is characterized by dark green or blue-green leaves with reddening along the veins.

Fast, luxuriant growth in nutrient-rich soils and hardly any growth at all in poor soils is characteristic of crop plants, especially modern varieties, and explains the universal use of fertilizers. Some wild plants, however, grow (albeit slowly) on extremely nutrient-poor soils without showing any deficiency symptoms, but they show little response to added fertilizer. Clearly, there are differences between plants in their ability to utilize abundant nutrients and to obtain scarce nutrients from soil: Sections 3.5 and 3.7 explore some of the reasons for these differences. Before this (Section 3.2) we examine the pathway of ion transport from roots to shoots, the metabolism of ions — how, if at all, they are altered after uptake — (Section 3.3), and the mechanisms of ion uptake in roots (Section 3.4). Finally (Section 3.8) the ways in which plants deal with unwanted or excess ions and the special problems posed by saline soils are considered.

3.2 LONG-DISTANCE TRANSPORT OF IONS

Ions can be absorbed over the whole surface of submerged plants (such as pondweeds) and of non-vascular plants (bryophytes and algae). For most vascular plants, however, ion uptake is largely confined to the roots, where ions are absorbed along with water from the dilute solution that surrounds soil particles. It is true that small amounts of ions can be absorbed from rainwater by leaves (**foliar feeding**, i.e. spraying leaves with a solution of fertilizer, depends on this absorption route); but, in general, the action starts in the roots. The question asked here is: how do ions move from the root to the growing parts of the shoot, where there is a high demand for mineral nutrients?

3.2.1 From roots to leaves

The journey occurs in two stages: (1) radial transport across the root from the epidermis to the xylem tissue in the middle; and (2) upwards to the shoot in the conducting cells of the xylem, which form part of the intricate network of veins. Figure 3.1 illustrates stage 1, with ions crossing the plasmalemma (the outer cell membrane) and entering the cytoplasm of epidermal cells. They then move across the cortex by diffusion and may equilibrate with reservoirs of ions accumulated in cell vacuoles.

◇ Ions may be transported across the cortex without moving out of the cytoplasm or crossing a plasmalemma. From your knowledge of plant cell structure and Figure 3.1, deduce how this is possible.

◆ Movement occurs from cell to cell via the *plasmodesmata* — the narrow channels containing cytoplasm and lined by the plasmalemma that cross cell walls (see Chapter 1). This is illustrated in Figure 3.1b.

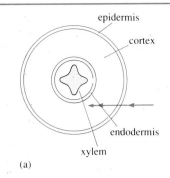

(a)

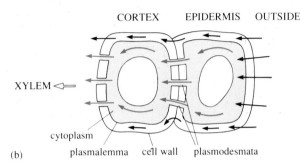

(b)

Figure 3.1 (a) A transverse section of a root showing the pathway of ion transport (green arrows) from epidermis to xylem. (b) A transverse section through cells of the root cortex showing pathways of ion movement in the cytoplasm (green arrows) and the cell walls (black arrows).

This continuum of living cells linked via plasmodesmata is termed the **symplast**, and movement in the way described is called symplastic transport (green arrows in Figure 3.1b). By contrast, the continuum of cell walls outside plasmalemmas is called the **apoplast** and much of the water taken up by roots and some ions (discussed in Section 3.3) move along this path (black arrows in Figure 3.1b). The distinction is important because the concentration of most ions is far higher inside living cells than in the soil solution or the cell walls and movement across the plasmalemma is usually 'uphill' against an electrochemical gradient. Such movement, therefore, requires an input of metabolic energy and plant roots expend a great deal of energy in absorbing ions from the soil. A notable exception here is calcium, whose concentration in most soils exceeds that in root cytoplasm by at least 1 000 fold: this will be discussed in more detail later.

Once ions (and water) reach the central xylem, a much more rapid kind of long-distance transport begins. It is neither in the symplast nor, strictly, in the apoplast but inside the dead, lignified vessels and tracheids (described in Chapter 1) and of a type described as **mass flow**. Another name for this is bulk flow, because the liquid and any substances dissolved or suspended in it move together 'in bulk'. It is caused by differences in pressure (a pressure gradient) that pull or push the column of liquid. For example, mass flow up a drinking straw is due to sucking and so lowering pressure at the top, whereas mass flow along an artery is primarily due to the action of a pump (the heart) that raises pressure and so pushes the blood from behind. Movement up the xylem is usually caused by lowering pressure at the top of the water column, brought about by the evaporation or *transpiration* of water from leaves (although in some circumstances a pushing force from below operates). How such a simple mechanism can produce a *transpiration stream* moving at velocities of about 3–4 metres per hour in trees is discussed in Chapter 5 where we discuss plant water relations. For now, the important point to grasp is that no metabolic energy is involved: the transpiration stream is entirely driven by evaporation from the shoot and ions are carried along willy-nilly.

On the way, ions may diffuse out of xylem vessels into living cells of the stem or trunk but the majority are delivered to leaves, where the xylem pathway ends in a network of fine veins. The system allows no 'targeting' of ions: more mineral nutrients may be delivered by the transpiration stream to a mature leaf, where the demand for them is quite low, than to young leaves at the stem apex, where the demand is very high. The usual way round this problem is to transfer surplus ions to the second long-distance transport system in plants — the phloem.

3.2.2 Transport of ions in phloem

Phloem serves primarily as a transport system for sugars and amino acids, but redistribution of inorganic ions is an important secondary function. Unlike xylem transport, phloem transport occurs in living cells (the sieve tubes, described in Chapter 1) and is 'demand-led'. Sugars move from areas where they are produced and in surplus to areas where they are used and 'in demand'. How this system works is described in detail in Chapter 4; the point here is that tissues with a high demand for sugars commonly have a high demand for mineral ions so, by entering sieve tubes, ions are carried along to the places where they are most needed. This means that surplus ions transported in the xylem to (say) a mature leaf may pass rapidly into the adjacent phloem and move to growing leaves, fruits, storage organs or back to the roots. Table 3.2 shows an example of such redistribution in young barley plants.

Table 3.2 Potassium turnover in the leaves of a young barley plant that was supplied with radioactively labelled potassium (^{42}K) for a brief period.

	Oldest leaf	Second leaf	Youngest emerged leaf
Initial dry weight of leaf/mg	20	13	1.8
Initial K content/μmol	46	26	2.9
Intake of K per day/μmol:			
^{42}K from roots	1.9	2.7	2.0
stable K from other tissues	−1.6	0.7	1.3
net intake	0.3	3.4	3.3

◇ From Table 3.2, what happens to most of the potassium entering the oldest leaf?

◆ It moves out (negative intake from other tissues) and passes to younger leaves, chiefly the youngest emerged leaf. This transfer occurs in the phloem.

Efficient lateral transport from xylem to phloem is often associated with the presence of special **transfer cells** that promote the active removal of ions from the xylem. The electron micrographs of xylem transfer cells in Plates 11 and 12 illustrate two important features: (i) the wall of the transfer cell adjacent to a xylem vessel is elaborately corrugated and (ii) there are large numbers of mitochondria adjacent to the corrugated wall (Plate 11). Feature (i) affords a huge surface area for uptake of ions from xylem and feature (ii) facilitates the supply of ATP necessary for the active transport of ions across the corrugated junction.

Transfer cells are not concerned solely with exported ions from xylem however. They occur in many situations where large quantities of ions or organic solutes move into or out of conducting cells or storage tissues. Figure 3.2 illustrates the complex transport pattern around xylem and phloem elements and there is further discussion of the roles of transfer cells in Chapter 4.

Frequently, ions and soluble organic molecules containing N and P also move out of the dying leaves of perennial plants before the leaves are shed, thus keeping them in the plant body for re-use. This elegant circulation and redistribution system does not, however, work equally well for all mineral nutrients. Table 3.3 shows the relative mobility in phloem of some essential nutrients and non-essential 'pollutants'.

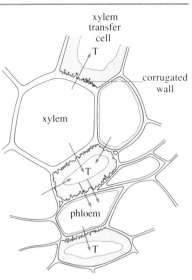

Table 3.3 Relative mobility of mineral elements in phloem (Transport is as inorganic forms unless indicated otherwise.)

Highly mobile	Intermediate mobility	Immobile
potassium	iron	calcium
nitrogen (as organic N)	manganese	strontium
caesium	zinc	barium
sodium	copper	boron
magnesium	molybdenum	lead
phosphorus		polonium
sulphur (mainly as organic S)		
chlorine		

Figure 3.2 The position and direction of transport (green arrows) into and out of transfer cells (T) adjacent to the xylem and phloem.

Among the former, calcium and boron stand out in having very little mobility in phloem and this has profound implications both for plant nutrition and for animals that eat plants.

◇ How will the concentrations of Ca and B change during the life of a leaf?

◆ They will increase steadily as delivery in the xylem continues with little export in the phloem. (This means that old leaves are the best source of Ca for people who don't eat animal products.)

◇ Why do you think that potato tubers have a very low Ca content?

◆ Because, being underground, they do not transpire and are largely bypassed by the transpiration stream. Thus they are supplied mainly through phloem transport.

The phloem immobility of lead and strontium may also have consequences for herbivores because these elements are often absorbed mainly through the leaves and so remain there to be consumed. Despite the growing use of lead-free petrol, Pb is still a major air pollutant from car exhaust fumes and the long-lived radioisotope ^{90}Sr is a major component of radioactive fall-out. Once they have entered leaves, these elements stay there until the leaf dies.

One reason for the differing mobility of ions in phloem is the ease with which they enter sieve tubes. Entry is not via plasmodesmata here but by movement across the plasmalemma, which is determined by the availability of ion-specific carriers. The necessary carriers are thought to be unavailable for phloem-immobile ions. For calcium and strontium there is a second reason why they are excluded from sieve tubes: energy-driven pumps in the plasmalemma remove these ions from cytoplasm (where their levels are always low) and deposit them in the cell walls. Furthermore, Ca, Sr and Pb could never be transported in quantity by sieve tubes, even if they could get in. At pH values above 7, these ions form insoluble phosphates which precipitate out of solution, and within sieve tubes there is both a high phosphate level and a pH of around 8!

Summary of Sections 3.1 and 3.2

1 In addition to the elements C, H and O, plants require at least 13 other essential elements for growth. Most of these are absorbed as ions from the soil and they are referred to as mineral nutrients.

2 Six mineral nutrients are required in large amounts (the macronutrients) and the remainder in relatively small amounts (the micronutrients). Both shortage and excessive absorption of any nutrient may cause reduced and abnormal growth. Shortages cause deficiency symptoms that are nutrient-specific.

3 For terrestrial plants, mineral nutrients are absorbed chiefly by roots from the soil solution. They move across the root cortex to the central xylem and up to the shoot in xylem vessels or tracheids.

4 For most nutrients, radial transport across the root cortex is in the symplast (cytoplasmic continuum) but some ions are transported in the apoplast (cell walls). Long-distance transport in xylem is by mass flow and nutrients are delivered in bulk to transpiring organs via the transpiration stream.

5 This transport system does not target mineral nutrients to growing organs where there is a high demand. Redistribution from older leaves to growing parts of the shoot, to fruits and storage organs, and back to roots occurs in the phloem.

6 The transfer of mineral nutrients from xylem to phloem is often facilitated by transfer cells.

7 Some ions, notably calcium and also boron, lead and strontium, are largely excluded from phloem and these phloem-immobile elements accumulate in older leaves.

Questions relating to Sections 3.1 and 3.2 are on p. 126.

3.3 METABOLISM OF IONS

Some mineral nutrients may remain in the plant in an unchanged state (e.g. K^+, Na^+, Cl^-) while others are incorporated directly into organic molecules after transport to their final destination (e.g. phosphate, iron, Ca^{2+}). This applies to animals as well as plants and is why supplementary minerals can be supplied as inorganic ions, for example as 'salt licks' to domestic animals or as 'mineral tablets' to humans. However, there are two major exceptions to this rule: the utilization of nitrogen and sulphur depends on chemical reduction and assimilation into organic molecules, i.e. they must be *metabolized*, just as carbon is metabolized when fixed in photosynthesis. The reduction reactions for nitrate and sulphate are energetically very expensive: only plants and certain micro-organisms do this and so are able to survive with entirely inorganic sources of nitrogen and sulphur. Other organisms must obtain pre-metabolized N and S in organic compounds, either from their food or from symbiotic micro-organisms in their guts. Because nitrogen metabolism is a key aspect of plant mineral nutrition and, furthermore, influences both the uptake and transport of N, we consider it in some detail (Section 3.3.1). Sulphur metabolism is discussed briefly in Section 3.3.2.

3.3.1 Nitrogen metabolism

Inorganic nitrogen is present in soils mainly as the nitrate ion (NO_3^-), unless conditions are acidic or waterlogged when ammonium ions (NH_4^+) predominate. Either of these ions can be taken up and used by plants but, whereas NO_3^- is often transported in xylem or stored in cell vacuoles, NH_4^+ cannot be, because it is toxic. The main reason for this toxicity is that high concentrations of NH_4^+ cause a lowering of cell pH: the equilibrium $NH_4^+ \rightleftharpoons NH_3 + H^+$ is far over to the right and so the concentration of H^+ rises. For cells to use nitrate, however, it must first be reduced to ammonia — one of the most important reduction reactions in plant metabolism: we consider this first.

Nitrate reduction

The overall reduction of nitrate to ammonia can be represented as:

$$NO_3^- + 8H^+ + 8e^- \longrightarrow NH_3 + 2H_2O + OH^- \tag{3.1}$$

◇ What will happen to cell pH when nitrate reduction is occurring?

▸ It will tend to *rise* because there is a net formation of hydroxyl (OH^-) ions.

The reduction occurs via two separate reactions, each catalysed by a different enzyme and occurring in a different cell compartment, as shown in Figure 3.3.

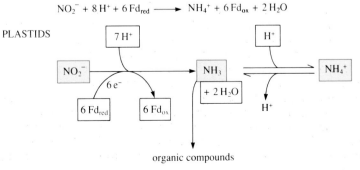

Step 1: nitrate reduction

$$NO_3^- + NADH + H^+ \longrightarrow NO_2^- + NAD^+ + H_2O$$

CYTOPLASM

Step 2 : nitrite reduction

$$NO_2^- + 8H^+ + 6Fd_{red} \longrightarrow NH_4^+ + 6Fd_{ox} + 2H_2O$$

PLASTIDS

organic compounds

Figure 3.3 The two steps in the reduction of nitrate to ammonia. (Fd_{red} and Fd_{ox} are the reduced and oxidized forms, respectively, of ferredoxin.)

Nitrate is first reduced to nitrite (NO_2^-) by the cytoplasmic enzyme **nitrate reductase**, a large multi-enzyme complex that requires molybdenum for activity. High levels of nitrate reductase are found in fast-growing plants — such as nettles and high-yielding arable crops — when they grow on nitrate-rich soils, whereas slow-growing heathers on acid, nitrate-poor soils contain very little of the enzyme.

Nitrite is then reduced to ammonia by the plastid-localized enzyme **nitrite reductase** and it is this step that requires the largest input of reducing power. There is about ten times more nitrite reductase than nitrate reductase, so nitrite ions never build up in plant cells — which is fortunate for us (as plant eaters) because nitrite can be converted to carcinogenic compounds in the human gut.

The NH_3 formed may be incorporated directly into organic compounds but, at physiological pH values, it combines with a proton, increasing cell pH even more, to give NH_4^+. The H^+ is released when NH_4^+ is metabolized and there is an important implication here for plants that take up NH_4^+ ions directly from the soil: NH_4^+ metabolism will result in a *net release* of H^+, so that cell pH is *reduced*. Cytoplasmic pH is under strict homeostatic control in all cells, so this tendency for nitrate metabolism to raise pH and ammonium metabolism to reduce it must be countered and we shall describe later how this is done.

Assimilation into organic compounds

The toxicity of NH_3 and NH_4^+ means that once nitrate has been fully reduced or ammonium ions absorbed from soil, rapid assimilation into organic compounds is essential. The main pathway is as follows:

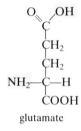

glutamate

glutamine

where Fd_{red} is reduced ferredoxin (i.e. reducing power) and Fd_{ox} is the oxidized form of ferredoxin (see Chapter 2, Sections 2.4.2 and 2.5.2).

In reaction (i) an amino group ($-NH_2$) from NH_3 is added to the amino acid glutamate, to give the amide, *glutamine*. The reaction uses ATP and is catalysed by the enzyme *glutamine synthetase* or GS. In reaction (ii) the 'extra' amino group of glutamine is transferred to the carboxylic acid, α-oxoglutarate, producing two molecules of glutamate, one of which is recycled to reaction (i). The enzyme *glutamate synthetase* (or in modern terminology, glutamine–oxoglutarate aminotransferase, GOGAT) and a source of reducing power are required for this reaction. Together, reactions (i) and (ii), the GS–GOGAT pathway, give a net production of one molecule of glutamate.

All the other amino acids and nitrogenous compounds in the plant can be synthesized from either glutamate or glutamine. Glutamine is also an important storage compound for nitrogen, abundant in storage organs such as potato tubers, carrots and radish, for example, and much of the N transported in xylem and phloem may be in this form. If you now look back at all the steps involved in nitrate reduction and nitrogen assimilation, three points emerge:

1 A great deal of energy, as ATP or (especially) reducing power, is required, as mentioned earlier.

2 There are potential problems for homeostatic pH control.

3 Carbon compounds, as glutamate and α-oxoglutarate, are necessary for assimilation.

To see how these points are resolved, we must consider where in a rooted green plant the various steps in nitrogen metabolism are carried out.

Where does it all happen?

It is not possible to generalize because the answer to this question varies for different plant species, as Figure 3.4 illustrates.

◇ From Figure 3.4, what are the main sites of nitrate reduction and nitrogen assimilation in (a) radish and (b) chickweed?

◆ Remember that xylem sap flows from roots to shoots, and so its composition reflects metabolic activity in the roots. The low nitrate level and high levels of amino acids and amides in the xylem sap of radish indicate that nitrate reduction and N assimilation occur mainly in the radish *roots*. Conversely, the high nitrate level and low levels of amino acids and amides in the xylem sap of chickweed indicate that nitrate reduction and N assimilation occur mainly in the *shoots* (presumably the leaves) of chickweed.

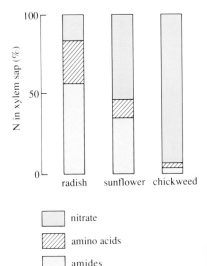

Figure 3.4 Relative amounts of nitrogen compounds in the xylem sap of radish (*Raphanus sativus*), sunflower (*Helianthus annuus*) and chickweed (*Stellaria media*). Plants were grown with their roots in sterile sand and watered with a sterile nutrient solution containing 10 mmol 1⁻¹ nitrate. Stems were then severed and xylem sap exuding at the cut surface was collected and assayed.

Consider first the situation where nitrogen metabolism occurs mainly in leaves, which is the norm for nitrate-rich soils. Photosynthesis ensures abundant supplies of ATP and reducing power (as reduced ferredoxin or NADPH) and of the necessary carbon skeletons, but regulation of cell pH poses a problem. The OH^- ions produced by nitrate reduction must be neutralized while, at the same time, maintaining the balance of positive and negative charges. Figure 3.5 illustrates how this can be achieved by the synthesis of organic acids.

Nitrate ions are delivered from the xylem with the negative charge balanced by K^+ ions. Nitrate reduction generates OH^- ions (Equation 3.1) that must

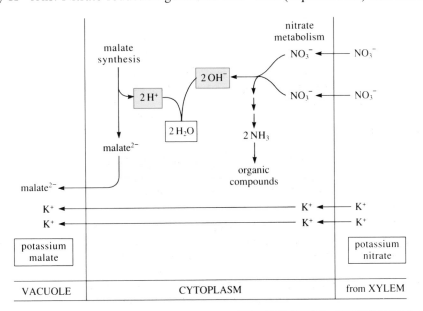

Figure 3.5 A model illustrating how electrical charge and pH are balanced during nitrate metabolism in leaves.

be balanced by H^+ ions if cell pH is not to rise. Malic acid is synthesized by the same reactions as in the C_4 pathway (Chapter 2, Section 2.5.3) and dissociates to give a malate anion (malate^{2-}) and two H^+ ions, which do the neutralizing job. This leaves potassium malate which can be stored in the cell vacuole or exported in the phloem.

Next consider the situation for nitrate reduction and assimilation in roots. Here, excess OH^- ions can be pumped out into the soil solution (or H^+ ions allowed to diffuse in) but obtaining the necessary energy and carbon compounds is a problem. They must be imported via the phloem, for example as sugars or as potassium malate which, as mentioned above, is produced by nitrate reduction in leaves. It appears that the balance between leaf and root nitrate reduction may be closely tied to malate traffic in phloem, with K^+ ions balancing electrical charge on the way down (phloem) and recirculating in xylem.

Similar import is required when NH_4^+ ions are taken up from the soil but, in this case, cell pH is maintained by pumping *out* the excess H^+ ions. Table 3.4 illustrates the consequences of this for soil pH.

Table 3.4 Effect of soil pH and supply of either ammonium or nitrate ions on pH of the soil adjacent to roots of young soya bean plants.

Initial soil pH	Final soil pH adjacent roots	
	NH_4^+ supplied	NO_3^- supplied
5.2	4.71	6.60
6.3	5.60	7.05
6.7	6.25	7.19

◇ From the data in Table 3.4, what is the effect on soil pH when plants take up (a) NH_4^+ and (b) NO_3^-?

◈ (a) The soil tends to become more acid (pH falls), especially close to the surface of roots. (This is one reason why the commonly used fertilizer ammonium sulphate increases soil acidity.)

(b) The soil tends to become more alkaline (pH rises).

3.3.2 Sulphur metabolism

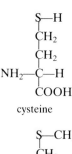

cysteine

methionine

The essential reaction of sulphur metabolism that plants carry out but animals cannot is the reduction of sulphate (SO_4^{2-}) to sulphide (S^{2-}), most of which is incorporated into amino acids such as cysteine or methionine. Many roots reduce sufficient sulphate to meet their own needs but the bulk of sulphate reduction is carried out in leaf chloroplasts after sulphate has been transported from roots to leaves in the xylem. It can be summarized as:

$$SO_4^{2-} + ATP + 8H^+ + 8e^- \longrightarrow S^{2-} + 4H_2O + AMP + PP_i \qquad (3.3)$$

◇ How does this reaction *differ* from the reduction of nitrate (Equation 3.1)?

◈ ATP is used in addition to a very large amount of reducing power and there is no net production of OH^-, so cell pH is unaffected.

Equation 3.3 summarizes a series of reactions which will not be described in detail. In effect the sulphate is reduced whilst attached to a series of carrier molecules, the first of which is ATP (with SO_4^{2-} replacing the last two phosphate groups in the side chain). In leaves, the reducing agent is reduced ferredoxin, as for nitrite reduction.

Summary of Section 3.3

1 The utilization of nitrate and sulphate by plants and certain prokaryotes involves reduction and assimilation into organic molecules.

2 This ion metabolism is energetically expensive and requires large amounts of ATP and reducing power. It may occur in roots but often occurs mainly in leaves.

3 Nitrate metabolism involves, first, reduction to ammonia and then assimilation into amino acids. Reduction occurs in two steps: (1) $NO_3^- \rightarrow NO_2^-$ (the rate-limiting step and catalysed by nitrate reductase) and (2) $NO_2^- \rightarrow NH_3$ (catalysed by nitrite reductase), the NH_3 being converted largely to NH_4^+ at cytoplasmic pH values. Reducing power is required and there is a net formation of OH^- ions.

4 Assimilation of NH_3/NH_4^+ requires ATP, reducing power and carbon skeletons (glutamate and α-oxoglutarate). The amide, glutamine, is formed first and subsequently the amino acid, glutamate.

5 Ammonium ions may be absorbed by roots and assimilated after loss of a proton ($NH_4^+ \rightleftharpoons NH_3 + H^+$). Utilization of NH_4^+ thus lowers pH, whereas utilization of NO_3^- raises pH (by producing OH^- ions).

6 pH homeostasis in cells requires that any excess OH^- or H^+ ions generated during nitrogen metabolism must be neutralized. This may be achieved in roots by pumping out H^+ ions into the soil solution (NH_4^+ assimilation) or pumping out OH^- ions (NO_3^- reduction). In leaves, excess OH^- ions from nitrate reduction may be neutralized by H^+ ions derived from malic acid, with K^+ ions balancing the charges on the malate anion. Potassium malate may be stored in the vacuole or exported in phloem.

7 Exported malate may be used in roots as a source of energy and carbon skeletons for nitrate reduction, thus coordinating nitrate reduction in leaves and roots.

8 Sulphate metabolism occurs mainly in leaves and involves the reduction of SO_4^{2-} to sulphide (S^{2-}), which is then assimilated into the amino acids cysteine or methionine.

Questions relating to Section 3.3 are on p. 126.

3.4 ION UPTAKE BY ROOTS

So far we have considered what happens to mineral ions once they have entered root cells — where they go and how they are metabolized. The question now is: how do essential ions enter root cells in the first place? Two points should be emphasized:

1 In most situations, ion uptake into roots is dependent on plant metabolism and requires an expenditure of energy, i.e. it is an *active* process. The only significant exceptions are for the uptake of borate, silicate and, sometimes, calcium, which are usually present at higher concentrations in the soil solution than in cytoplasm and appear to enter roots passively. This also happens in saline soils, where essential ions such as chloride are present at high concentrations in the soil solution.

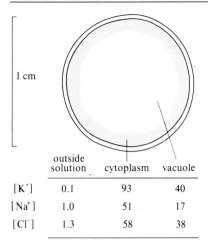

	outside solution	cytoplasm	vacuole
[K⁺]	0.1	93	40
[Na⁺]	1.0	51	17
[Cl⁻]	1.3	58	38

Figure 3.6 A giant algal cell (*Hydrodictyon*) showing values for ion concentrations, measured as mmol 1^{-1}, in different cell compartments and in the external solution.

2 Ion uptake is highly *selective*, so that, for example, an essential ion such as K^+ is taken up much more readily than the similar but non-essential Na^+ ion. This principle of **ion selectivity** is illustrated in Figure 3.6 for giant cells of the alga *Hydrodictyon*. The size of these cells makes them easy to study using techniques such as ion-selective microelectrodes, where electrodes that measure the concentration of specific ions can be inserted into different cell compartments.

◇ Which ion is accumulated most 'strongly' by *Hydrodictyon*? What is the increase in concentration of this ion in the cytoplasm relative to the exterior?

◆ K^+, which shows a 930-fold increase in concentration between outside solution and cytoplasm, far greater than the increase for either Na^+ or Cl^-.

The existence of transport proteins (carriers), highly specific for particular ions, explains the selective uptake of ions. To explain active transport, however, we need to consider the nature of the driving force that moves ions into roots.

3.4.1 Mechanisms of active uptake

Active transport is a fundamental process in all living cells, essential for maintaining cellular homeostasis and, in animals, for activities such as digestion, excretion and osmoregulation*. The essence of ion uptake into plant roots is summarized in Figure 3.7. A proton-pumping ATPase in the plasmalemma pumps H^+ *out* of cells, resulting in a gradient of hydrogen ion concentration (or ΔpH) and a gradient of electrical charge (a potential difference, p.d., alternatively described as an electrical driving force, ΔE, or membrane potential). These gradients then *drive* the uptake of other ions, whose transport is coupled to H^+ movements by their ion-specific transport proteins.

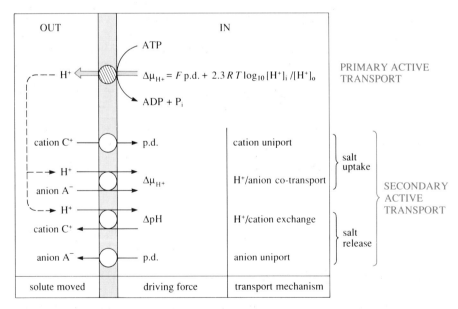

Figure 3.7 Secondary active transport of ions by root cells driven by the primary active transport of protons by an H^+-ATPase (hatched circle). p.d. = potential difference (gradient of electrical charge); $\Delta\mu_{H^+}$ = the electrochemical gradient; ΔpH = the pH (or H^+ concentration) gradient.

*These activities are described in Chapters 8, 11 and 12, respectively, of Michael Stewart (ed.) (1991) *Animal Physiology*, Hodder & Stoughton Ltd in association with The Open University (S203, *Biology: Form and Function*, Book 3). Active transport is discussed more fully in Norman Cohen (ed.) (1991) *Cell Structure, Function and Metabolism* (Book 2 in the same series).

Because it is directly linked to ATP hydrolysis, proton movement of this type is described as **primary active transport** and this applies also to proton pumping linked to photosynthetic or respiratory electron transport. Combining the electrical and concentration components of the proton gradient, the overall electrochemical potential gradient for protons, $\Delta\mu_{H^+}$ (where μ is the greek letter mu), can be expressed as

$$\Delta\mu_{H^+} = F\,\text{p.d.} + 2.3RT\log_{10}[H^+]_i / [H^+]_o \tag{3.4}$$

where F is the Faraday constant, R the universal gas constant, T absolute temperature and the subscripts i and o for proton concentrations refer to inside and outside root cells. $\Delta\mu_{H^+}$ is equivalent to the proton-motive force, a term that is more often used when talking about ATP synthesis — not ion uptake, as here.

Ion uptake into roots that is driven by $\Delta\mu_{H^+}$ is described as **secondary active transport**, and it may operate in two ways (illustrated in the lower part of Figure 3.7). The membrane potential created by proton pumping has an inside-negative/outside-positive distribution of electrical charge; this drives cations into root cells via their transport proteins, 'attracted' by the negatively charged interior and down *their* electrochemical potential gradient. This system is described as **cation uniport** and it is driven by the p.d. component of $\Delta\mu_{H^+}$.

At the same time, protons tend to diffuse back into cells down their electrochemical potential gradient, $\Delta\mu_{H^+}$, and if this flow is coupled by suitable transport proteins to the movement of other solutes, including anions, they can enter the cell *up* a gradient of both concentration and electrical charge. The system shown in Figure 3.7 for anions is described as *H^+/anion co-transport* although, since solutes such as amino acids and sugars can enter cells in this way, a more general term is **H^+/solute co-transport**. It is driven by $\Delta\mu_{H^+}$.

For completeness (and because you will meet it later in this chapter), we have also included in Figure 3.7 two systems of secondary active transport by which cations and anions can be moved *out* of cells. For cation export, protons moving back into a cell are exchanged for cations moving out. The cations are now moving *up* a gradient of electrical charge and the driving force is, in effect, the pH gradient. This is described as **H^+/cation exchange**. It is used, for example, to get rid of unwanted sodium ions in some plants. Anion export works in the same way as (but in the opposite direction to) the cation uniport system above. It is similarly driven by the p.d. and is called **anion uniport**.

The only other ions that may undergo primary active transport in vascular plants are Ca^{2+} and Cl^- but this involves pumping ions *out* of the cytoplasm and not in. In most other organisms, however, primary active transport of sodium ions occurs (the sodium pump) and, in animals, it is the electrochemical potential gradient of Na^+, $\Delta\mu_{Na^+}$, rather than $\Delta\mu_{H^+}$, that drives the uptake of other ions.

◇ Using Equation 3.4, write out the equation for $\Delta\mu_{Na^+}$, the electrochemical potential difference for Na^+ set up by the sodium pump.

◆ $\Delta\mu_{Na^+} = F\,\text{p.d.} + 2.3RT\log_{10}[Na^+]_i / [Na^+]_o$

In vascular plants, however, it is proton pumping that powers ion uptake into roots — and many other transport processes. In marked contrast to animals, there seems to be no primary active Na^+ transport and secondary active transport is never linked to $\Delta\mu_{Na^+}$ in vascular plants.

Electron transport

Earlier in this discussion about active transport we mentioned that primary ion pumping may be driven not only by ATP but also by electron transport — as occurs in photosynthetic and respiratory electron transport. There is some evidence that proton pumping out of plant cells, including roots, may be linked to an electron transport chain in the plasmalemma, which is commonly referred to as the **plasmalemma redox chain**. What the carriers are in this chain and how much it contributes to $\Delta\mu_{H^+}$ is still unknown, but it may turn out to be quite important.

A plasmalemma redox chain may also be involved in the uptake of iron (a topic discussed more fully in Section 3.5.3). In most soils, iron is present as Fe^{3+}, the Fe(III) form, but it is taken up by roots as the Fe(II) form, Fe^{2+}. What happens is that a reducing agent is released at the root surface and converts Fe^{3+} to Fe^{2+}.

◇ How might a plasmalemma redox chain be involved here?

◈ Fe^{3+} could serve as the terminal electron acceptor for the chain, becoming reduced when it does so.

A model indicating how both proton secretion from roots and iron reduction may be linked to a plasmalemma redox chain is shown in Figure 3.8.

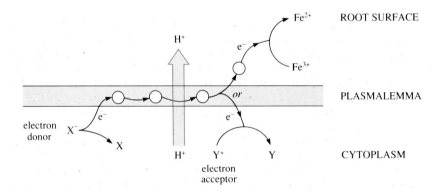

Figure 3.8 A model for a plasmalemma redox chain in root cells linked to proton secretion and iron reduction. Reduction of Fe^{3+} is shown as a possible alternative to reduction of an intracellular substrate, Y^+.

3.4.2 Uptake into the xylem

Once ions have crossed the plasmalemma and entered cells near the root surface, some remain and are used there but the majority move on; they cross the cortex in the symplast (as described in Section 3.2) and eventually enter the lignified conducting elements of xylem. There is still uncertainty about whether ions diffuse passively into xylem vessels via transport proteins (facilitated diffusion) or enter by secondary active transport driven by proton pumping from adjacent xylem parenchyma cells (see Chapter 1). Figure 3.9 illustrates the situation envisaged with a passive xylem entry or *one-pump model*; the critical requirement for this model is that transport proteins allowing ion *exit* must be more abundant in the xylem parenchyma than at the root periphery.

Active secretion from the xylem parenchyma could be especially important when ion concentrations in xylem sap approached those in root cell cytoplasm, which is most likely to happen when the transpiration stream is virtually static.

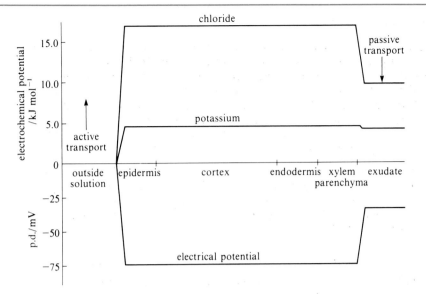

Figure 3.9 A 'one-pump' model of radial ion transport summarizing the forces acting on K^+ and Cl^- ions moving from the epidermis to the xylem in maize roots. Ions were collected in the xylem exudate. It is assumed that secondary active uptake occurs in the epidermis and that movement across the root is passive. p.d. = the electrical potential difference across a membrane relative to the outside solution.

Note: Active transport occurs when movement is *up* the electrochemical gradient and passive transport when it is *down* the gradient.

◇ In what situations will the transpiration stream be static?

◆ When stomatal pores are closed (e.g. at night) so that little transpiration occurs or, in deciduous plants, before the leaves expand in spring. The bleeding xylem sap from a vine stem cut in spring contains high levels of ions.

How active secretion of ions into the xylem might work is still not resolved, although H^+/cation exchange and anion uniport is one possibility (Figure 3.7). Figure 3.10 illustrates the situation with a *two-pump model* for ion uptake and radial transport in roots. Whichever model is nearer the truth, the nature of the ion-specific transport proteins involved in xylem influx has implications for the control of transport. There is evidence that shoots 'signal' their ion requirements to roots and that the movement of ions into root xylem is adjusted accordingly; plant hormones (or growth regulators) moving from shoots to roots in the phloem are likely candidates as signals. Control points close to the xylem, and regulated by substances released from nearby phloem, would provide a sensitive system for coordinating the movement of ions from roots to shoots, and transport proteins in the stele appear to be especially sensitive to chemical control.

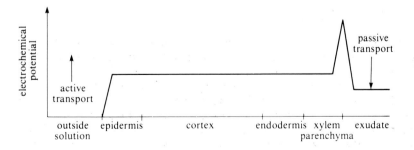

Figure 3.10 A 'two-pump' model for the radial transport of ions across the root cortex and into the xylem.

Calcium: the special case

Calcium ions are a notable exception to the pattern of uptake just described, i.e. active uptake, symplastic transport to the xylem, active secretion or passive diffusion into xylem vessels. The basic reason is because the cytoplasmic concentration of Ca^{2+} is always low (1 μmol 1^{-1} or less) and powerful homeostatic mechanisms operate to keep it low: for example, Ca^{2+}-ATPases pump calcium out of the cytoplasm and into the cell wall (a primary active transport system) and H^+/Ca^{2+} exchange causes movement of Ca^{2+} into the vacuole (a secondary active transport system driven by proton pumps on the vacuolar membrane). This means that very little Ca^{2+} moves across the root cortex in the symplast and instead it moves passively by diffusion along the cell wall or apoplast route. However, there is a problem here in the form of the *endodermis* (Chapter 1).

◇ From your knowledge of endodermal structure from Chapter 1, why will the endodermis pose a problem for apoplastic transport?

◆ The endodermal cell wall develops a waterproof (suberized) band, the *Casparian strip*, running right round the cell (Figure 3.11a and b); ions cannot diffuse across this band.

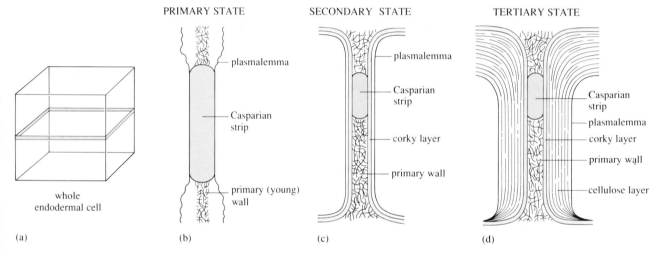

PRIMARY STATE   SECONDARY STATE   TERTIARY STATE

whole endodermal cell

plasmalemma — Casparian strip — primary (young) wall

plasmalemma — Casparian strip — corky layer — primary wall

Casparian strip — plasmalemma — corky layer — primary wall — cellulose layer

(a)   (b)   (c)   (d)

Figure 3.11 (a) Diagram of a whole endodermal cell with the Casparian strip shown in green. (b)–(d) Endodermal cell walls of a monocotyledon in cross-section: (b) primary state; (c) secondary state; (d) tertiary state. In the secondary state a corky (suberin) layer is deposited over the whole inner surface of the wall. In the tertiary state a thick cellulose layer is deposited on the inner wall surface and the whole wall may become lignified.

In roots of monocotyledons, the endodermis develops, in addition to the Casparian strip, secondary and tertiary layers of waterproof thickening, as shown in Figure 3.11c and d; these block the apoplastic route even more effectively. So how does calcium cross the endodermis? There are only two possibilities: either (i) Ca^{2+} enters endodermal cytoplasm on the cortical side and is pumped out into the cell wall on the other side — a kind of symplast loop which could operate only *before* secondary wall thickening developed; or (ii) Ca^{2+} moves across in the apoplast in the few regions where there is no 'waterproofing' of the endodermal wall. This last applies close to the root tip, where secondary roots grow out through the endodermis, and at scattered 'passage' cells along the root. The question is still not fully resolved, although the consensus is that Ca^{2+} transport *is* largely apoplastic. Consistent with this view, Ca^{2+} uptake and transport to xylem is relatively insensitive to inhibitors of respiration compared with ions transported in the symplast.

◇ In what way are the data in Figure 3.12 *difficult* to reconcile with the view that transport is purely apoplastic?

◆ In these roots, Ca^{2+} is transported most readily to the xylem about 8 mm behind the tip, i.e. not at the extreme tip but in a region where the Casparian strip is fully developed. This suggests symplastic transport may be involved. However, secondary thickening of the endodermis clearly restricts the passage of Ca^{2+} and there is a close correlation between decreasing Ca^{2+} transport and increased secondary thickening.

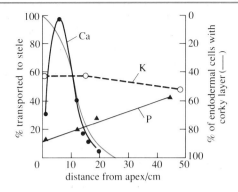

Figure 3.12 The transport of Ca^{2+}, K^+ and phosphate across roots and into the xylem for a monocot species in relation to distance along the root and the development of thickening in the endodermis. Black lines show ion transport to xylem as a percentage of total ions taken up by a 3.5 mm length of root. The green line shows the proportion of cells that had corky layers (secondary state thickening) in the endodermis.

With the probable exception of calcium, therefore, the uptake of ions into roots requires a supply of energy, proton-pumping ATPases (primary active pumps), and ion-specific carriers (for secondary active transport). The number and distribution of these carriers will determine largely which ions get in where. Transport to the shoot requires the same sorts of factors operating in the xylem parenchyma adjacent to conducting cells. It does not follow, however, that what is *required* for uptake necessarily *determines* the rate at which a whole plant absorbs mineral ions from the soil. This depends on numerous other factors which are described in the next two sections.

Summary of Section 3.4

1 Ion uptake into roots is highly selective because movement across the plasmalemma occurs mainly via ion-specific transport proteins or carriers. The types, number and distribution of these carriers determine to a large extent which ions are taken up in particular places.

2 For the majority of ions, uptake into roots also requires metabolic energy (ATP). Uptake is by secondary active transport driven by the electrochemical gradient set up by proton pumping *out* of root cells.

3 Proton export is by primary active transport, where ATP is used directly by a H^+-ATPase to pump H^+ ions across the plasmalemma. The gradient of electrical charge (potential difference, p.d.) and of hydrogen ion concentration together make up the proton-motive force, alternatively called the electrochemical potential gradient for protons, $\Delta\mu_{H^+}$ (Equation 3.4).

4 Cation uptake is driven by the electrical gradient (p.d.) in cation uniport systems; anion uptake is coupled to the re-entry of protons into the cell (H^+-anion co-transport) driven by $\Delta\mu_{H^+}$.

5 Cations and anions can be moved *out* of cells, by H^+/cation exchange and anion uniport, respectively. The former is, in effect, driven by ΔpH and the latter by the p.d.

6 A still undefined electron transport chain in the plasmalemma (the plasmalemma redox chain) may also bring about proton export and contribute to $\Delta\mu_{H^+}$. It also appears to facilitate the uptake of iron by reducing Fe^{3+} to Fe^{2+} at the root surface.

7 Following energy-linked uptake into the epidermis or outer cortex of the root, ions diffuse in the symplast to the xylem. It is possible that efflux from xylem parenchyma into the dead xylem cells is also energy-linked — the two-pump model of radial transport — but passive entry (the one-pump model) is also feasible.

8 Calcium ions do not behave in this way, because $[Ca^{2+}]$ in the cytoplasm is usually much *lower* than in the soil solution. Ca^{2+} diffuses across the root in

the apoplast up to the endodermis, where the Casparian strip blocks further movement. Entry to the stele and access to xylem may be either in regions where there is no Casparian strip or by a symplast loop through endodermal cells: the question is still unresolved.

Question 1 (*Objectives 3.1, 3.2 and 3.4*) Classify statements (a)–(d) as true or false and, if false, explain why.

(a) The majority of ions move from the soil to plant shoots without entering a living cell.

(b) Calcium accumulates in older leaves because it cannot enter the phloem.

(c) Mineral nutrients move from roots to leaves by mass flow in the xylem: most can then pass into the phloem via transfer cells and circulate to other parts of the plant.

(d) Most cations accumulate in root cells by primary active transport but anions accumulate by a system of secondary active transport described as H^+/anion co-transport.

Question 2 (*Objectives 3.1 and 3.3*) Each of the following enzymes: glutamine synthetase, nitrite reductase, glutamate synthetase, nitrate reductase, catalyses a reaction that occurs during nitrate reduction and incorporation into amino acids in plants. List the enzymes to show the appropriate sequence of reactions, indicate the nature of the reaction catalysed and whether any of the following are required: reducing power, ATP, organic substrate.

Question 3 (*Objectives 3.1, 3.2, 3.3 and 3.4*) Suggest explanations for each of the statements (a)–(c).

(a) On boron-deficient soils, the older leaves may be healthy while younger leaves show symptoms of boron deficiency.

(b) Supplying nitrate fertilizers to crops growing on acid soils is better for the *soil* than supplying ammonium fertilizers.

(c) It is possible that active ion uptake from the soil into the roots could be powered by a proton gradient in the absence of proton-pumping ATPases in the plasmalemmas.

Question 4 (*Objectives 3.1, 3.4 and 3.9*) A segment 1.5–2.0 cm from the root tips of barley seedlings was stripped of epidermal and cortical cells. This procedure disrupts the endodermal cells but leaves the rest of the stele undamaged. Radioactively labelled phosphate ions were supplied to the stripped segments and to the same region of intact control roots and their uptake and transport to the shoot compared (Table 3.5).

(a) Describe the different patterns of uptake and translocation for the stripped and intact root segments.

(b) Explain the differences in these patterns.

Table 3.5 Phosphate uptake by intact and stripped roots of barley. (For Question 4)

Uptake period/hours	State of segment	Phosphate per segment/pmol		
		in the segment	transported to shoot	total uptake
0–0.25	intact	3.4	0.3	3.7
	stripped	0.3	3.5	4.0
0–4	intact	21.0	2.5	24.0
	stripped	0.3	54.0	54.5

3.5 NUTRIENT FORAGING: ACQUIRING MINERAL IONS

When plants have plenty of water and light, the supply of mineral nutrients is commonly rate-limiting for growth (Figure 3.13). This is why fertilizers are applied to crops and why 'good' crop varieties have come to be regarded as those which show the biggest increase in yield for a given amount of fertilizer, i.e. they absorb and use effectively a lot of nutrients. How do they do this? From reading Section 3.4 you might reasonably conclude that the most important factors are the numbers and properties of proton pumps and ion carriers in the roots, but Figure 3.14, from Clarkson (1985), indicates that this is not necessarily true.

Figure 3.13 The effect of depriving barley plants of certain major nutrients: A, control; B, deprived of K; C deprived of N; D, deprived of P.

Figure 3.14 shows the predicted change in phosphorus uptake by a soya bean plant when various root and soil variables are changed. The conditions apply to a plant growing in nutrient-rich soil. Phosphate uptake and the values of all variables were actually measured for one set of conditions and this is the point defined as 1 on the change ratio axis. P uptake was then calculated assuming either an increase or a decrease in a particular variable (increase or decrease in the change ratio), all other variables remaining constant. Doubling of a variable increases the change ratio from 1 to 2, for example, and halving a variable reduces it from 1 to 0.5.

◇ What variable influences most strongly the rate of P uptake?

◆ The rate of root elongation, k: P uptake approximately doubles (from 0.25 to 0.5) when this factor is doubled (change ratio increases from 1 to 2).

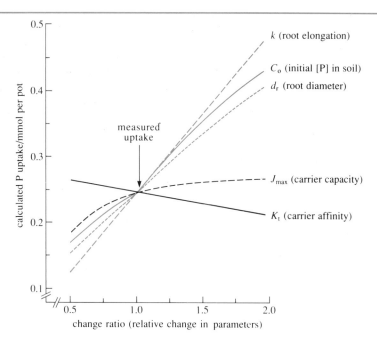

Figure 3.14 Effect of changing different variables on the predicted uptake of phosphorus by pot-grown soya beans on a nutrient-rich soil. K_t and J_{max} are the affinity and transport capacity, respectively, of the phosphate transport protein; C_o is the initial concentration of phosphate in the soil; k is the rate of root elongation; and d_r is the root diameter. Predicted (i.e. calculated) P uptake correlates with real measurements for value 1.0 on the change ratio scale.

Root diameter (d_r) and the initial concentration of P in the soil solution (C_o) likewise have large effects but, in contrast, changes in ion carriers — their affinity for phosphate (K_t) and transport capacity (J_{max}) — have hardly any effect at all. This means that properties of the root system (growth rate and root diameter) and ion concentration in the soil solution are more important for the uptake of phosphorus than properties of the ion-specific carriers. It must be emphasized that this situation applies only to *nutrient-rich* soils: there is good evidence that a more efficient ion transport system significantly increases ion uptake from nutrient-poor soils. Here, however, we shall concentrate on the other factors that influence the ability of plants to obtain nutrients from soil, which is often described as **nutrient foraging**.

Plants can influence nutrient foraging in three main ways:

(i) by changes in the growth or geometry of the roots, which relates to elongation rate, k, and root diameter, d_r, in Figure 3.14;

(ii) by changing the concentration or availability of ions in the soil, which relates to C_o in Figure 3.14;

(iii) by forming an association with (or even capturing) other organisms that supply mineral nutrients.

(i) and (ii), which relate to properties of a plant's own roots, are considered in this section and (iii) is discussed in Section 3.6.

3.5.1 Root growth and geometry

Root elongation and nutrient depletion zones

Roots continue to grow throughout the growing season by cell division in the meristem and elongation of the cells produced and, as Figure 3.14 shows, the rate of root elongation has a powerful influence on ion uptake. To understand why, consider what happens when a root extends into 'new' soil. At first, ion uptake occurs at a rate that depends mainly on the concentration of ions in the soil solution. This lowers ion concentrations adjacent to the root surface, setting up a concentration gradient down which ions diffuse towards the root. However, uptake proceeds much faster than diffusion so that a region that is depleted of ions — the **nutrient depletion zone** — gradually extends out around the root, as illustrated in Figure 3.15a. This zone can be regarded as 'exhausted' soil and the rate of ion uptake falls considerably. This applies even if the root continues to take up water at a constant rate, because ions and water diffuse independently of each other. The width of the depletion zone shows little increase after about 5 days but varies dramatically for different ions (Figure 3.15b). This variation reflects the different mobilities or rates of diffusion of ions through the soil, which depends on the strength of binding interactions between ions and electrically charged components of soil (clay and organic matter).

◇ From Figure 3.15 what can be deduced about the properties of nitrate and phosphate ions?

◆ The depletion zone of nitrate is wider than that of phosphate (in fact, nitrate is 1 000 times more mobile than phosphate). This is because nitrate binds very weakly to soil components whereas phosphate binds much more strongly.

For maximum ion uptake, therefore, it is essential that roots grow continuously into new areas of unexploited soil. This is particularly important for relatively immobile ions such as phosphate and zinc because the volume of soil depleted by a single root is so small. It is also important that one root does not grow into the depletion zone created by another — and this is one reason why the geometry of the root system matters.

The geometry of root systems and individual roots

If root density is high, depletion zones overlap and roots compete for nutrients. This can happen at quite low densities for mobile ions such as nitrate and sulphate, which have wide depletion zones. Therefore, if these nutrients are limiting growth, then even, wide spacing between root branches is desirable in terms of plant economy, i.e. extracting maximum nutrients from a given volume of soil for the minimum dry weight of roots. Another desirable feature from this point of view is a root system with many fine branches, as in the fibrous roots of grasses, rather than with few thick branches, as you often find in tree seedlings, for example.

◇ From the data in Figure 3.16, work out for (a) and (b): (i) the total areas of root cross-section (giving a measure of plant 'expenditure' on roots); (ii) the total areas of depletion zones (giving a measure of nutrient uptake); (iii) relative nutrient uptake per unit root area (a measure of 'root economy').
Note: the area of a circle is πr^2 (where r is the radius and π (pi) a value of 3.14).

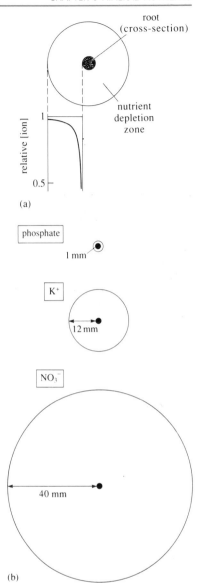

(a)

(b)

Figure 3.15 (a) Relative ion concentration in the nutrient depletion zone of a root. (b) Sizes of nutrient depletion zones for three ions.

width of all depletion zones = 1 mm

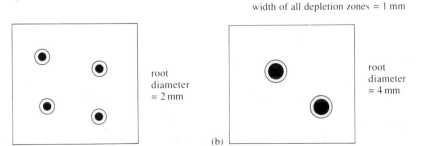

(a)

(b)

root diameter = 2 mm

root diameter = 4 mm

Figure 3.16 Identical areas containing (in cross section) (a) four roots of diameter 2 mm and (b) two roots of diameter 4 mm. All roots have a depletion zone that is 1 mm wide.

◆ (i) Total root area in (a) is $4 \times \pi \times 1^2 = 12.6 \, mm^2$ and in (b) is $2 \times \pi \times 2^2 = 25.1 \, mm^2$.

(ii) Total areas of depletion zones are obtained from (area of roots + depletion zones) − (area of roots). For (a) this is

$$(4 \times \pi \times 2^2) - 12.6 = 50.2 - 12.6 = 37.6 \, mm^2;$$

for (b) it is

$$(2 \times \pi \times 3^2) - 25.1 = 56.5 - 25.1 = 31.4 \, mm^2.$$

(iii) For (a) relative uptake per unit root area is $\dfrac{37.6}{12.6} = 3.0$

and for (b) it is $\dfrac{31.4}{25.1} = 1.25$

so it is more than twice as efficient to have four roots of diameter 2 mm than to have two roots of diameter 4 mm!

In reality, root systems are inevitably a compromise. If immobile nutrients such as phosphate are growth-limiting, then closely packed root branches may be necessary to satisfy plant demand — even though many of these branches are of little use for nitrate uptake because depletion zones for nitrate overlap.

The degree of root branching may also vary if the distribution of soil nutrients is uneven, as shown in Figure 3.17. How roots 'sense' nutrient-rich patches in soil is unknown but this capacity for local variation in geometry greatly increases the foraging capacity of root systems.

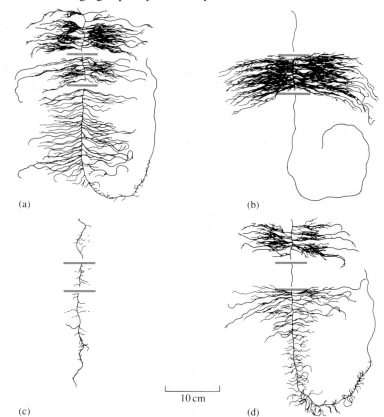

(a)

(b)

(c)

10 cm

(d)

Figure 3.17 Changes in the amount of lateral branching along one major root of barley, induced to grow through three successive zones, each containing either 'high' (1.0 mmol 1^{-1}) or 'low' (0.01 mmol 1^{-1}) nitrate. The central zone is marked by two horizontal green bars. The locations of 'high' and 'low' nitrate were: (a) high, high, high; (b) low, high, low; (c) low, low, low; (d) high, low, high. The experiment was carried out over 26 days. (Results of M. C. Drew)

A mechanism by which individual roots increase their foraging power is through the development of root hairs (Figure 3.18). These fine extensions from single epidermal cells, usually from 0.5 to 8 mm long, increase the absorption area of roots but, since they grow initially *within* the depletion zone, their effectiveness increases sharply when they extend beyond this zone.

◇ For which ion(s) are root hairs most likely to extend beyond the initial depletion zone?

◆ Ions such as phosphate, which have a very narrow depletion zone (Figure 3.15b), or possibly, for long root hairs, ions such as potassium, or ammonium which also binds quite strongly to soil particles.

It follows that root hairs are most effective in the absorption of relatively immobile nutrients, such as phosphate, whose concentration near the epidermal surface is low. However, phosphate uptake is influenced far more by another very common mechanism: formation of a symbiotic association between roots and a fungal partner. In this association, which is called a *mycorrhiza*, fungal hyphae extend outwards from roots and act like extra long, nutrient-absorbing root hairs. Mycorrhizas are discussed in Section 3.6.

3.5.2 Increasing the availability of mineral ions in soil

Soil pH and nutrient availability

Plants often show deficiency symptoms for a particular nutrient when soil analysis reveals a high content of that nutrient. The problem here is that the nutrient, although present in soil, is *unavailable for uptake* by plant roots. There are several reasons for this: ions may bind to charged particles of clay or organic matter; certain combinations of ions precipitate out of solution as insoluble salts — phosphates of iron or aluminium, for example; and elements such as iron and phosphorus occur in various ionic forms, only one of which is taken up readily by roots. All these availability factors are influenced by one important soil property: the *pH of the soil solution*. Figure 3.19 shows how the availability of nutrients varies with the pH of the soil.

(a)

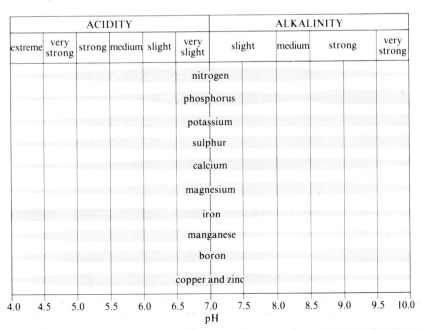

	ACIDITY					ALKALINITY			
extreme	very strong	strong	medium	slight	very slight	slight	medium	strong	very strong

nitrogen
phosphorus
potassium
sulphur
calcium
magnesium
iron
manganese
boron
copper and zinc

pH: 4.0 4.5 5.0 5.5 6.0 6.5 7.0 7.5 8.0 8.5 9.0 9.5 10.0

(b)

Figure 3.18 (a) Root tip of a radish seedling showing the development of root hairs. (b) Diagram of epidermal cells in longitudinal section to show a single root hair growing as an extension of a single cell.

Figure 3.19 The relationship between the pH of the soil and the availability of mineral nutrients: the wider the bar, the greater the availability.

The distribution of many plants is strongly influenced by soil pH. Lime-lovers or **calcicoles**, for example, grow well only on calcium-rich soils of alkaline pH and obtain sufficient iron, manganese, boron, copper and zinc there despite the low availability of these ions at high pH. The poor growth of calcicoles on acid soils may be either because of nutrient shortages — sometimes the major nutrients are not readily available under these conditions — or because of mineral *toxicity*.

◇ Which ions are most likely to be available at levels toxic to calcicoles in soils of strong to medium acidity?

◆ Iron, manganese, boron, copper and zinc, whose availability is very high in this pH range (Figure 3.19). Iron toxicity is the most commonly encountered but high levels of aluminium (not shown in Figure 3.19) may also cause toxicity in acid soils. The limited agricultural potential of acid soils is said to be due largely to the presence of aluminium.

In contrast to calcicoles, lime-haters or **calcifuges** are usually restricted to acid soils. They include species such as heather (*Calluna vulgaris*), *Rhododendron* spp. and bilberry (*Vaccinum myrtillus*). If such plants are grown on alkaline soils, they commonly show severe yellowing which is called **lime chlorosis** and is caused by a deficiency of iron. They may also be affected by calcium toxicity. These examples of calcicoles and calcifuges illustrate, therefore, the varying ability of plants to obtain adequate soil nutrients when availability is low and to avoid toxicity problems when availability is high. One reason for this variation is the degree to which plant roots actually modify the soil and influence the availability of mineral ions.

Modification of nutrient availability by roots

Two general mechanisms have been identified by which roots influence the local availability of soil nutrients. The first involves lowering pH at the root surface by active secretion of protons and the second involves secretion of mucilage that glues together soil particles and improves root–soil contact.

◇ From Figure 3.19 identify one major nutrient and one micronutrient whose availability in a moderately alkaline soil could be improved by acid secretion.

◆ Phosphorus is the obvious major nutrient and the micronutrient mentioned earlier as most commonly in short supply in such soils is iron. Availability of the micronutrients manganese, boron, copper and zinc would also be improved.

Rape (*Brassica napus*) is one of the species that is particularly effective in obtaining phosphate from P-deficient soil. When all available P has been absorbed, acid secretion is induced which solubilizes insoluble phosphates, increasing by up to 10-fold the uptake of P. Solubilization of unavailable iron and manganese has also been demonstrated and roots can be thought of as 'mining' these minerals, making the ions available for uptake.

Mucilage release, which occurs from surface cells near root tips, influences nutrient availability indirectly by affecting soil structure. By gumming together roots and soil particles, the intimate root–soil contact necessary for nutrient uptake is maintained. In anaerobic mud, however, root mucilage may have a quite different function related to protection against ion toxicity. Reduced and soluble iron (Fe^{2+}) is often present at toxic levels in anaerobic mud but, in plants such as rice, oxygen diffusing into the mud from the air-filled *aerenchyma* tissue of the roots (see Chapter 1) causes oxidation to Fe^{3+}. This forms an insoluble layer of reddish oxide/hydroxides which is stabilized by mucilage to give a protective sheath that blocks the access of Fe^{2+} ions (Figure 3.20).

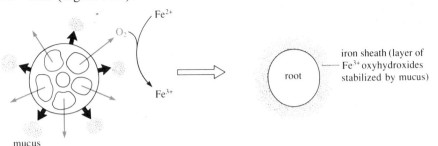

Figure 3.20 Development of an iron sheath around the root of a plant in waterlogged soil or mud. Oxygen diffuses from the large air-spaces (aerenchyma tissue) in the root into the soil and mucus is secreted from the root surface.

3.5.3 Iron-efficient sunflowers: a case study in nutrient foraging

To get an overall view of how roots forage for nutrients it is useful to consider what happens in sunflowers (*Helianthus annuus*) when iron is in short supply. Sunflowers have been classed as **iron-efficient plants** because, when their growth medium is or becomes deficient in available iron, remarkable changes occur in the roots that increase greatly the capacity for iron uptake. Similar behaviour is thought to explain the ability of calcicoles to avoid iron deficiency on soils of high pH. The changes induced in sunflower roots by iron deficiency are summarized in Figure 3.21.

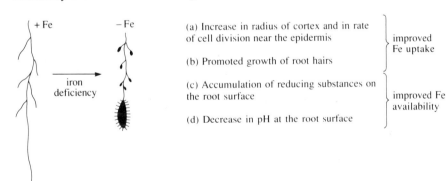

Figure 3.21 Changes in root morphology and physiology in sunflowers when grown at low levels of iron. The net result of (a)–(d) is increased mobilization of sparingly soluble Fe^{3+} compounds at the root surface.

Changes in root morphology, (a) and (b), increase by up to 70-fold the capacity for Fe^{2+} uptake at the root tips; whilst biochemical changes (release of reducing agents and protons, (c) and (d) make available for uptake the insoluble Fe^{3+} compounds at the root surface.

◇ Recall a biochemical system that may supply both protons and reducing power to the root surface.

◆ The plasmalemma redox chain (Section 3.4.1).

Linked with the acidification and increase in reducing power at the root surface there is also a 70-fold increase in the transfer of iron from root cortex to xylem and its transport to shoots. However, this super-efficient state with respect to iron uptake and transport is not sustained indefinitely, as the data in Table 3.6 show.

Table 3.6 Effect of growing sunflower plants with low and high levels of available iron for one to six days on subsequent accumulation of iron by roots and its translocation to shoots.

Iron accumulation and translocation were determined by supplying plants for 10 hours with radioactively labelled iron, ^{59}Fe, (at an intermediate concentration) and measuring ^{59}Fe in roots and shoots. Values are ^{59}Fe content in nmol g^{-1} dry wt.

Growth conditions		\multicolumn{6}{c}{Days of growth}
		1	2	3	4	5	6
low [Fe]	shoots	6	636	1 329	6	45	786
	roots	266	2 502	1 491	261	1 294	1 759
high [Fe]	shoots	5	3	3	3	12	3
	roots	353	265	282	280	293	267

In this experiment, iron-efficient behaviour was induced on the second day after transferring plants to low-iron solutions but persisted only until day 3. On day 4, iron uptake and transport to shoots declined markedly but increased again on days 5 and 6. These dramatic fluctuations contrast sharply with the constant rates of uptake and transport in the plants cultured in high-iron solutions.

What controls this complex response to iron shortage is still not understood but there is certainly a strong genetic element: certain strains of tomatoes and soya beans, for example, are much more iron-efficient than others. However, this is not always related to the efficiency of foraging and uptake from the soil: the critical step may be transport of iron from roots to shoots. Iron entry into root xylem involves binding to citrate, an organic acid, and oxidation of Fe^{2+} to Fe^{3+} and, in some iron-inefficient tomato varieties, this xylem entry step seems to be rate-limiting.

Inducible increases in uptake efficiency also occur for other ions, such as phosphate, and an attractive alternative to applying very high levels of fertilizers is the production of 'nutrient-efficient' crop strains by selection or genetic engineering. However, for wild plants growing on nutrient-poor soils, partnerships with micro-organisms are often far more important for nutrient acquisition than the uptake efficiency and nutrient foraging of the plants' own roots.

3.6 OTHER WAYS OF ACQUIRING NUTRIENTS: MICROBIAL SYMBIOSES AND CARNIVORY

Symbiosis means 'living together' and there are two symbioses that play key roles in plant nutrient relationships: (1) **mycorrhizas** (from the Greek *mukes*, a mushroom, and *rhiza*, a root), involving roots and fungi; and (2) nitrogen-fixing symbioses, involving (mainly) roots and prokaryotes (bacteria or actinomycetes). Both commonly involve a transfer of carbon compounds from the plant host to the microbial partner and a transfer of mineral nutrients from micro-organism to host, so they are mutualistic, i.e. they benefit both partners. In (2), the nutrient is reduced nitrogen (NH_3/NH_4^+) but in (1) several nutrients may be involved, although phosphate is usually the principal one.

One or other of these microbial symbioses occurs in the majority of plants but, in contrast, **carnivory** — literally, eating animals (usually insects) — is confined to a few specialized species. Here, extra mineral nutrients are obtained by trapping and digesting small invertebrates and we describe in Section 3.6.3 some of the bizarre mechanisms used.

3.6.1 Mycorrhizas

A feature of mutualistic mycorrhizas is that some fungal hyphae are intimately associated with the root and some extend out into the soil rather like an extended network of root hairs (Figure 3.22a). The soil hyphae do indeed function like very long root hairs, absorbing mineral nutrients and passing some back to the root. Figure 3.22 illustrates an external or *ecto*mycorrhiza, so-called because hyphae do not penetrate within host cells. This type occurs only in woody plants of cool climates, including trees such as birch, beech and pine: the tremendous effect of ectomycorrhizas on the growth and nutrient content of seedling pine is shown in Table 3.7. The primary effect of the ectomycorrhizal fungus in this instance was on P uptake, phosphate being in limiting supply for growth; once the phosphate limitation was removed, growth and the uptake of N and K were greatly increased. When particular species of toadstools (fungal fruiting bodies) are found always associated with certain tree species, the fungi are nearly always ectomycorrhizal partners.

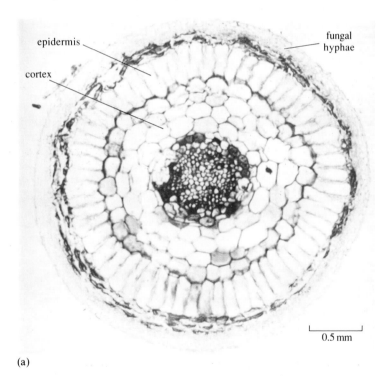

(a)

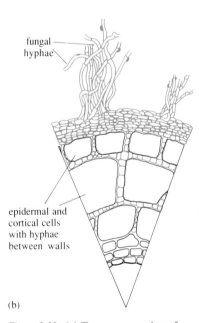

(b)

Figure 3.22 (a) Transverse section of an ectomycorrhizal root of beech (*Fagus sylvatica*). Fungal hyphae form an outer sheath from which they extend into the soil; they also penetrate between the radially elongated epidermal cells and the cortical cells. (b) Partial transverse section of an ectomycorrhizal pine root (*Pinus maritima*) showing fungal hyphae (green).

Table 3.7 The growth and nutrient content of pine seedlings (*Pinus elliotti*) with and without ectomycorrhizal infection.

	Growth/ g dry wt per plant	Macronutrients /mg per plant		
		N	P	K
no mycorrhiza	0.26	2.4	0.22	2.27
+ ectomycorrhiza	2.81	38.9	3.90	35.6

At least six other kinds of mycorrhizas occur but they are all internal types (*endo*mycorrhizas) where fungal hyphae enter root cells. These can be formed in almost all higher plants, the nettle family (Urticaceae) and cabbage family (Cruciferae) being two of the few exceptions. In general, the lower the level of soil nutrients (particularly phosphate), the greater the dependence of the plant on its fungal partner for obtaining nutrients, and thus for growth and survival. This is why the 'cost' to host plants, in terms of supplying carbohydrates to support mycorrhizal fungi, is fully justified when soil nutrients are in short supply. In some ectomycorrhizas, 15% or more of host net production passes to the fungus. However, in nutrient-rich soils, the benefit to host plants is less certain and the 'cost' of mycorrhizas may even reduce growth. With the emphasis in agriculture shifting towards 'low input' systems, which means less use of fertilizer, ensuring that crops have the most effective mycorrhizal partners is receiving more attention, particularly in the tropics. In tropical America, for example, over 40% of soils are infertile and acidic, and inoculating soil with particular endomycorrhizas has improved growth dramatically for many crops, including tea and coffee.

3.6.2 Nitrogen-fixing symbioses

The biggest reservoir of nitrogen on Earth is nitrogen gas in the atmosphere (which is 80% N_2) but, unfortunately, green plants cannot use it!

◇ Recall the forms of nitrogen that green plants can use.

◆ Nitrate and/or ammonium ions (Sections 3.1 and 3.3).

Certain prokaryotes, however, can 'fix' nitrogen gas, by reducing it to ammonia and some of these **nitrogen fixers** form symbiotic associations with plants. On roots of peas, beans and other members of the legume family there are pinkish nodules (Figure 3.23) inside which live N-fixing bacteria of the genus *Rhizobium*. The pink colour is due to the presence of an oxygen binding protein (a kind of haemoglobin) that maintains very low levels of O_2 in the nodule; this is essential because N fixation is an anaerobic process and would be poisoned by O_2. The bacteria depend for their energy supply entirely on carbon compounds from the host, so — as in mycorrhizas — there is a substantial 'cost' to the host. The pay-off is that most of the ammonia produced by N fixation passes into the host root where it is assimilated into amino acids and amides.

Some woody plants, including alder (*Alnus* spp.), sea buckthorn (*Hippophae rhamnoides*) and sweet gale (*Myrica gale*), have root nodules containing N-fixing actinomycetes (also prokaryotes). And in *Sphagnum* moss, cycads (gymnosperms) and some liverworts (bryophytes) N-fixing cyanobacteria live symbiotically in pouches within green tissues. In nearly all these examples the host plant often grows on nutrient-poor soils where there is a particular shortage of nitrogen — acid, waterlogged conditions or free-draining sands, for example. However, on agricultural soils, legume crops such as clover (*Trifolium* spp.) have long been used as 'living fertilizers' because when they die and decay the N content of the soil is increased: improvements may be possible by inoculating soil with particularly 'good' strains of *Rhizobium*. Another aim for the future is to produce N-fixing *cereal* crops by genetic engineering, either by introducing bacterial genes (and there has been some progress here) or by introducing legume genes necessary for establishing a root nodule symbiosis (which appears to be difficult, perhaps impossible). The aim is high grain yields with low inputs of nitrogenous fertilizer.

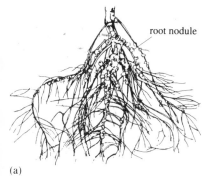

(a)

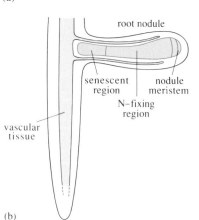

(b)

Figure 3.23 (a) Soya bean roots showing root nodules. (b) Diagram of a root plus nodule in longitudinal section.

3.6.3 Carnivorous plants

In six families of angiosperms, species have evolved that possess complex and often weird structures which trap animals: examples include pitcher plants (e.g. *Nepenthes* and *Sarracenia* spp.), fly traps (e.g. *Dionaea* spp.), sundews with sticky hairs (*Drosera* spp.) and bladderworts (*Utricularia* spp.); some of these trapping mechanisms are shown in Figure 3.24. The soft tissues of trapped animals are broken down to small, soluble molecules which can be absorbed by the plant, in much the same way that animals digest and absorb food.

(b)

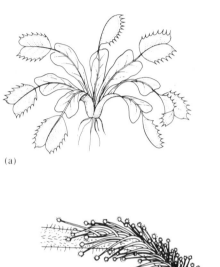

(a)

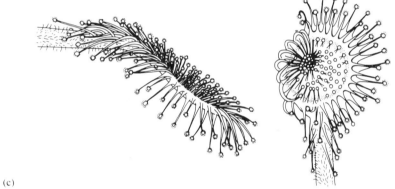

(c)

Figure 3.24 Trapping mechanisms of insectivorous plants. (a) The Venus' fly trap (*Dionaea muscipula*); when an insect touches one of the sensitive hairs, cells near to the midrib 'hinge' change shape, causing rapid inward movement of the two halves of the leaf. The insect is trapped inside. (b) The pitcher of *Nepenthes*, a pitcher plant; the pitcher is a modified leaf and contains a digestive fluid. Insects attracted by secretions near the rim slide down the slippery inner surface. (c) Leaves of a sundew (*Drosera* sp.); an alighting insect is held by the sticky secretion produced by glandular hairs, which are then stimulated to curl inwards and release digestive enzymes.

Botanists argued for many years about *why* this carnivorous habit had evolved because all the species are green autotrophs and obtain energy through photosynthesis. The favourite hypothesis was that digested animals provided essential mineral nutrients but until the 1970s there was little experimental evidence to support this. Table 3.8 shows data about the value of insect trapping for a sundew.

Table 3.8 The effect of feeding a sundew (*Drosera binata*) with insects (*Drosophila* sp.) in the presence and absence of inorganic nitrogen.
Plants were grown in sand and watered with an inorganic nutrient solution with or without nitrogen and containing all other essential elements. Values for dry weight increase were normalized for full supplies (= 100%) and are to the nearest 5%.

Supply of inorganic N to roots	Supply of insects	Dry wt. increase (%)
+	+	100
+	−	65
−	+	145
−	−	50

◇ What do the data in Table 3.8 suggest?

◆ *Drosera binata* appears to obtain a high proportion of its nitrogen from trapped insects, even when inorganic N is supplied to roots in a nutrient medium. The N from digested insects must be in organic form (amino acids etc.) and line 3 of Table 3.8 suggests that this is the 'preferred' form of N — which is highly unusual for a green plant. Similar experiments with other species of sundew indicate that supplies of sulphur and phosphorus may also be increased by carnivory.

Sundews grow in acid bogs where nitrogen is nearly always in short supply so it is reasonably certain that insects supply most of the plant's nitrogen in the wild. Similarly many of the other carnivorous species grow in very nutrient-poor habitats and, unlike neighbouring species, all have poorly developed root systems and form neither mycorrhizas nor nitrogen-fixing symbioses. Carnivory, therefore, appears to be an alternative strategy for obtaining nutrients that are in short supply.

Summary of Sections 3.5 and 3.6

1 The ability of plants to obtain nutrients from soil (nutrient foraging) may be influenced by: (i) root growth and geometry; (ii) the supply of energy and the properties of ion transport proteins; (iii) the concentration and availability of ions in the soil; (iv) the supply of nutrients from symbiotic partners or 'prey'.

2 (i) and (iii) above are usually of greatest importance in nutrient-rich soils and (ii), (iv) and possibly (iii) in nutrient-poor soils.

3 Continuous elongation of roots is essential for maintaining high rates of ion uptake because the soil adjacent to the root surface becomes depleted of nutrients.

4 The diameter of nutrient depletion zones varies for different ions, depending on their mobility in the soil, and the optimal root geometry for uptake depends on the nutrient in shortest supply. Relevant aspects of geometry are the thickness and degree of branching of roots and the number and length of root hairs.

5 The availability of ions for uptake is strongly influenced by soil pH. Roots may acidify the soil locally by proton secretion, thus increasing the availability of e.g. phosphate and iron. Secretion of mucilage facilitates ion uptake by maintaining root–soil contact and a mucilage sheath also protects roots from taking up toxic amounts of Fe^{2+} ions in anaerobic mud.

6 Changes that occur in sunflower roots when the supply of iron becomes growth-limiting, illustrate how nutrient foraging can be enhanced in this iron-efficient species. Some changes influence the uptake of iron and others influence iron availability.

7 The formation of root–fungus associations (mycorrhizas) is a widespread mechanism that improves the supply of phosphate to plants. Improved supply of nitrogen on nutrient-poor soils is achieved for certain species by association with nitrogen-fixing prokaryotes. These symbiotic associations involve a significant 'cost' to the host plant, which provides all the carbon compounds required by microbial partners.

8 Carnivory is an alternative and highly specialized strategy for obtaining mineral elements in nutrient-poor conditions.

Questions relating to Sections 3.5 and 3.6 are on pp. 145–146.

3.7 REGULATION OF ION ACCUMULATION

The last two sections described various mechanisms by which a plant can acquire more nutrients: improve foraging efficiency of its own roots, support a microbial forager or capture animal prey. However, this does not answer another question posed at the beginning of this chapter: why, if there is plenty of light and water and suitable temperatures, do species vary so widely in their response to high levels of mineral nutrients? Some, like crop plants / the nettle (*Urtica dioica*), grow at a prodigious rate and can be described as **nutrient-responsive**; whilst others, like wild thyme (*Thymus praecox*) or mat-grass (*Nardus stricta*), found on moorlands, show very little increase in growth and can be described as **nutrient-insensitive**.

3.7.1 The importance of growth rate

At one level there is a simple answer to this question. The rate at which plants absorb nutrients is largely determined by their 'demand' for nutrients and this, in turn, depends on their rate of growth. The maximum potential growth rate is genetically determined and varies between species. Nutrient-responsive seedlings of nettle, for example, have maximum growth rates of about 2.3 g dry weight per g dry weight per week — more than three times greater than the rates for nutrient-insensitive seedlings of thyme or mat-grass. This principle was applied to breed the modern high-yielding crop varieties, particularly those of maize and rice that are the basis of the so-called **green revolution**. Varieties were selected with very high potential growth rates, where growth of the grain was the main feature. Because of this high 'demand', these crops take up large amounts of nutrients and produce high yields — but *only if* supplied with a great deal of N, P and K in fertilizers. What the farmer needs to know is the minimum quantity of fertilizer necessary to sustain maximum growth, the philosophy being to modify nutrient supply to match plant demand.

This coupling between growth rate and the accumulation of major nutrients raises some interesting questions about control and coordination. Invariably, for example, the growth rate of shoots is highest when they are young and slows down progressively until senescence or dormancy occurs. For nutrient-responsive plants there is a corresponding fall in the net uptake of major nutrients such that concentrations of N, P and K in the tissues remain approximately constant. But how do the uptake systems in the roots 'know' and respond to what is happening in the shoots? Some evidence suggests that plant growth regulators carry the information from shoots to roots; alternatively, feedback control by the levels of inorganic ions or their organic metabolites in the root cytoplasm may operate. There is no clear answer yet.

Nutrient-responsive plants

Figure 3.25 summarizes in a simplified way how the concentration of nitrogen in tissues, $[N]_t$, net nitrogen accumulation and growth rate are related to external levels of N, $[N]_{ext}$, for a typical nutrient-responsive plant.

◇ How does $[N]_t$ change when growth rate is limited by the supply of nitrogen and $[N]_{ext}$ is increased?

◈ When $[N]_{ext}$ is very low and plants are showing symptoms of N deficiency, $[N]_t$ rises as $[N]_{ext}$ is increased. $[N]_t$ then reaches a plateau — a steady-state level — and remains constant, even though growth is N-limited.

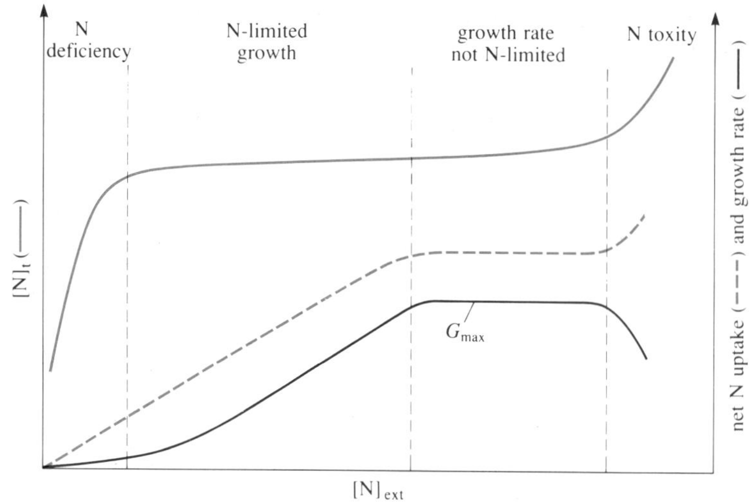

Figure 3.25 Effect of changing the external concentration of nitrate $[N]_{ext}$, on nitrogen concentration in the tissues, $[N]_t$, net nitrate uptake and growth rate.

This steady-state level of nitrogen in tissues is maintained over a wide range of external N levels and appears to be under homeostatic control. The changes that can occur if the supply of N or other major nutrient falls below the level needed to sustain maximum growth rate are:

(i) The number or 'effectiveness' of specific ion transport proteins could increase so that uptake of those ions per unit area of root surface increases. Effectiveness can increase either because J_{max} (maximum rate of transport) rises or because K_t (affinity of the carrier for its substrate) falls.

(ii) The form of the root system may change; it may become more branched (as described in Section 3.5.1) and grow relatively more than the shoots. Modification of the soil by roots (Section 3.5.2) may also increase nutrient supply.

If these changes sufficiently improve the supply of nutrients, then the maximum potential growth rate (G_{max}) can be restored; if this is not possible, growth rate falls — but the steady-state, internal concentration of nutrients remains constant.

To retain this constant $[N]_t$ when the supply of major nutrients increases requires a general damping down of ion accumulation. In the case of nitrate, ion influx may fall or the *efflux* from roots may increase although, for other nutrient ions, decreased influx is usually the only mechanism. In the longer term, a decrease in root growth relative to shoots occurs. These plants rarely indulge in 'luxury' uptake of major nutrients, taking up more than they need when $[N]_{ext}$ is high and storing the excess, although luxury uptake of phosphorus, does occur in some species. Such nutrient-responsive plants are typical of rich soils; they grow rapidly under these conditions and make few provisions for 'hard times', when soil nutrient levels fall.

Nutrient-insensitive plants

These plants, which are typical of nutrient-poor soils, differ from the nutrient responders primarily in their intrinsically low growth rate. In addition, the nutrient-insensitive plants respond less flexibly to a change in external nutrient levels: growth rate, the ratio of roots to shoots and the overall capacity for accumulating major nutrients such as phosphorus change relatively little.

◇ Given this information, which species in Figure 3.26 is nutrient-insensitive and which is nutrient-responsive?

◈ *C. crassiuscula* is nutrient-insensitive because its root:shoot ratio shows little change with increasing phosphate supply whereas the ratio for *C. pallens*, which is nutrient-responsive, falls markedly.

A consequence of this low flexibility coupled with low growth rate is that when supplied with high levels of major nutrients, tissue concentrations of nutrients rise much more in nutrient-insensitive than in nutrient-responsive plants (Table 3.9). This luxury uptake may be advantageous in the wild, where scarce nutrients are often released into the soil in pulses at certain times of year (usually autumn). But it becomes a liability on fertile soils because herbivores tend to graze selectively on the more nutritious nutrient-insensitive plants and tissue concentrations of nutrients may even rise to levels that are toxic to the plant. This latter is the price paid by nutrient-insensitive plants for their ability to grow on infertile soils. It is common in biology to find that adaptations have a 'cost': the advantages they confer in one situation almost inevitably mean that the organism is worse off in other situations.

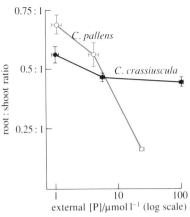

Figure 3.26 Mean ratio of root to shoot dry weight for *Chionochloa pallens* and *C. crassiuscula* seedlings grown at different external phosphate levels.

Table 3.9 Phosphorus accumulation in shoots, shoot dry weight and [P] in shoots for seedlings of two alpine grasses from New Zealand, *Chionochloa pallens* and *C. crassiuscula*, grown for three months in solutions of different phosphate concentration.

Phosphorus supply/μmol 1^{-1}	*C. pallens* shoot			*C. crassiuscula* shoot		
	total P/μmol	dry wt/mg	[P]/μmol g^{-1} dry wt	total P/μmol	dry wt/mg	[P]/μmol g^{-1} dry wt
1	0.60	30	20.0	0.31	20	15.3
10	29.6	700	42.3	13.0	175	74.3
100	247	1 800	137	75.6	275	275

All plants have a potential problem regarding toxic levels of ions in their tissues with respect to, for example, calcium, micronutrients and certain non-essential ions such as sodium and aluminium. Ways round this problem are described next.

3.8 DISPOSING OF EXCESS IONS

Any ion accumulated to excess in the cytoplasm can be damaging. For major plant nutrients — N, K and usually P and S — this is a rare event because, in the short term, cell vacuoles provide a temporary store and buffer and, in the long term, growth rate matches nutrient supply (except for nutrient-insensitive species on rich soils). For other nutrients, however, there is no such matching of supply and demand and a wide array of strategies have evolved that serve to exclude unwanted ions from the cytoplasm. Usually, these strategies involve one or more of three basic processes:

(i) pumping excess ions out of the roots;

(ii) storing the excess, sometimes converted to a non-toxic form, in the vacuoles;

(iii) secreting excess onto the surface of leaves.

Sections 3.8.2 and 3.8.3 describe two examples relating to calcium ions and saline soils, in both of which these processes are used. There are other mechanisms, however, one of which — formation of an iron sheath in waterlogged soils preventing excess intake of iron — was described in Section 3.5.2. Another 'special case' relating to the disposal of sulphate in the oceans is described below.

3.8.1 Sulphur disposal in marine phytoplankton

Seawater is rich in sulphate and some unicellular marine algae, phytoplankton, absorb more than they need. The excess is disposed of in a way that now appears to have potentially widespread environmental consequences. The phytoplankton incorporate excess sulphur into volatile organic compounds which are released into the water and then escape into the atmosphere. Here they may be oxidized to sulphur dioxide which contributes to the problem of 'acid rain'. The volatile organic compounds also act as condensation nuclei, causing water vapour to condense to form clouds and the general effect of clouds is to keep the Earth cooler, because they prevent solar radiation from reaching the surface. Massive increases in algal production have been recorded in the late 1980s in areas such as the North Sea, mainly because of nutrient enrichment from inflowing rivers; but a possible beneficial result is that increased cloud formation could counteract the **greenhouse effect** (global warming due to increases in CO_2 and other 'greenhouse gases'), which is arguably the most severe environmental problem of the next 100 years.

3.8.2 Calcium disposal

In all living cells the concentration of calcium in the cytoplasm is low — usually in the range of 0.01 to 1 μmol l^{-1}, which may be more than 1 000 times lower than in the soil solution. Since cell membranes are permeable to Ca^{2+}, there is a very strong tendency for Ca^{2+} to diffuse into root cells with the inevitable consequence that too much Ca^{2+} enters the cytoplasm. The excess must be removed and as a first line of defence, there are Ca^{2+}-ATPases that pump calcium out into the soil solution or the cell walls by a primary active transport system (Section 3.4.1). Another disposal mechanism is sequestration of Ca^{2+} in cell vacuoles by means of a secondary active transport system (H^+/Ca^{2+} exchange; see Figure 3.7).

◇ Recall from sections 3.4.2 and 3.2.2 reasons why these mechanisms would not exclude Ca^{2+} from shoots.

◈ First, Ca^{2+} ions move to root xylem (and thence to shoots) mainly via the cell wall or apoplastic route; so Ca^{2+} exclusion from root cells is of no relevance. Second, once Ca^{2+} ions reach a particular shoot organ in the transpiration stream, they cannot move out again because very little Ca^{2+} is transported in phloem (Section 3.2.2).

The usual mechanism for disposing of excess Ca in shoots is to store it in cell vacuoles. In some species such as Dog's Mercury (*Mercurialis perennis*) and *Rheum* sp. (e.g. rhubarb) it is present mainly as insoluble crystals of calcium oxalate (see Figure 3.27).

In other species, many of which are CAM plants (Chapter 2) or calcicoles growing on lime-rich soils, vacuoles contain soluble calcium malate and Ca^{2+}, therefore, contributes significantly to osmotic pressure. Other species pump Ca^{2+} out of leaf cells and, in the brassicas, it may then be precipitated as $CaCO_3$ in the cell walls. Some members of the saxifrage family excrete

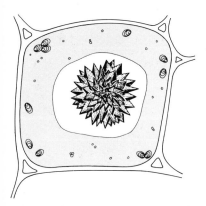

Figure 3.27 Calcium oxalate crystals in the vacuole of a leaf cell of *Rheum* sp.

$CaCO_3$ onto the leaf surface, where it forms a characteristic powdery deposit. The general rule is that the older the leaf the more calcium it contains and if all means of calcium disposal are exhausted, the leaf dies and is shed.

3.8.3 Saline soils and halophytes

Most species of vascular plants cannot survive on saline soils. These are soils with high concentrations of Na^+ ions, which are usually accompanied by Cl^-, as in salt marshes, or sometimes, in saline deserts, by bicarbonate (HCO_3^-) or sulphate (SO_4^{2-}) ions. Species that can survive in these environments are described as **halophytes** (from the Greek *hals*, salt and *phyton*, plant).

Saline conditions present two kinds of problems for plants, one related to the uptake of water (discussed in Chapter 5) and the other in connection with ion relationships, which hinges mainly on the need to keep Na^+ and the accompanying anion out of the cytoplasm.

◇ From the early part of this section, suggest a first line of defence against this last problem.

◆ To transport Na^+ out of root cells into the soil.

This is actually done by non-halophytes tolerant of low salinity and seems to be their sole defence (e.g. the edible tomato, *Lycoperiscum esculentum*; Figure 3.28). Most halophytes, however, do it to only a limited degree, as you can see for *L. cheesmanii* in Figure 3.28; they transport NaCl out of the cytoplasm but there is far too much and it would require too much energy to transport it into the soil. Considerable amounts pass into cell vacuoles, where Na^+ replaces K^+ as the major cation responsible for maintaining osmotic pressure and this is one of the chief characteristics of halophytes. Their cells have high osmotic pressures due to high concentrations of sodium salts in the vacuoles and special organic solutes in the cytoplasm (discussed in Chapter 5). For roots, this seems to be all that happens and when vacuoles can absorb no more salt, the excess passes to shoots where there are several strategies for dealing with excess salt.

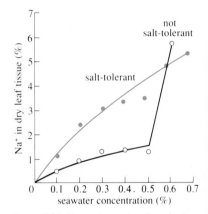

Figure 3.28 Percentage of sodium in dry leaf tissue as seawater was added to the nutrient solution for two species of tomato, one of which was salt-tolerant (*Lycopersicum cheesmanii*) while the other was not (the edible tomato, *L. esculentum*).

Dealing with excess salts in leaves

Simply looking at and tasting plants give useful clues about the strategies for salt disposal:

(i) Some plants, such as the glassworts (Figure 3.29), develop **succulence**, having thick, fleshy leaves with a high surface area : volume ratio. The high osmotic pressure in young cells leads to an influx of water that generates high turgor pressure; young cell walls stretch more than usual and large-celled 'succulent' tissues are formed. Salts in the vacuoles are diluted by this process so that, although leaves contain large amounts of salts, the salt *concentrations* are not especially high.

(ii) Some plants taste very salty if chewed and the older the leaf, the saltier it tastes. These **salt accumulators** do nothing more than store salt in leaf cell vacuoles until the concentration is so high that the leaf can absorb no more and dies. Herbaceous seablite (*Suaeda maritima*) and sea arrow-grass (*Triglochin maritima*), both salt-marsh species, are of this type.

(iii) Other plants, such as the tamarisk bush, some mangrove trees and sea-lavender (*Limonium* spp.), taste salty if licked and leaves may be encrusted with salt. Here the excess salts are secreted onto the leaf surface through special **salt glands** (Figure 3.30). How these remarkable glands work

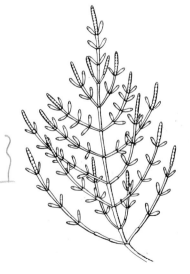

Figure 3.29 A typical halophyte with succulent leaves, *Salicornia europaea* (glasswort), which grows on the seaward edges of salt-marshes.

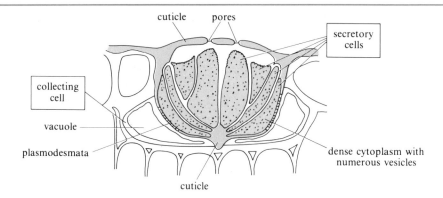

Figure 3.30 A salt gland of sea-lavender (*Limonium latifolium*).

is still not certain but what seems to happen is that Na^+ and Cl^- ions accumulate in the vacuoles of collecting cells, sometimes to ten times the concentration in seawater! Ions then move passively via plasmodesmata into the complex of secretory cells, which is largely surrounded by an impermeable cuticle except for a few surface pores. The secretory cells have the dense cytoplasm and corrugated outer wall characteristic of transfer cells (Section 3.2.2) and, in some way (probably by active transport at the plasmalemma), secrete salt onto the leaf surface. Where salt-secreting plants grow in areas of high humidity, as in the desert of northern Chile, they become covered with a film of liquid as the salt absorbs water hygroscopically.

The desert saltbushes (*Atriplex* spp.) from Australia taste salty for a slightly different reason: the leaf surfaces are covered with **salt hairs** where salt is accumulated (Figure 3.31). Each hair consists of two cells, a lower stalk cell

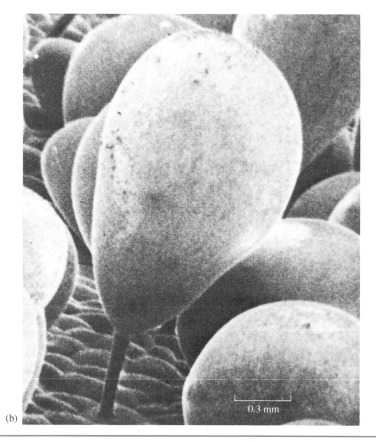

Figure 3.31 (a) Diagram to show the structure of a salt hair. (b) A scanning electron micrograph of salt hairs on the leaf surface of a salt-bush (*Atriplex spongiosa*).

which is much like the secretory cells of the salt gland; and an upper bladder cell where salt accumulates in a huge vacuole. The bladder eventually bursts or falls off but in some species new bladders then develop.

Much of the research on halophytes is aimed at developing salt-tolerant crops because saline soils are already a serious agricultural problem over vast areas of semi-arid land — and the problem is getting worse. The primary cause is irrigation: water used for irrigation always contains some dissolved salts and more are dissolved as the water percolates through the soil. Most of these salts are left behind in the surface layers of the soil as the water evaporates or is taken up by plants. So the surface layers become progressively more saline with repeated irrigation, a process termed **salinization**. Other than washing out this surface salt by applying huge amounts of freshwater (and where is the water to come from?), the only hope for salinized areas is to plant crops tolerant of salty conditions and control very carefully any further irrigation.

Summary of Sections 3.7 and 3.8

1 Based on their growth response to high levels of major nutrients (chiefly N, P and K), plants can be classified broadly as nutrient-responsive or nutrient-insensitive.

2 Nutrient-responsive plants usually grow on fertile, nutrient-rich soils and the accumulation of major nutrients is coupled closely to growth rate, which is intrinsically high. Tissue concentrations of major nutrients, especially nitrogen, are maintained at a constant level when the external supply of nutrients alters, by flexible changes in growth rate and in the mechanisms for acquiring nutrients.

3 Nutrient-insensitive plants usually grow on infertile, nutrient-poor soils. Growth rates are intrinsically low and are not closely coupled to nutrient supply. If the supply of nutrients increases, luxury uptake in excess of immediate requirements may occur and tissue concentrations rise, sometimes to toxic levels.

4 When plants accumulate excess ions, three general mechanisms are used for disposal.

(i) Excess ions may be actively secreted by roots (as happens for NaCl on slightly saline soils).

(ii) They may be stored in cell vacuoles (e.g. Ca^{2+} as calcium oxalate in the leaves of some species; NaCl in the salt-accumulating halophytes).

(iii) They may be secreted to the outside of shoots (e.g. Ca^{2+} as calcium carbonate in some saxifrages; NaCl via salt glands in some halophytes). Secretion into salt hairs that are subsequently shed is a variation of this mechanism. Another variation occurs in certain marine phytoplankton, which secrete excess sulphur as a volatile organic compound.

Question 5 (*Objectives 3.5 and 3.6*) For plants growing in a nutrient-rich soil (a) which single item from (i)–(vii) would cause the largest increase in the uptake of nitrate; (b) which items (any number) would increase the uptake of phosphorus?

(i) an increase in the number of specific transport molecules ('carriers') in root cell membranes

(ii) an increase in the number of lateral root branches

(iii) an increase in the rate of elongation for all roots

(iv) an increase in the number and length of root hairs

(v) formation of a mycorrhizal association

(vi) formation of nitrogen-fixing root nodules

(vii) proton secretion and acidification of the soil around roots.

Question 6 (*Objectives 3.1 and 3.5*) (a) Define briefly an iron-efficient plant. (b) What are the two possible explanations for iron-inefficiency in a plant?

Question 7 (*Objectives 3.1, 3.6 and 3.7*) If the sole objective were to increase the yields of arable crops, which of (i)–(v) would probably be most helpful for (a) a modern, intensive farm in southeast England and (b) a peasant holding, farming at subsistence level, in tropical Brazil?

(i) Breed crops with greater phosphate efficiency.

(ii) Breed crops with higher maximum potential growth rates.

(iii) Engineer crops that are able to fix their own nitrogen.

(iv) Find an endomycorrhizal fungus that could be inoculated into the soil and would increase phosphate uptake.

(v) Develop a species of insectivorous plant as a crop species.

Question 8 (*Objectives 3.1 and 3.7*) Plants of a nutrient-efficient type were cultured in conditions of slight potassium deficiency, all other nutrients being in non-limiting supply. If the supply of K^+ were increased, what would you expect to happen to (a) plant growth rate (increase in dry weight per unit dry weight per day); (b) the net uptake rate of nitrate; (c) the concentrations of K^+ and N in the leaves?

Question 9 (*Objectives 3.1 and 3.8*) Classify statements (a)–(f) as true or false.

(a) A nutrient-insensitive species is one that does not show toxicity symptoms when the external concentration of any essential ion is very high.

(b) Pumping calcium ions out of roots into the soil is one of the strategies for disposing of excess Ca^{2+} in plants.

(c) Preventing excess NaCl from reaching leaves is one of the main strategies employed by halophytes.

(d) If very thirsty, the sap squeezed out of succulent halophytes would be a reasonable source of fluids.

(e) Plants with a high growth rate are more likely to suffer from copper toxicity on a copper-rich soil than are plants with a low growth rate.

(f) Active transport of Na^+ and Cl^- ions occurs in salt glands both into collecting cells and out of secretory cells.

OBJECTIVES FOR CHAPTER 3

Now that you have completed this chapter you should be able to:

3.1 Define and use, or recognize definitions and applications of, each of the terms printed in **bold** in the text. (*Questions 1, 2, 3, 4, 6, 7, 8 and 9*)

3.2 Describe or recognize correct descriptions of the pathways of ion transport into and within plants and indicate ways in which the pathway for calcium and boron differs from those of other ions. (*Questions 1 and 3*)

3.3 Summarize the main features of (a) the metabolism of nitrate and sulphate in plants and (b) the assimilation of inorganic nitrogen into organic compounds. Explain how and why the metabolism and assimilation of nitrogen may influence soil pH. (*Questions 2 and 3*)

3.4 List the processes involved in the transport and accumulation of mineral nutrients (including calcium) within plants. (*Questions 1, 3 and 4*)

3.5 Explain how the availability and acquisition of mineral ions may be influenced by (a) soil pH, (b) root growth and geometry and (c) root secretions. (*Questions 5 and 6*)

3.6 Give examples which illustrate how heterotrophs may increase the supply of mineral nutrients to plants with reference to mycorrhizal and nitrogen-fixing symbioses and carnivory. (*Questions 5 and 7*)

3.7 Explain how nutrient accumulation is regulated in nutrient-responsive plants and how these plants differ from nutrient-insensitive species. (*Questions 7 and 8*)

3.8 Describe and give examples of the mechanisms by which plants exclude or dispose of excess ions. (*Question 9*)

3.9 Interpret data about ion transport or ion relations in ways that are consistent with principles given in this chapter. (*Question 4*)

4.1 TRANSPORT IN PHLOEM: AN INTRODUCTION

To function properly, a plant must constantly move substances around from one organ to another. Ions and water taken up by roots move up to the shoots and you saw in Chapter 3 that this occurs in the xylem, with recirculation of ions from leaves occurring in phloem. Similarly, the products of photosynthesis (**assimilates**) must be transported from the organs where they are synthesized or stored to the places where they are used, and this too occurs in phloem — indeed, assimilate transport is generally regarded as the primary function of phloem. Xylem and phloem provide the plumbing system of plants that allows long-distance transport between organs and in this chapter we shall be concentrating on phloem.

The green arrows in Figure 4.1 show the directions in which organic substances move in phloem during the development of a pea plant (*Pisum sativum*).

Movement may be up, down or in both directions along a stem so, clearly, phloem provides a very flexible transport system.

It is also a system that can operate over very long distances (it may be 30 m or more between roots and the nearest leaves in some trees) and move enormous quantities of material at high rates. These last properties are illustrated most vividly in certain tropical monocot plants — species of palms, *Agave* and *Yucca*. For centuries the inflorescence stalks (bearing the flower heads) of these plants have been tapped to obtain a sweet, sugary fluid (called toddy in India) that is often fermented and distilled to produce alcoholic

Figure 4.1 The different transport routes (green arrows) for organic solutes during the development of a pea plant. (a) A young seedling. (b) A rapidly growing young plant. (c) A branch bearing a fruit.

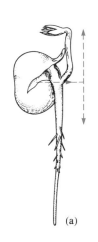

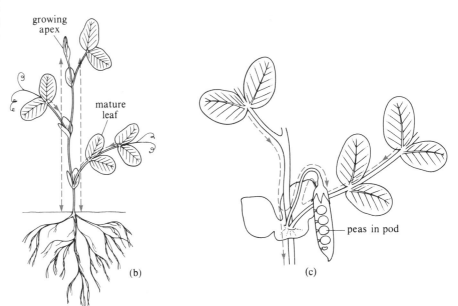

beverages such as arrack (in Asia) or tequila (in Mexico). The sugary fluid is now known to derive from phloem and contains 15 to 19% solids, mostly sucrose; amounts ranging from three to an astonishing 45 litres per plant per day have been collected. The name given to this mass transfer of solutes in phloem is **translocation**. This simply means the movement of soluble substances from one part of a plant to another and can be applied to movement of ions in xylem; usually, however, the term is restricted to phloem transport.

Unravelling the story of translocation has occupied plant physiologists for over 100 years. Not until 1928 was it certain that phloem is the only tissue involved and not until the late 1970s was a consensus reached about the mechanism of transport. The focus of research now (1990) is the control and coordination of the system so that supply of materials from source organs meets the demand by sink organs. Transport from **source to sink** is a key feature of phloem transport. Aspects of all these topics — how phloem is studied, the mechanism of transport and source-to-sink relationships — will be considered in this chapter. We begin, however, by looking at phloem structure because there are fewer better illustrations of how knowledge of structure provides the essential foundation for an understanding of function.

4.2 PHLOEM STRUCTURE

Unlike xylem, where the majority of cells are dead and lignified, phloem consists of living cells, although lignified phloem fibres are present in some species. There are three types of cells: *sieve tube elements, companion cells* and *phloem parenchyma*. The phloem parenchyma serves as 'packing' but it may also be modified to form transfer cells (Chapter 3, Section 3.2.2) and is then involved in the local transfer of material from xylem to phloem. It is the first two that are intimately involved in the long-distance transport of solutes and Figure 4.2 shows their main features. We describe them in more detail in the next two sections.

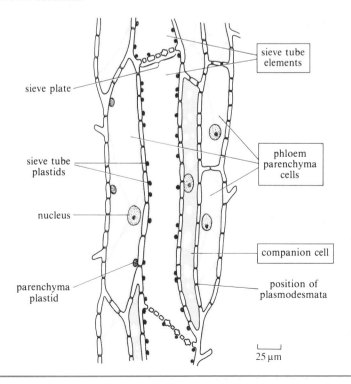

Figure 4.2 A longitudinal section of phloem tissue from the stem of *Nicotiana tabacum* (tobacco).

4.2.1 Sieve tube elements

Sieve tube elements are highly specialized cells through which translocation actually takes place. In angiosperms, they link up with each other via end-walls to form long conduits called **sieve tubes**.

◇ Recall from Chapter 1 the characteristic feature of end-walls that facilitates the linking up of sieve-tube elements.

◐ Large pores; the perforated end-walls are called **sieve plates** (see Figure 4.2).

Material moving along sieve tubes obviously has to pass through these pores, and you will see later that the state of the pores — whether they are clear 'holes' or whether they contain cytoplasmic or other plugging material — is a crucial factor in assessing the mechanism of transport.

During their development, sieve tube elements undergo several degradative changes: they lose their nuclei and ribosomes; the membrane round the vacuole (the tonoplast) breaks down; mitochondria and plastids become disorganized internally: you might say that the cell is specialized to the point of death! However, the plasmalemma of sieve tube elements remains intact, and large amounts of a special protein are formed. This **P-protein** (P = phloem) exists as fibrils composed of numerous subunits and is dispersed through the cell lumen. Some of the earlier hypotheses about the mechanism of phloem transport assigned a special role to P-protein, suggesting that it acted like muscle fibres and pumped fluid along. But it must be noted that not all species appear to form it: ferns, gymnosperms and many monocotyledons, for example, do not.

In essence, the sieve tube is a living, pipe-like structure. It has a cellulose cell wall (often very thick), a cell membrane (the plasmalemma) and watery cytoplasm containing a few peripheral organelles but large amounts of P-protein. In order to relate structure to function, however, it is essential to know about the ultrastructure of sieve tubes, in particular how the P-protein is organized at the sieve plates — and here is where electron microscopists encountered a (still) insoluble problem. The reasons for this relate to a universal feature of sieve-tubes: their contents are under *very high pressure* (or high *turgor*). The membrane-bound protoplast of a sieve tube thus exerts a large force on the cell wall, much as the air in a highly inflated balloon presses against the rubber 'wall' of the balloon.

◇ What observation described in Section 4.1 indicates high pressure in sieve tubes?

◐ The rapid exudation of 'phloem sap' in certain monocots (palms etc.). Assuming that the sap derives from sieve tubes, then the contents must have been under high pressure.

When preparing phloem material for electron microscopy, there are two main reasons why high turgor gives rise to artefacts. The first is a direct result of high turgor because whenever the tissue is cut there is an immediate *surging* of cell contents as pressure is released. No matter how carefully material is prepared, it seems impossible to rule out the possibility that the cell contents have been disorganized by surging. Very rapid freezing before cutting is the method used to minimize this problem.

The second reason arises because plants have a defence mechanism to prevent 'bleeding' to death if the phloem is nibbled into or cut: they deposit massive amounts of an insoluble polysaccharide called **callose** (a β-1, 3-linked glucose polymer or glucan) that plugs the wound. To maintain sap flow, the cut surface of bleeding palms described earlier has to be shaved every day to remove callose plugs. Callose deposition occurs very readily indeed (even rough handling may cause it), and material prepared for electron microscopy nearly always contains some callose, often blocking parts of the sieve plates. It is difficult to know whether this is another artefact or whether sieve tubes really do contain small amounts of callose at all times.

Given these problems, it is not surprising that different methods of preparation and different investigators produce widely different pictures of sieve tube structure.

Two contrasting images, one in which the sieve-plate pores appear free of all obstruction (has everything 'surged' out of the way?) and the other in which P-protein and callose occupy most of the pores (has P-protein surged into pores and callose been deposited?) are typical: compare Figure 4.13a with j and k (p. 162). Years of research and debate resulted from these doubts but the consensus now (for reasons described later) is that most pores are probably free of obstructions.

Sieve tube structure: summary

1 We know that sieve tubes are composed of living cells (sieve tube elements) which have a very high turgor pressure and are interconnected via perforated sieve plates.

2 We do not know how P-protein is arranged inside the cells — whether the strands are arranged randomly or grouped to form fibrils; nor is it known if individual strands run through the sieve plate pores, linking one element with the next.

3 We do not know with certainty the state of the pores in sieve plates: whether they are free of obstructions or occluded by P-protein or callose. The former is usually assumed.

4.2.2 Companion cells

As their name implies, and as Figure 4.2 shows, **companion cells** are close associates of sieve tube elements. The two types of cells arise in pairs from the same parent cell, and the companion cell remains alive only as long as its, usually much larger, sister sieve tube does. Structurally, companion cells appear less specialized than sieve tube elements: they retain a nucleus and have numerous ribosomes and organelles — usually taken to indicate a high level of metabolic activity. There are also many plasmodesmata connecting each companion cell with its adjacent sieve tube element and this high level of intercommunication supports the view that companion cells 'service' and help to maintain the integrity of sieve tube elements.

◇ Why are mature sieve tube elements likely to have a special need for help of this kind?

◈ Because they lack both nucleus and ribosomes and so cannot synthesize proteins (Section 4.2.1); any replacement P-protein, enzymes (or even substrates) for cell maintenance would have to come from other cells.

Another role, which companion cells sometimes share with phloem parenchyma cells, is in the active loading of sugars into, and possibly out of, sieve tubes. You can think of a companion cell plus sieve tube element as a functional unit. In some plants the companion cells in the smallest leaf veins, which are the most active sites of solute loading into sieve tubes, become specialized to form **transfer cells**. These were considered in Chapter 3, Section 3.2.2 (in connection with their role in the circulation of ions from xylem to phloem). Now you can see that they function also in the transfer of organic solutes from one cell to another. This operates not only for phloem but also in glands, reproductive structures and in symbiotic (including host–parasite) associations. This wider role of transfer cells is considered briefly in the next section.

4.2.3 The wider role of transfer cells

Transfer cells are specialized parenchyma cells that seem to function as collectors of both organic and inorganic solutes which are then transferred to specific, neighbouring cells. Structural features relevant to this role are:

1 A greatly increased surface area because of complex ingrowths of the cell wall, which may be all over the surface or in localized regions and which are followed by convolutions of the plasmalemma.

2 Numerous plasmodesmata linking up with adjacent living cells, to which or from which transfer is thought to occur.

3 Dense cytoplasm containing numerous mitochondria, which indicates a high level of metabolic activity.

These features are illustrated in Plates 11, 12 and 13. The general principles of how transfer cells work is that *active* transport of solutes (ions or organic molecules) occurs wherever wall ingrowths are found; movement may be into or out of the transfer cell. Passive movement to or from adjacent living cells can occur symplastically via plasmodesmata. In vascular tissue, transfer cells often function like a lens, scavenging solutes from the apoplast (cell walls) and focusing them into the phloem (Plate 14 and Figure 4.3).

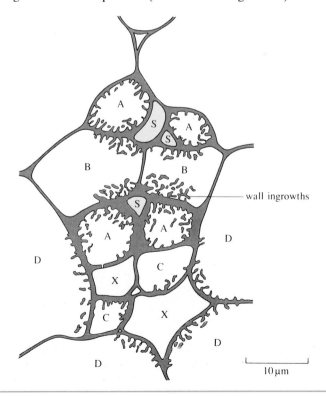

Figure 4.3 Transverse section through a minor leaf vein of the C_4 plant *Anacyclus pyrethrum*. Four types of vascular transfer cell are shown: A, modified companion cells, and B, modified phloem parenchyma cells, both associated with sieve tube elements (S); C, modified xylem parenchyma, and D, modified bundle sheath cells, both associated with xylem (X). Note the characteristic wall ingrowths in transfer cells.

wall ingrowths

10 μm

Transfer cells develop in nearly every situation where solutes are diverted to (or collected by) particular, localized groups of cells. Examples include: the junction between parent plant (supplying nutrients) and developing sporophyte* offspring (Figure 4.4a and b and Plate 13); glands that secrete salt or nectar to the exterior (Figure 4.4c and d); and the epidermis of submerged leaves which absorb ions (Figure 4.4e). Other examples occur when parasites or mutualistic partners invade the tissues (Figure 4.5); A and B in this figure illustrate how the presence of a symbiont can *induce* the differentiation of transfer cells; in A they facilitate transfer of nitrogen compounds from a nitrogen-fixing root nodule to the xylem and transfer of sugars from phloem to nodule, and in B they facilitate transfer from vascular cells to a root gall which feeds a nematode worm. An electron micrograph showing transfer cells in a nitrogen-fixing root nodule is shown in Plate 15.

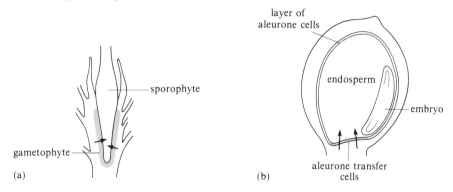

(a)

(b)

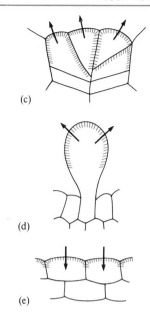

(c)

(d)

(e)

Figure 4.4 (a) and (b): the position of transfer cells (green shading) in reproductive structures. (a) At the junction between sporophyte and gametophyte in a moss. (b) Aleurone transfer cells in a grass seed; these transfer nutrients from the parent plant to the endosperm (food-storing tissue). (c)–(e): the positioning of wall ingrowths (green lines) in transfer cells from specialized plant organs. (c) A salt gland that secretes salt to the exterior. (d) A nectary, which secretes nectar to the exterior. (e) Cells from the surface of a submerged leaf in a water plant that absorbs ions from the surrounding medium. Arrows show the direction of transport.

In general, transfer cells are agents for the *short-distance* transport of solutes and they are especially prominent whenever there is no cytoplasmic continuity (via plasmodesmata) between tissues, as at the xylem boundary and the boundaries between host and parasite or gametophyte and sporophyte. Since they occur in all land plants from bryophytes to angiosperms, transfer cells are clearly of very ancient origin.

Figure 4.5 The relationship between vascular transfer cells and the conducting elements of xylem and phloem for four symbiotic associations A–D. The position of wall ingrowths (green) and the presumed directions of transport (grey arrows) are shown. A, a cell in a legume root nodule where symbiotic bacteria fix nitrogen; nitrogen compounds move out to the root xylem and carbon compounds move in from the phloem via several intervening cells. B, giant cell of a root gall caused by nematodes, which feed on the gall tissue. C, a projection from a special absorptive cell of the stem parasite, *Cuscuta* (dodder). D, a similar projection in the root parasite, *Orobanche* (broomrape).

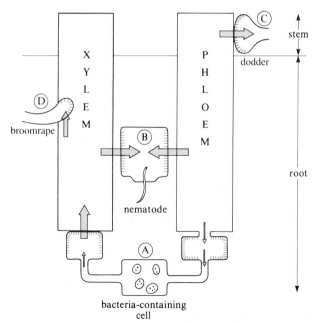

*Sporophyte and gametophyte generations of plants are described in Chapter 1 of Caroline M. Pond (ed.) (1990) *Diversity of Organisms*, Hodder & Stoughton Ltd in association with The Open University (S203, *Biology: Form and Function*, Book 1).

Summary of Sections 4.1 and 4.2

1 Transport in phloem (translocation) involves the movement of ions and assimilates (sugars and other products of photosynthesis) from sources to sinks.

2 Phloem contains three types of living cells: sieve tube elements, companion cells and phloem parenchyma.

3 Sieve tube elements link up to form sieve tubes, through which translocation occurs. The perforated end walls form sieve plates and the interior (lumen) usually contains fibrillar P-protein and a few organelles but no nucleus. Turgor (internal hydrostatic pressure) is very high.

4 The internal structure of functioning sieve tubes is very difficult to study because surging and callose deposition cause artefacts when preparing material for microscopy. Consequently, it is not known how P-protein is arranged nor the degree of blockage of sieve plate pores.

5 Companion cells derive from the same parent cells as sieve tube elements but have dense cytoplasm, many organelles and a nucleus. They function in the loading, unloading and maintenance of sieve tubes, a companion cell plus sieve tube element forming a functional unit. Companion cells, and sometimes phloem parenchyma, may become specialized as transfer cells.

6 Transfer cells are characterized by wall ingrowths, numerous plasmodesmata and dense cytoplasm with many mitochondria.

7 Transfer cells may be involved in both the loading and unloading of sieve tubes, acting like a lens to concentrate solutes released into the apoplast. They also occur in a wide range of situations where intense, short-distance transfer of solutes occurs between two tissues, especially if the tissues are not linked by plasmodesmata. Active transport into or out of transfer cells occurs where there are wall ingrowths and passive movement occurs via plasmodesmata.

Questions relating to Sections 4.1 and 4.2 are on pp. 164–165.

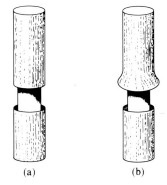

Figure 4.6 Malpighi's experiment. (a) Girdling a woody shoot by removing the bark leaves the xylem intact. (b) A few weeks after girdling, the tissue on the upper side has swollen.

(a)　　　　(b)

4.3 EXPERIMENTAL STUDIES OF THE PHLOEM

Now that you know something about phloem structure and before going on to consider how the phloem functions, it is useful to examine the kinds of experiments used to investigate the tissue and some of the data they provide about, for example, rates of translocation and the composition of phloem sap. The earliest experiments aimed simply to show that some tissue, other than xylem tissue, was involved in the long-distance transport of sugars. In 1679, for example, the Italian Malpighi (of kidney Malpighian tubule fame) removed a ring of bark and a ring of soft tissues external to the wood of a tree (a procedure called **ring barking** or **girdling**) and observed that 'nutrient material' accumulated and caused swelling above, but not below the ring (see Figure 4.6).

The accumulation occurred only in the light when green leaves were present above the ring, so Malpighi concluded that nutrients were being 'elaborated' by leaves in the light and transported down the stem. As most sap — that is the transpiration stream in xylem — was thought to rise up the stem, this experiment provided the first indication that a special tissue was involved in the long-distance transport of nutrients out of leaves and down the stem. Not until this century, however, long after phloem tissue had been observed down the microscope, was it shown convincingly that Malpighi's special tissue was phloem.

4.3.1 Evidence that organic solutes move in phloem

We describe below two of the main techniques that have been used to study phloem transport and provide evidence that sugar translocation occurs in phloem sieve tubes.

Girdling

These experiments are similar in principle to those of Malpighi just described and usually depend on careful surgery combined with chemical analysis of tissues or the use of isotopic markers. In 1933, for example, Schumacher carried out experiments which demonstrated that phloem rather than xylem is the transport tissue for organic solutes. He used the leaf stalk of *Pelargonium*, and after dissecting away all the outer tissues to leave a single, large vascular bundle (with the phloem on one side and the xylem on the other) connecting stalk and leaf (Figure 4.7), he monitored the loss of organic nitrogen compounds from yellowing leaves that had been kept in the dark. Translocation of these substances continued after surgery but stopped when phloem tissue (zone A in Figure 4.7) was removed or damaged, the level of nitrogen in the leaf remaining constant.

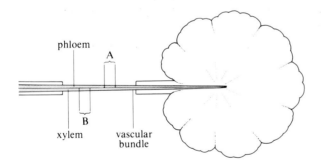

Figure 4.7 Schumacher's procedure with a leaf of *Pelargonium*. Xylem and phloem are exposed along a length of the leaf stalk.

◇ Predict the effect of removing zone B instead of zone A (Figure 4.7).

◈ Initially there should be no effect since the first experiment shows that there is no translocation of organic nitrogen into the leaf in xylem. However, xylem supplies the leaf with water so that eventually nitrogen loss would slow down as the leaf became short of water for phloem transport.

An experiment performed by Rabideau and Burr in 1945 showed that phloem transport may occur both up and down a stem. They combined stem girdling with use of the heavy isotope of carbon, supplied as $^{13}CO_2$, as a marker (the more convenient radioactive isotope, ^{14}C, was not available). They supplied $^{13}CO_2$ to a single leaf on a bean plant kept in the light, and subsequently monitored the distribution of ^{13}C throughout the plant. Before supplying $^{13}CO_2$, some plants ·were girdled at either or both of positions X and Y shown in Figure 4.8.

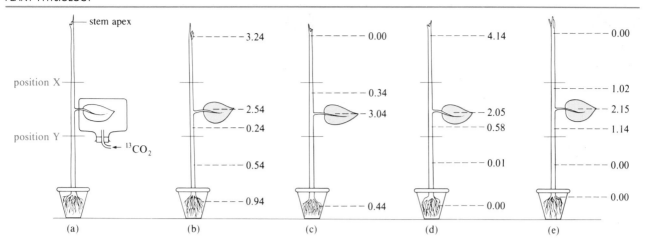

Figure 4.8 Rabideau's and Burr's experiment (1945) using French bean plants (*Phaseolus vulgaris*). (a) The system for supplying $^{13}CO_2$ to a leaf, and the positions where the stem was girdled in some plants (X and Y). (b)–(e) The relative distribution of ^{13}C after fixation of $^{13}CO_2$ by the leaf, and translocation with and without girdling at X and Y.

◇ Look at the data in Figure 4.8b–e showing the distribution of ^{13}C and decide whether the girdles were applied at X alone, Y alone, both positions X and Y or neither. *After checking your answer, mark the positions on Figure 4.8.*

◆ For the plant shown in Figure 4.8b, no girdles were applied; the marker moved both up and down the stem but mainly upwards. For Figures 4.8c, d and e, the positions of girdles were at X alone, Y alone and at both X and Y, respectively.

Autoradiography

The two experiments just described show that phloem tissue is involved in translocation but they do *not* show exactly which cells are used for long-distance movement. The earliest evidence that sieve tubes are key elements came almost accidentally when Schumacher was experimenting with *Pelargonium*. He was trying to follow the movement of dyes in phloem and found that although the dye eosin entered phloem, it did not move, and nor for that matter did anything else — translocation stopped completely! The reason, it turned out, was that eosin causes the deposition of callose, which blocks sieve tubes but no *other* phloem cells.

Since the 1950s, a standard technique in translocation studies has been the localization of radioactively labelled sugars by *autoradiography**. A single leaf is usually supplied with $^{14}CO_2$ in the light and, after allowing a suitable period for carbon to be fixed and exported, stem tissue is rapidly frozen and dehydrated before sectioning. For autoradiography the sections are placed on a photographic emulsion in the dark, the position of radioisotopes being revealed as dark silver grains on the developed emulsion. Figure 4.9a shows clearly the localization of silver grains over the phloem and Figure 4.9b shows the more precise location of grains, mainly over sieve tubes with some over the smaller companion cells.

*The principle of autoradiography is explained in more detail in Chapter 2 of Norman Cohen (ed.) (1991) *Cell Structure, Function and Metabolism*, Hodder & Stoughton Ltd in association with The Open University (S203, *Biology: Form and Function*, Book 2).

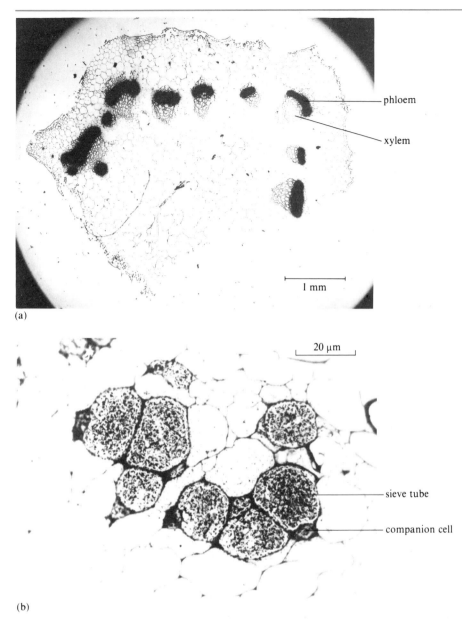

Figure 4.9 Autoradiographic evidence that assimilates from leaves are transported in phloem. (a) A transverse section of sugar beet leaf stalk 30 minutes after the assimilation of $^{14}CO_2$ by the leaf. The blackened phloem marks the path of translocation. (b) Localization of ^{14}C-labelled assimilates in the sieve tubes and companion cells of *Ipomoea hederacea* (morning glory). A transverse section of stem is shown, 10 cm below a single leaf that was supplied with $^{14}CO_2$ for 6 hours.

4.3.2 Studies of translocation rates and the composition of phloem sap

Dry weight gain

A simple way of estimating solute flux through phloem is to monitor the gain in weight of a sink organ, such as a fruit or tuber. Usually, the increase in size of several sink organs is measured over a period of days or weeks, with occasional samples removed so that size increase can be related to dry weight increase. Then the organs are harvested and the stems that supplied them with assimilates are sectioned and the cross-sectional areas of phloem or, for the highest precision, just sieve tubes are determined. The average gain in weight over a given time divided by the average cross-sectional area of phloem or sieve tubes allows you to calculate the **specific mass transfer** (or **SMT**) with units of weight gain per unit area per unit time. There are many inaccuracies associated with this method; for example sink organs use some of the imported assimilate for respiration, and it is also very difficult to measure the cross-sectional area of phloem accurately. This method has, however, allowed useful comparisons to be made between species.

Collection of phloem exudates

To obtain information about *what* is transported in phloem, you need to know the composition of sieve tube contents. Two methods have been used to do this: collecting and analysing bleeding sap directly from phloem and sampling sieve tubes by means of aphids. Both methods depend on the high pressure within sieve tubes, which means that sap exudes if the sieve tube is punctured. Sap collection from certain monocots was mentioned in Section 4.1 but the amount of exudate obtained from these plants is quite exceptional. In most species, the cut phloem is rapidly sealed by P-protein or callose deposition and only in a few species can sufficient exudate be obtained for chemical analysis. There is also the possibility that exudate is contaminated by materials from cut cells adjacent to sieve tubes.

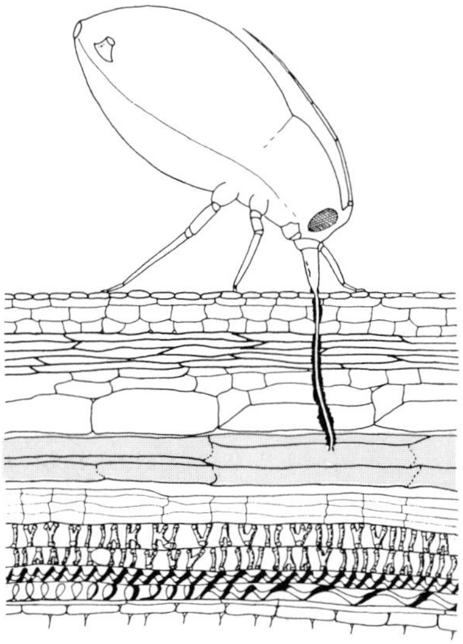

An elegant and more precise way of tapping phloem involves the use of sap-sucking aphids, the **aphid stylet technique** (Figure 4.10). These insects feed by inserting their tubular mouthparts (or stylets) into a *single* sieve tube so that sugary sap surges up them — the aphid does not even have to suck. If a feeding aphid is anaesthetized (to prevent it from withdrawing its stylets) and the body severed from the stylets, the contents of the sieve tube continue to exude from the cut stylets, often for several days, and can be collected by a micropipette for analysis. Aphids pass saliva into sieve tubes and this is thought to contain substances that inhibit callose deposition for a considerable time. It may also contain substances that cause growth distortion in the plant (leaf curling or galls induced by aphids are quite common) and which influence translocation by increasing the 'sink strength' of the aphid (discussed in Section 4.5). All this means that stylet exudate cannot be assumed to be identical with normal sieve tube contents –– but it is probably close to this ideal and provides the best sampling technique so far available for individual sieve tubes. The main drawbacks are that it is technically difficult to do and produces only very small samples of sap. Nevertheless, it is a powerful technique allowing not only sap analysis but also, when combined with the use of radioactive tracers (described below), measurement of translocation rate and the effects on it of treatments such as cooling and water stress; by gluing a tiny manometer — a device for measuring pressure — to exuding stylets, it is even possible to measure the turgor of sieve tubes.

Figure 4.10 Diagram of an aphid in feeding position with its stylets inserted into a sieve tube element. (Phloem tissue is shaded green.)

Use of radioactive tracers

Nearly all recent work on translocation involves the use of radioactive isotope labels for sugars, phosphate or potassium (all of which are components of phloem sap). The velocity of transport, for example, can be measured by introducing labelled material into the phloem (by feeding it to a leaf or injecting it into a stem) and monitoring the time taken for the label to travel a known distance. The monitoring may be done either with a sensitive Geiger counter placed on the outside of the plant or by using aphid stylets.

4.3.3 Facts about phloem transport

From the battery of techniques just described, we can now consider some of the information obtained about phloem transport.

Phloem sap

Compared with xylem sap from a transpiring plant, **phloem sap** is a much more concentrated solution of considerably higher pH. Xylem sap is typically a dilute, slightly acid solution (about 0.01% (w/v) solids and pH 6) whereas phloem sap contains 15–30% (w/v) dissolved solids and has a pH in the range 7.2–8.5. Usually, more than 90% by weight of the phloem solutes are soluble carbohydrates and, although some species contain large quantities of 'unusual' sugars such as raffinose (a trisaccharide) or mannitol (a hexose sugar alcohol), in the majority of plants the most abundant sugar is the disaccharide, sucrose (Chapter 2, Section 2.5.1). This generally comprises about half the solute molecules in phloem sap, as you can see from Figure 4.11, which shows the composition of phloem exudate obtained from the castor bean plant (*Ricinus*) by cutting and bleeding the stem.

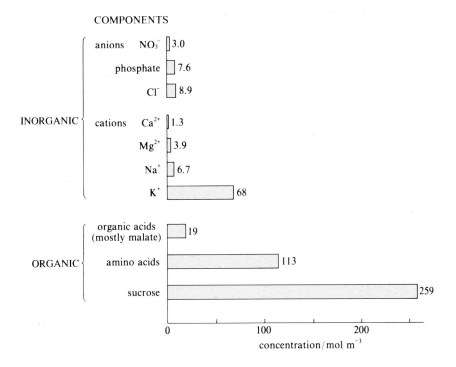

Figure 4.11 The average composition of phloem sap from *Ricinus* (castor bean plant).

Such concentrations of sucrose are considerably greater than those in surrounding tissues.

◇ What does this concentration difference suggest about the mechanism by which sucrose enters sieve tubes?

● Because sucrose entry is against (or 'up') a concentration gradient, it probably requires an energy input, i.e. sucrose is loaded actively into the sieve tubes by some kind of active transport. This is discussed further in Section 4.5.

Other points to notice from Figure 4.11 are that amino acids (and amides) are the next most common solutes after sucrose and that the commonest ion present is potassium.

One feature of phloem sap deserves special mention because it is highly relevant to the mechanism of transport, discussed in the next section. The concentration of sugar in sieve tubes is always higher close to the source, where it is produced or released, than it is at the sink end of the pathway, where it moves out of the phloem. So there is a gradient of sugar concentration from source to sink along sieve tubes.

Rates of transport in phloem

The 'rate of transport' is an umbrella term covering two different parameters: the **velocity of transport** (distance travelled per unit time) and the solute flux, defined earlier as SMT. These parameters are interrelated because

$$\text{SMT} \ (\text{g cm}^{-2}\,\text{h}^{-1}) = \text{sap concentration} \ (\text{g cm}^{-3}) \times \text{average velocity} \ (\text{cm h}^{-1}).$$

SMT matters to a plant because the rate at which an organ can grow depends to a large extent on the rate at which energy and raw materials are delivered to it by the phloem. Measured values for SMT are mostly in the range 0.5–$6.3 \text{ g cm}^{-2}\,\text{h}^{-1}$ but for the cut inflorescence stalk of the *Arenga* palm an astounding $99 \text{ g cm}^{-2}\,\text{h}^{-1}$ has been recorded. Over a period of 130 days, the amount of sucrose in bleeding sap from this palm amounted to 252 kilograms!

Transport velocities for sugar are usually in the range 10–100 cm h^{-1} but for short periods, and often in experimental situations, maximum velocities of 500–$1\,500 \text{ cm h}^{-1}$ are quite common. The *Arenga* palm once again holds the record: 7.2 *metres* per hour. To give you some idea of what these velocities mean, it has been calculated that for young willow stems sampled by aphid stylets and with a transport velocity of 120 cm h^{-1}, 235 sieve elements are emptied every minute. Similar calculations for other species tapped by aphid stylets suggest that the punctured sieve tube empties and refills between three and 28 times per second. The rate at which sugar molecules move by diffusion is slower by a factor of around 10^5 than the rate of movement in sieve tubes; so whatever the mechanism of phloem transport, it is *not* simple diffusion. We consider next what the mechanism is thought to be.

Summary of Section 4.3

1 Early experiments which demonstrated that translocation occurs in phloem, that it can occur both up and down a stem between sources and sinks and that sieve tubes are probably the main pathway, involved the use of stem girdling, isotopic markers to label translocated material and autoradiography.

2 There are two measures of translocation rate: specific mass transfer (SMT) and sap velocity. SMT = [sap] × velocity.

3 Estimates of SMT can be obtained from the rate of dry weight gain in sink organs and the cross-sectional area of phloem or sieve tubes. Alternatively, knowledge of sap concentration and velocity can be used to estimate SMT.

4 The composition and concentration of phloem sap can be determined after sap collection from cut stems or sampling sap from single sieve tubes by means of aphid stylets. Phloem sap is an alkaline solution with 15–30% (w/v) solids, of which more than 90% are sugars, usually sucrose. Organic nitrogen compounds are the next most abundant solute and K^+ the most common ion. Sap concentration is higher close to sources than close to sinks.

5 Sap velocity and much other information about translocation can be obtained by using radioactive tracers, sometimes in combination with the aphid stylet technique. Velocities are commonly in the range 10–100 cm h^{-1} but often reach 500–$1\,500 \text{ cm h}^{-1}$. Simple diffusion could not account for velocities of this order.

Questions relating to Section 4.3 are on pp. 164–165.

4.4 THE MECHANISM OF TRANSPORT IN PHLOEM

Until about 1980, few topics in plant physiology aroused so much controversy as the mechanism of phloem transport: what force drives solutes along sieve tubes? A key reason was the lack of precise information about the structure of sieve tubes (Section 4.2) but, although there is still uncertainty about this, nearly all research workers have accepted one model of phloem transport: the **pressure flow hypothesis**. We shall first describe this hypothesis and then consider why, for more than 50 years, many scientists did not accept it.

4.4.1 The pressure flow hypothesis

The German physiologist Münch first proposed this hypothesis in its simplest form in 1930. Since then it has been elaborated somewhat but remains a delightfully simple concept, which is this:

The driving force for phloem transport is a gradient of hydrostatic pressure within sieve tubes: sieve tubes at the source end of a pathway have a higher turgor than those at the sink end.

The key point to grasp is how a pressure gradient in the phloem might be generated and this is illustrated by the model in Figure 4.12. In its current form, the pressure flow hypothesis suggests that sugars and other solutes are actively transported into sieve tubes at the source end; in Figure 4.12 this would be equivalent to adding dry solutes to chamber A. Water then enters by osmosis generating a high turgor pressure, which causes mass flow of water and solutes along the sieve tubes (the tube linking chambers A and B in Figure 4.12). There is relatively little movement of solutes out of sieve tubes along the pathway but, at the sink end, solutes move actively or passively into the surrounding tissues. Water moves out either in response to the pressure from the source (i.e. by reverse osmosis, as happens when water from the blood enters the Bowman's capsule of mammalian kidneys*) or, if there are high solute concentrations in the cell walls (apoplast), by osmosis, first into the cell walls and then into the sink cells. It then either moves back to the source, e.g. in the xylem from roots to shoots, or remains and contributes to growth (cell expansion) of the sink — think of a swelling fruit or growing shoot tip. Consider next the implications of this hypothesis and how they stand up to experimental testing.

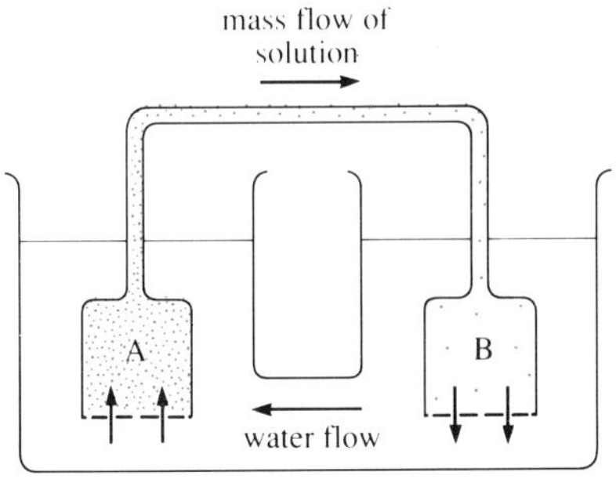

Figure 4.12 A model to illustrate the concept of pressure flow in phloem. There is a higher concentration of solutes in the source (A) than in the sink (B). Water from the surrounding compartment enters A by osmosis across the semipermeable membrane (dashed line), generating a hydrostatic pressure (turgor) which causes mass flow of solution along the connecting tube to B. Because the osmotic pressure at A is higher than at B, water is forced out across the semipermeable membrane at B by reverse osmosis.

4.4.2 Predictions and testing of the pressure flow hypothesis

The pressure flow hypothesis predicts that:

1 *There should be a pressure gradient along sieve tubes*. We mentioned in Section 4.3.2 that gradients of sugar concentration have been measured in sieve tubes but, for reasons that will become apparent when you have studied Chapter 5, this does not necessarily imply the existence of a pressure gradient. Few direct measurements of sieve tube turgor have been made because they are very difficult to do. However in a species of oak, phloem turgor at different heights was measured with a fine needle and pressure gradients of the order of 0.03–0.05 MPa per metre were found. The critical question is whether gradients of this size are sufficient to drive phloem sap at the kinds of velocities measured.

*The structure and function of kidneys are described in Chapter 11 of Michael Stewart (ed.) (1991) *Animal Physiology*, Hodder & Stoughton Ltd in association with The Open University (S203, *Biology: Form and Function*, Book 3).

The size of gradient necessary can be calculated from Poiseuille's law*, provided you know the resistance to flow of the sieve tubes — which depends on the dimensions and frequency of sieve plate pores *and the extent to which pores and sieve tube lumina are blocked* by P-protein or callose. If there is no blockage at all, then a 10% (w/v) sugar solution (that is, 10 g sugar per 100 cm³ solution) travelling at 100 cm h⁻¹ would require a pressure gradient of 0.027 MPa m⁻¹. This is a reasonable figure, well within the range of measured values cited above. But evidence from electron micrographs indicates that obstructions *are* present and this dramatically increases the necessary pressure gradient. Calculations by Weatherley in 1972, for example, showed that if P-protein filaments are no less than 200 nm apart and spread out uniformly through the sieve tube lumen and across pores, then a gradient of 12 MPa m⁻¹ would be required to move a 10% sugar solution at 100 cm h⁻¹. If the filaments were grouped into bundles as they crossed the lumen, the necessary gradient fell to 1.1 MPa m⁻¹, but even this lower value is hopelessly unphysiological: such high pressure gradients do not exist in the phloem. Weatherley calculated that with a physiologically acceptable pressure gradient of around 0.1 MPa m⁻¹, not more than 1% of the cross-sectional area of the pores should be occupied by filaments, and filaments in the lumen should be arranged either as a small number of tightly packed bundles or round the edge of the cell.

Figure 4.13 shows the range of images of phloem obtained with the electron microscope using various fixation techniques.

◇ Four of the arrangements shown in Figure 4.13 are definitely incompatible with the pressure flow hypothesis. Which four do you think these are?

◈ The last four, (i) to (l), in which considerable numbers of filaments block the pores. Arrangements (a) to (d) are definitely and (f) to (h) probably compatible with pressure flow; that is, they offer a sufficiently low resistance to flow.

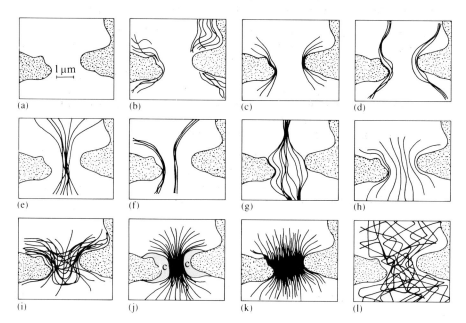

Figure 4.13 Arrangements of P-protein filaments in sieve pores (seen in section) that have been observed in the electron microscope. Material labelled c in (j) represents callose lining the pore and compressing filaments in it.

* Poiseuille's law was described in Chapter 5 of *Animal Physiology*, in relation to blood flow. It states that the rate of flow of liquid (J) in a rigid tube equals the pressure gradient, ΔP, divided by the resistance to flow, R, i.e. $J = \Delta P/R$, where R equals the fourth power of the radius (r) of the tube divided by the viscosity of the liquid, η, and the tube length, l, i.e. $R = r^4/\eta l$.

It was the frequent occurrence of electron microscope images like (i) to (l) in Figure 4.13 that led many plant physiologists to reject the pressure flow hypothesis. There must, they argued, be another mechanism, another force causing flow. But you know from Section 4.2 that artefacts due to surging and callose deposition arise very easily during the preparation of phloem for electron microscopy and it is now widely accepted that blocked sieve tubes are not the normal, functioning state. If this is true, then the main objection to the pressure flow hypothesis is removed but there is always the possibility that new techniques will show unequivocally that functioning sieve tubes are partially blocked, and the hypothesis would have to be rejected once again. However, the pressure flow hypothesis also leads to other predictions to which objections were also raised.

A second prediction of the pressure flow hypothesis is that:

2 *Water and solutes should move together in the same direction and at the same rate down sieve tubes.* This applies to any movement by mass flow (Chapter 3) and means, for example, that if ^{14}C-labelled sugars and tritiated water $^{3}H_2O$ (incorporating the tritium isotope ^{3}H) are introduced simultaneously into phloem, they should move at the same rate along it. Some experiments found that they did not, with sugars moving much faster than water, but there is a good explanation for this. Sieve tubes are permeable to water, which may exchange rapidly with surrounding tissues and hence produce an *apparent* slowing down of tritiated water.

Another apparent problem was **bidirectional flow** — simultaneous translocation in opposite directions. Some workers claimed that because sieve tubes in the same vascular bundle could be shown to transport in different directions, this invalidated the pressure flow hypothesis. In fact it does not. You know from Figure 4.8 that a single source may transport material both up and down a stem. However, if you study Figure 4.14 you can see that vascular bundles branch and anastomose in quite complicated ways along the stem, so that sieve tubes in one vascular bundle may quite reasonably be transporting in opposite directions without any violation of pressure flow principles.

Much more serious was the claim that bidirectional flow could occur in a *single* sieve tube (Figure 4.15): this really is incompatible with pressure flow. But the experiments purporting to show it were carried out with maturing leaves that change from a sink to a source as they mature. This change is associated with a reversal of flow in the sieve tubes and gave a spurious impression of bidirectional flow. True bidirectional flow occurring at the same time in a single sieve tube has never been demonstrated, so the pressure flow hypothesis cannot be rejected on this account.

At present, therefore, the pressure flow hypothesis is consistent with all the known facts about phloem transport. The only doubts about its validity hinge on the disputed evidence about sieve tube structure, but otherwise it has withstood rigorous testing and is widely accepted.

Summary of Section 4.4

1 The pressure flow hypothesis is widely accepted as the mechanism of phloem transport. It suggests that transport is a mass flow phenomenon driven by a gradient of hydrostatic pressure (turgor) between sources and sinks.

2 Two important predictions of this hypothesis are that (a) sieve tube turgor should be highest close to sources and decrease towards sinks; and (b) water and all solutes should move together at the same velocity and in the same direction.

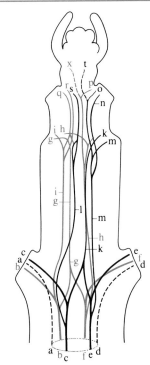

Figure 4.14 Anastomosing vascular bundles in a stem of *Clematis vitalba*. The letters should help you to follow the course of each individual bundle.

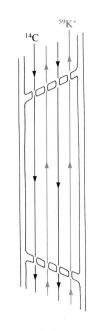

Figure 4.15 A diagram to illustrate bidirectional flow in a single sieve tube; ^{14}C-labelled sugars move down and $^{59}K^+$ ions upwards. *Note that this kind of bidirectional flow has never been shown to occur.*

3 Measurements of pressure gradients in sieve tubes support prediction (a) *provided* it is assumed that P-protein and callose cause no significant blockage of the sieve tube lumen and sieve plate pores, respectively.

4 Bidirectional flow at the same time and in a single sieve tube has never been demonstrated conclusively. Thus prediction (b) has not been disproved.

Question 1 (*Objectives 4.1, 4.2 and 4.4*) Which of the following statements are *true* and which are *false*? Qualify your answers where appropriate.

(a) Translocation is the transport of sugars from shoots to roots.

(b) A sieve tube is a unit of translocation composed of interlinked sieve tube elements.

(c) The major components of sieve tube cytoplasm are P-protein and callose.

(d) The convoluted surfaces of transfer cells are probably involved in the active uptake from or active secretion into the apoplast.

(e) Source-to-sink transport can be defined as the movement of sugars in the phloem from leaves to roots.

Question 2 (*Objective 4.3*) Explain which of techniques (i) and (ii) you would consider more appropriate as preparation for an ultrastructural examination of phloem sieve tubes in as 'natural' a state as possible.

(i) Remove a segment of stem tissue from a species with vascular bundles lying close to the surface; puncture the sieve tubes to release the pressure and then fix the tissue very slowly and gently before embedding and sectioning it for EM observation.

(ii) Take an intact plant of a species (such as certain water-lilies) where there is a single central vascular bundle in the stems or leaf stalks; very gently dissect away the cortex to expose a length of vascular bundle and then freeze this segment very rapidly before removing and fracturing it for EM observation.

Question 3 (*Objectives 4.5 and 4.6*) Exuding stylets were obtained from aphids feeding on the roots of two plants. For plant A, $^{14}CO_2$ was supplied to the lowest leaf on the stem and ^{14}C-labelled material (identified as sucrose on analysis) was subsequently collected in stylet exudates. For plant B, $^{14}CO_2$ was supplied to the uppermost leaf but when exudate was subsequently collected it was found to be unlabelled. The plants had no side branches and no flowers or fruits. Which of (i) to (iv) are valid conclusions from these data?

(i) Assimilates move in the phloem sieve tubes.

(ii) Phloem transport can be bidirectional in a stem.

(iii) Sinks are supplied with assimilates by the sources that are nearest to them.

(iv) Sucrose is the major component of phloem sap.

Question 4 (*Objectives 4.1 and 4.6*) (a) If the specific mass transfer (SMT) into a sink alters over a short period, what two factors might have caused the change?

(b) If you knew the velocity of transport in the stem phloem of a specific plant, how could you use the aphid stylet technique to estimate the SMT to the roots?

Question 5 (*Objectives 4.1 and 4.7*) Which of (a)–(d) are true statements about or predictions of the pressure flow hypothesis?

(a) The pressure flow hypothesis states that sugars move in phloem from regions of higher concentration to those of lower concentration.

(b) Amino acids can be transported up a sieve tube in which sucrose is being transported down.

(c) Cooling a small length of stem between a source and a sink should not affect SMT to the sink.

(d) An increase in sieve tube turgor close to a sink would slow down the rate of translocation to that sink.

4.5 ASSIMILATE PARTITIONING AND THE CONTROL OF TRANSLOCATION

Now that you know something about the pathway and the mechanism of phloem transport, we can turn to the wider issue of overall control: what determines how much of a particular substance — sucrose or organic nitrogen, for example — goes to a particular sink? If you compare crop plants with their wild ancestors, the harvestable organs — storage tissue, fruit or seed — are always much bigger in the crop plant and a major reason is that a higher proportion of the plants' resources are translocated to these organs. You can say that the harvestable organs act as *'stronger' sinks*. But what does this mean? We shall explore this question and the related topics of phloem loading and unloading in this section.

4.5.1 The 'strength' of sinks

When potato tubers start to grow, a high proportion of assimilates is diverted to them — they have a high **sink strength** — and the main reason seems to be that they are areas of intense metabolic activity. They have high rates of cell division and expansion, respiration, and biosynthesis and, consequently, a high demand for assimilates. What determines demand is really a question for developmental biologists because you might just as well ask why a particular organ begins to grow at a particular stage of development. It has been suggested that growth regulators (plant hormones) affect the strength of a sink, and certainly strong sinks are often regions with high levels of certain regulators that are known to stimulate cell division and cell expansion.

◇ If substances such as plant hormones were shown to control assimilate demand and hence sink strength, how might this knowledge be useful for agriculture?

◆ By applying the substances to harvestable organs it might be possible to stimulate sink activity, increasing growth and hence crop yields.

As yet, no such trigger substance has been found and it now appears that, besides metabolic activity and demand, another factor may be of considerable importance in determining sink strength: this factor is the capacity to unload solutes from the phloem and transport them to the sink.

4.5.2 Phloem unloading

Early in the life of a leaf it is a sink organ and solutes are unloaded mainly from the smallest veins, which are described as **minor veins**. As the leaf

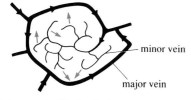

(a) young leaf (unloading)

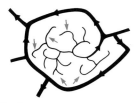

(b) mature leaf (loading)

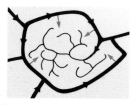

(c) mature, shaded leaf
(unloading from major veins only)

Figure 4.16 Phloem loading or unloading in the minor veins of a leaf. Green arrows show the direction of solute transport. (a) Unloading in a young leaf that acts as a sink. (b) Loading in a mature leaf that acts as a source. (c) Unloading from major veins only in a shaded mature leaf; no unloading occurs from minor veins.

Figure 4.17 (a) Diagrammatic section through a maize kernel illustrating the phloem termination in the stalk and assimilate movement (green arrows) to the endosperm (food-storing tissue) through the placenta region and the endosperm transfer cells. (b) As for (a) but with endosperm and embryo removed and replaced by trap solution or agar, as is done in the empty seed-coat technique. (c) Diagram to show the pathway of assimilate unloading in a maize kernel. Arrows within cells indicate probable symplastic intercellular transport via plasmodesmata (shown as gaps in the cell walls). Arrows within cell walls indicate apoplastic movement. Outline arrows (entering endosperm transfer cells) represent what is thought to be active transport.

matures it gradually changes to being a source organ and solutes are now loaded for export into the minor veins. But if a mature leaf is shaded, so that it is unable to photosynthesize and it becomes once again a sink organ, the minor veins do not switch from loading to unloading: they seem to have undergone an irreversible developmental change that allows them to load but not to unload, although solutes *are* translocated into the darkened leaf and there is limited unloading from the larger veins (Figure 4.16). Possibly the 'refusal' to unload from minor veins in mature leaves is a device that protects the plant from supporting 'parasitic' leaves which have become accidentally covered up or heavily shaded during development.

This kind of evidence suggests, therefore, that phloem unloading is not simply an innate capacity of any organ but a capacity that is developmentally programmed and involves specific structures or processes. Clearly, this capacity must be an essential attribute of strong sinks.

In the early 1980s an ingenious method was developed to study phloem unloading into seeds, the **empty seed coat technique**, and this has provided much information about the process. It has been used with plants such as legumes and maize (*Zea mays*), which have fairly large seeds, and is illustrated for maize in Figure 4.17a and b together with the presumed pathway of phloem unloading–seed loading (Figure 4.17c).

The empty seed coat technique depends on the fact that between the maternally-derived seed coat and the seed (embryo plus food-storing tissues, which are called endosperm in maize) there are no plasmodesmata. Because mother and offspring are not cytoplasmically linked, it is possible to remove the seed tissues without damaging maternal tissues and these continue to deliver solutes to a 'trapping solution' or agar trap at much the same rate as to the seed. By supplying inhibitors or varying pH or solute concentration in the trap, which is in contact with maternal tissues, their effects on solute delivery and thus, indirectly on phloem unloading, can be studied.

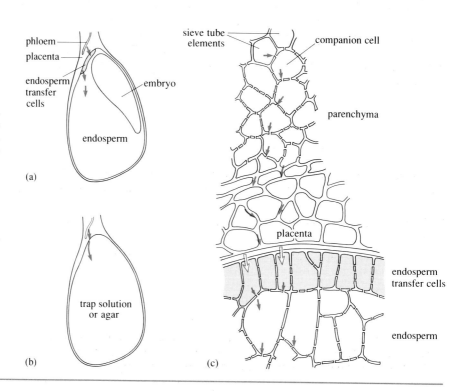

Consider first the presumed pathway of solute entry into maize seeds (Figure 4.17c). Solutes move out of sieve tubes probably via plasmodesmata and move symplastically through companion cells and phloem parenchyma.

◇ If this unloading route occurs also in the minor veins of leaves, what might prevent phloem unloading in mature leaves which had been converted to sink organs by shading?

◈ Blockage or disruption of plasmodesmata.

Solutes then enter the apoplast (cell walls) of cells in the placenta and finally are accumulated (probably actively) by an outer layer of transfer cells in the endosperm of the seed. Organic solutes entering the endosperm are rapidly converted to insoluble products — starch or protein, for example — so a steep concentration gradient exists between seed and maternal phloem, with the endosperm transfer cells acting like a booster station in the seed (Figure 4.18); since passive diffusion rates are proportional to concentration gradients, the steepness of these gradients influences strongly the rate of transport into seeds.

Maintaining the overall gradient is a function of seed metabolic activity and this is where the 'demand' element of sink strength is involved. Equally important, however, is the development and maintenance of the unloading pathway from phloem to seed and any interruption of this pathway will reduce assimilate entry to the sink. In certain mutants of barley, for example, shrunken grains are produced because the cells equivalent to the placenta in maize degenerate early in development and no further assimilates are transported to the endosperm.

In some species, active transport of assimilates appears to occur somewhere along the unloading pathway in maternal tissues. Thus in soya bean (*Glycine max*, a legume), supplying a fungal toxin that stimulates H^+-ATPases (see Chapter 3, Figure 3.7) will stimulate the unloading of ^{14}C-labelled assimilates into empty seed coats. In maize, however, there is no such effect and only passive movement seems to occur from phloem to endosperm transfer cells. Both the pathway and the processes involved in unloading may vary between species and probably between different organs on one plant, but phloem unloading is undoubtedly one of the keys to understanding the control of translocation.

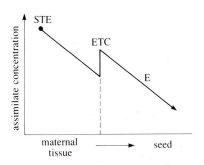

Figure 4.18 The concentration gradient of assimilates in maize from sieve tube elements (STE) in maternal tissue to seed endosperm (E). Note the rise in concentration in the endosperm transfer cells (ETC) of the seed, where assimilates are thought to be accumulated actively.

4.5.3 Feedback from sink to source

One reason why it is important to know how sinks function is that their activity may have a considerable effect on sources. Consider the following data:

1 When potato tubers begin to form, the rate of photosynthesis in the leaves increases.

2 If young tubers are removed from a potato plant, photosynthesis in the leaves declines.

◇ What do these data imply about the relationship between sources and sinks?

◈ They imply that there is a *feedback* of information from sink to source; increased demand at the sink can stimulate activity at the source, and decreased demand at the sink can inhibit source activity.

In both situations there are accompanying changes in the rate of phloem transport between sources and sinks so that *source–phloem–sink represents a highly integrated system*. To underline this point, look at the data in Figure 4.19.

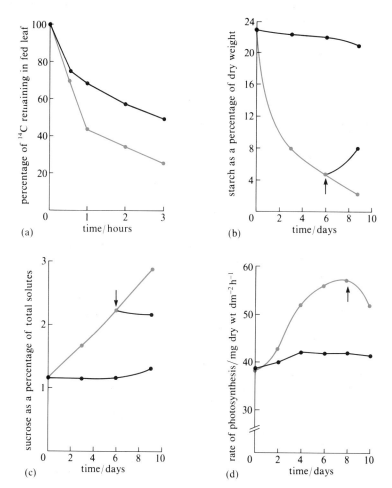

Figure 4.19 Measurements relating to a source — a single leaf of soya bean — when either all other leaves were shaded to prevent photosynthesis (green lines) or, as a control, were left unshaded (black lines). Arrows show when shading was removed from some plants. Changes with time in: (a) the rate of translocation out of the source leaf, measured as ^{14}C remaining after feeding $^{14}CO_2$ to the source leaf, (b) the concentration of starch, (c) the concentration of sucrose in the source leaf and (d) the rate of leaf photosynthesis. (Note the different time-scale for (a).)

This shows the results of an experiment in which changes were monitored in a single source leaf on a soya bean plant when all the other leaves were shaded; other leaves were left unshaded in the controls. Notice the sequence and time-scale of events in the source leaf. The first thing to happen (within 1–2 hours of shading other leaves) is a decrease in the amount of ^{14}C-labelled assimilate in the source leaf, i.e. an increase in the rate of phloem loading and transport (Figure 4.19a).

◇ What appears to happen next?

◆ A decrease in the amount of starch stored in the source leaf (Figure 4.19b), closely followed by (and possibly at the same time as) an increase in the level of sucrose (Figure 4.19c).

Later, about three days after shading, the rate of photosynthesis in the source leaf increases (Figure 4.19d). In this experiment the 'demand' of sinks on the source leaf was artificially increased and it shows clearly that activity at sources and sinks and the flux of solutes in the phloem are closely integrated. The experiment also shows that phloem loading is particularly sensitive to changes in sink demand, so we shall examine this process in more detail.

4.5.4 Phloem loading

We mentioned earlier that **phloem loading** is likely to be an active process because the concentration of sucrose in sieve tubes is so much higher than in other cells. Over the last 20 years, a mechanism for active phloem loading has been identified and this is illustrated in Figure 4.20.

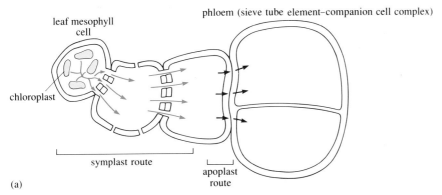

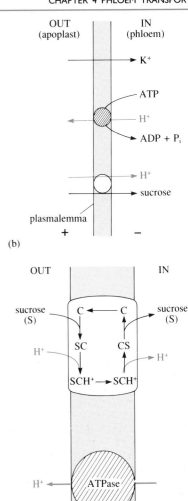

Figure 4.20 Phloem loading. (a) Apoplastic unloading into the phloem of a leaf. Assimilates move in the symplast from leaf mesophyll cells to the sieve tube element–companion cell complex (either type of cell), which they enter via the apoplast (cell walls).

(b) Sucrose–proton co-transport in phloem loading. The primary active efflux of H^+ ions occurs via the ATPase (hatched circle) and secondary active transport of sucrose with H^+ occurs via a sucrose transport molecule (white circle). Secondary active influx of K^+ may also occur.

(c) A model of how sucrose–proton co-transport may work. A membrane carrier protein (C) in the sucrose transport complex binds sucrose (S) and a proton (H^+) at the outer surface, changes conformation and moves the sucrose and proton to the inner surface, where they are released.

The suggestion is that sucrose from source cells (leaf mesophyll in Figure 4.20) passes via plasmodesmata along a symplast route until it reaches the phloem, where it crosses a cell membrane and enters the apoplast. The precise phloem cell where this apoplast transfer occurs is not certain and probably varies: it could be a companion cell or sieve tube element or even a phloem parenchyma cell (not shown in Figure 4.20). However, apoplast transfer is a critical step because it is from the apoplast that active transport occurs and sucrose is accumulated in phloem cells (Figure 4.20b).

◇ Is the sucrose accumulated by primary or secondary active transport (Chapter 3, Section 3.4.1)?

◆ Secondary active transport: an electrochemical gradient produced by primary active pumping of protons *out* of the phloem drives sucrose plus protons *into* phloem via specific transport molecules.

This process is described as **sucrose–proton co-transport** and a model of how it might operate is shown in Figure 4.20c. The proton gradient can also drive K^+ into phloem and is used for active loading of amino acids and amides, although by separate transport molecules to that for sucrose.

◇ If sucrose and other phloem solutes are loaded actively into companion cells or phloem parenchyma, by what route and mechanism might they reach the sieve tubes?

◆ Possibly by passive movement in the symplast, entering sieve tubes via plasmodesmata; or they might diffuse passively across the cell wall and membrane or even enter sieve tubes by secondary active transport. At present the answer is not known.

There are also doubts that active transport via the apoplast is the *universal* mechanism of phloem loading: some research workers believe that an entirely symplastic entry route is possible, but the mechanisms involved, e.g. to move sucrose against a concentration gradient, are not clear. Bearing in mind these doubts, you can accept active transport from the apoplast as a consensus view of phloem loading and it is certainly very useful in explaining, for example, the great specificity of phloem transport.

◇ How does it explain why sucrose is the only sugar translocated in many plants?

◈ Presumably because of the high specificity of the sucrose transport molecule and the absence of transport molecules for other sugars.

It also allows more precise questions to be asked about control of the *rate* of phloem loading and about 'communication' between sources and sinks. The supply of ATP, sucrose and, possibly, K^+ could all influence the rate of active loading but such control by the source is likely to be very coarse. Fine control, where loading responds rapidly to a change in sink demand, must depend mainly on conditions inside sieve tubes. Work with castor beans (*Ricinus*) suggests that sieve tube *turgor* is an important factor. Thus, cutting into a stem caused an outflow of phloem sap and reduced turgor (hydrostatic pressure) to near zero — simulating a very strong sink. The result was a rapid increase in phloem loading at the sources, which was explained as a response to lowered turgor in source phloem. In effect, a pressure wave travelling from sink to sources acted as a signal. How pressure signals are sensed or amplified at sources so that phloem loading is influenced is still not known, although there are plenty of speculative ideas. Pressure signalling would, however, provide a simple means of communicating sink demand to sources.

The general message emerging from studies of source activity and phloem loading is that these processes *respond* to changes in sink activity but do not usually control it. Sink activity is, therefore, seen as the final limiting factor that determines translocation rate and the ability of a sink to off-load sucrose and to absorb the assimilates will ultimately determine its growth rate.

Summary of Section 4.5

1 The activity of source and sink organs is closely integrated and the distribution of solutes between different sinks is tightly controlled.

2 The major controlling factor appears to be sink strength, a combination of sink metabolic activity ('demand') and the capacity to unload solutes from sieve tubes.

3 The capacity to unload phloem is lost by leaves as they mature and it develops in sink organs. Anatomical studies combined with the empty seed coat technique suggest that unloading occurs into the symplast and is probably passive. As solutes move to the sink cells which actually use them, transfer to the apoplast and active transport (often involving transfer cells) may occur. The pathway of unloading and the processes involved seem to vary between organs and species.

4 There is positive feedback between sinks and sources, an increase in sink strength leading to an increase in source activity and phloem loading.

5 Phloem loading is thought to involve active transport from the apoplast either directly into sieve tubes or indirectly via adjacent companion cells or phloem parenchyma. Secondary active transport of sugars, K^+ and amino compounds occurs, driven by a primary proton gradient. The mechanism for

sugars is described as sucrose–proton co-transport and involves a sucrose transport molecule; this mechanism explains the high specificity of phloem loading.

6   Pressure signalling is one hypothesis about sink-to-source communication. If sink strength changes, affecting the unloading and turgor of sieve tubes, this information might be transmitted to sources as a pressure wave, which influences source activity and phloem loading.

Question 6 (*Objective 4.9*) When applied to a leaf, a substance X inhibited sucrose loading but had no effect on proton extrusion or the uptake of K^+ by the phloem. Using Figure 4.20 as a guide, suggest two possible sites of action of substance X. *Note*: X did not inhibit sucrose synthesis.

Question 7 (*Objective 4.8*) When tomato plants are partly defoliated, there is often no decrease in the total yield of fruit. Select from (i) to (v) items that help to explain this observation.

(i) negative feedback between sink and source

(ii) positive feedback between sink and source

(iii) increased flux of solutes from sources to sink

(iv) increased sink activity

(v) increased source activity.

Question 8 (*Objective 4.8*) The respiration rate (R) of a sink organ can be taken as a measure of sink metabolic activity. Figure 4.21b shows how R for a single pod on a soya bean plant (Figure 4.21a) varies when the stem was girdled at the positions marked. The pod was darkened so that it could not carry out photosynthesis. Figure 4.21c shows how gross photosynthesis (GP) varies in the source leaf after girdling. For the control plant, R and GP were measured for a darkened pod and a leaf in the same positions but no girdling was carried out. The results in Figure 4.21 are the average from nine separate plants measured simultaneously.

(a) Describe and explain the effects of girdling on R and NP in the experimental plants.

(b) Do these data suggest that source activity and phloem loading are the major factors controlling translocation rate and sink activity?

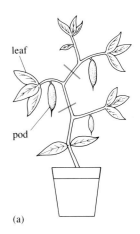

(a)

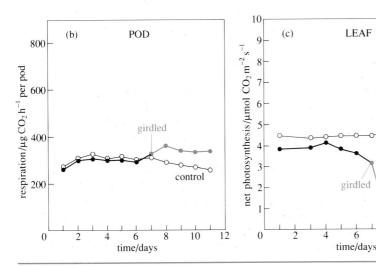

Figure 4.21 The effect of girdling one node of a soya bean plant so as to isolate a single trifoliate leaf (source) and one pod (sink). (a) Diagram of the plant, with girdle positions shown as green lines. (b) Effect of girdling on day 7 on the respiration rate of the isolated pod. (c) Effect of girdling on day 7 on the photosynthetic rate of the isolated source leaf.

OBJECTIVES FOR CHAPTER 4

Now that you have completed this chapter, you should be able to:

4.1 Define and use, or recognize definitions and applications of, each of the terms printed in **bold** in the text. (*Questions 1, 4 and 5*)

4.2 Describe the structure and arrangement of cells in phloem and relate structure to function. (*Question 1*)

4.3 Give two reasons why the ultrastructure of phloem is difficult to investigate, and assess critically the methods suggested for doing this. (*Question 2*)

4.4 Relate the structure of transfer cells to their supposed function and describe at least three situations where these cells occur. (*Question 1*)

4.5 Cite and evaluate evidence showing that assimilates move in phloem. (*Question 3*)

4.6 Describe the kinds of experimental techniques used to study phloem transport and interpret data about phloem transport in ways that are consistent with principles given in the chapter. (*Questions 3 and 4*)

4.7 Describe the pressure flow hypothesis, and cite or recognize data that support or contradict it. (*Question 5*)

4.8 Explain the nature and consequences of source–sink–phloem interactions. (*Questions 7 and 8*)

4.9 Describe possible mechanisms of loading and unloading the phloem, and predict or explain the effects of specified factors on the rates of these processes. (*Question 6*)

PLANTS AND WATER

5.1 INTRODUCTION

Water makes up, on average, 90–95% by weight of soft plant tissues and, since land plants cannot move about to find a drink, the ways in which they obtain water, distribute it between tissues and conserve it are vital aspects of their functioning. This chapter, therefore, is about plant water relationships.

An important point to grasp about water in vascular land plants is that only a tiny proportion of that taken up by the roots (typically 2–5%) is usually retained in the tissues and used for growth or chemical reactions. The rest evaporates from leaves, a process called **transpiration**; rapidly transpiring leaves can easily lose their own weight in water every *hour*. From general knowledge and earlier chapters in this book (bearing in mind that ions and sugars move in solution) you already know quite a lot about the paths of water movement in land plants.

◇ Describe the pathway of water movement in a vascular land plant from point of entry to point of loss.

◆ Water enters from the soil via the roots, moves across the root cortex to the xylem (as ions do, Chapter 3), passes from roots to shoots in the transpiration stream, and is then lost from leaves as water vapour, mainly through the stomatal pores. This pathway is illustrated in Figure 5.1. However, you should also appreciate from Chapter 4 that some water recirculates from leaves back to roots or other organs, moving in the *phloem*.

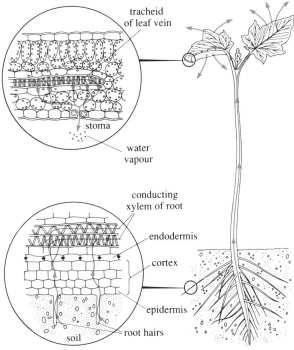

tracheid of leaf vein

stoma

water vapour

conducting xylem of root

endodermis

cortex

epidermis

root hairs

soil

Figure 5.1 The passage of water through a plant from roots to leaves. Green arrows show the pathway of water in the whole plant (right) and details of the pathway in roots and leaves are shown in the large circles (left).

The first question addressed in this chapter is: *why* does water follow this pathway — what are the *driving forces* that move water to the tops of tall trees (Sections 5.2 and 5.3)? We then ask what determines how fast and how much water moves in different parts of a plant, especially the rate of water loss from leaves (Sections 5.4 and 5.5), and what controls the overall flux of water between soil, plant and atmosphere (Section 5.6). Finally, in Section 5.7, we consider the various mechanisms that enable plants to survive in situations of water stress — drought, saline soils and freezing temperatures. As a background to the rest of the chapter, we discuss first certain properties of water that are particularly relevant to plant water relations.

5.1.1 Properties of water

Water is a very unusual compound. Unlike hydrides of non-metals (e.g. foul-smelling hydrogen sulphide, H_2S, or ammonia, NH_3), which are gases at room temperature, water is liquid. The reason for this and for most of the other unusual properties of water is that its molecules are *polar* and form *hydrogen bonds* with each other (Figure 5.2); this greatly raises the melting and boiling temperatures.

Figure 5.2 Water molecules. (a) Overall representation of one molecule with the two hydrogen atoms (grey) attached to the central oxygen atom (green). (b) Average distribution of electrical charge, showing net positive charges on both H atoms and a net negative charge on the O atom. (c) Hydrogen bonds (arrowed) arise through attraction of the negative side of one water molecule to the positive side of another.

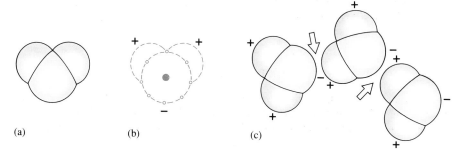

(a) (b) (c)

The tendency of water molecules to stick together because of hydrogen bonding is described as **cohesion**. This is why thin columns of water can move to the top of a tree in xylem vessels without the column rupturing; cohesion gives the column a very high tensile strength. Water molecules are also attracted to other polar molecules and therefore they wet solid surfaces such as glass or cell walls and form *hydration shells* around ions and around macromolecules such as proteins.

Cohesion of water molecules means that an unusually large amount of energy is required to cause evaporation and this is why transpiration from leaves (or the surface of an animal) has an important cooling effect. Furthermore, the packing of water molecules in the liquid state through hydrogen bonds is actually more efficient (more molecules per unit volume) than in the solid state (ice). This is why water expands when it freezes, resulting in phenomena such as burst water pipes and, in plants, the risk of tissue damage if cell water freezes. Many more examples of the relevance of water properties to plant water relations will become apparent as you read this chapter.

5.2 THE MOVEMENT OF WATER AND WATER POTENTIAL

A general point about the movement of water in plants is that it is entirely a *passive* process. There are no pumps equivalent to an animal heart and, as in animals, no transport molecules for the primary or secondary active transport of water. Active transport of solute molecules, e.g. ions or sugars, may give rise to conditions that cause water movement — but this is still regarded as a passive process. The question addressed in this section, therefore, is: what forces cause water to move in a particular direction in plants? Before answering it we need to consider briefly the two types of water movement that occur.

5.2.1 Mass flow and diffusion

Mass flow (Chapters 3 and 4) is the movement of water and solute molecules, together and in one direction, because of differences in pressure.

◇ Name two situations in plants where mass flow occurs.

◖ In the conducting elements of xylem, i.e. the transpiration stream (Chapter 3), and in the sieve tubes, where translocation occurs (Chapter 4).

Mass flow, therefore, is the type of water movement found in long-distance transport systems. By contrast, water movement between or into living cells or through the soil is usually by diffusion: water molecules move in all directions but more of them go in one particular direction and hence there is a net flux in that direction. The driving forces involved are considered next.

5.2.2 Water movement into living cells

Water may enter living cells by *osmosis*, i.e. it moves across a semipermeable membrane (the plasmalemma) from a region of lower to one of higher solute concentration. The plasmalemma allows free diffusion of water but obstructs the passage of solute molecules; if solute concentration (which gives rise to **osmotic pressure**, Π) is higher inside the cell than outside, then water moves into the cell. *Why* does water move in this way? Remember that solute molecules attract the polar molecules of water; this causes more attraction and stronger hydrogen bonding between the *remaining* water molecules which are then less able to diffuse freely — their *chemical potential* or free energy per mole is reduced, and the higher the solute concentration, the greater the reduction. So inside the cell, where osmotic pressure is high, the potential for water molecules to diffuse out is relatively low and osmosis occurs — a net movement of water *into* the cell.

For animal cells, this is usually taken as the end of the story. The difference in osmotic pressure across the cell membrane describes completely the driving force for water movement and this force can be written as $\Delta\Pi$ (Greek *delta*, meaning difference between, and *pi*, osmotic pressure). For plant cells, however, the story is not ended: another force, **turgor pressure**, which *opposes* $\Delta\Pi$ must be taken into account. Turgor pressure (which is given the symbol P) arises because plant cells are surrounded by a relatively rigid cell wall that restricts the swelling of the protoplast (the membrane-bound unit) as water enters by osmosis. It can be defined as the pressure (above atmospheric pressure) exerted by the protoplast on the cell wall or, equally, the pressure exerted by the cell wall on the protoplast — the two are numerically identical. When water is under pressure, this increases the tendency of the water molecules to move apart, i.e. intermolecular attraction is decreased, so free energy increases and the molecules diffuse more freely.

Compare Figure 5.3a (representing an animal cell) with 5.3b (a plant cell). Both cells have the same initial osmotic pressure, Π_1, which is greater than

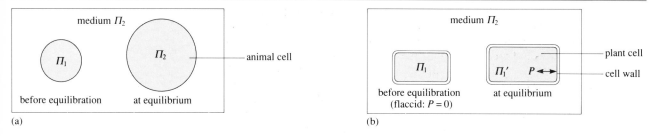

Figure 5.3 A comparison of (a) animal and (b) plant cells to illustrate the origin of turgor pressure in plant cells.

that outside, Π_2. Water enters by osmosis, diluting the cell contents, and the animal cell swells until it reaches equilibrium with the outside and osmotic pressure has fallen to Π_2 (assuming, of course, that it does not burst before reaching this point). The plant cell, however, stops swelling when its osmotic pressure is only Π'_1, well below Π_2. At this equilibrium point the cell wall has stretched as far as it can and the turgor pressure, which increased during water influx, is at a maximum — it is roughly equivalent to the point when no more air can be pumped into a strong tyre — and the plant cell is described as being *fully turgid*. So water movement into a living plant cell depends not only on the initial osmotic pressure, Π, but also on turgor pressure, P.

Plant physiologists and soil scientists use a single term, *water potential*, to describe the free energy of water in a system — its tendency or potential to move relative to another, defined system. We consider next the meaning and usage of water potential.

5.2.3 Water potential and what affects it

The concept of **water potential**, Ψ (pronounced *psi*) was first used in plant water relationships in 1960 by Ralph Slatyer, an Australian. Although based on the free energy ('tendency to move') of water in a system, a mathematical sleight of hand was used to convert the units of water potential from those of energy to those of *pressure*, e.g. pascals (Pa) or megapascals (MPa). Moreover, water potential is a *relative* measure: just as altitude on land is measured relative to sea-level, so water potential in a system is relative to that of a *standard* — pure water at atmospheric pressure and the same temperature as the system being studied. The water potential of pure water under these conditions is arbitrarily *set at zero* and this is an important point to remember when doing calculations about water potential.

Another key point to remember at this stage is that differences in water potential, $\Delta\Psi$, constitute the driving force for water movement and *water always moves down gradients of water potential*. This applies to all kinds of systems, not just plants, which is one reason why water potential is such a useful concept: it can be applied in all situations — living cells, dead cells or soil, for example — in contrast to osmosis, which occurs only when semipermeable membranes are present. When $\Delta\Psi$ between two systems is zero, the systems are in equilibrium and there is no net movement of water between them.

Now consider what happens if a living plant cell that is not fully turgid is placed in pure water.

◇ (a) In what direction will water move? (b) Is the water potential inside the cell above or below that outside?

◆ (a) Water will move *into* the cell because of osmosis. (b) Because water moves *down* gradients of water potential, Ψ inside the cell must be *lower* than that outside.

However, Ψ outside the cell (pure water) is, by definition, zero; so it follows that the water potential of the cell has a *negative* value. This is an important principle: with rare exceptions, *all living cells have a negative water potential*. The reason for this is that solute molecules, as explained earlier, lower the free energy of water — and hence water potential. The amount by which Ψ is lowered depends on solute concentration which, of course, is equivalent in pressure terms to the osmotic pressure, Π. We can say that Π is one factor that determines Ψ and, for animal cells, it is the only factor, i.e. $\Psi = -\Pi$, which is why animal physiologists do not use the water potential concept!

◇ From the previous section, what other factor influences Ψ in living plant cells?

◆ Turgor pressure, P.

As explained in Section 5.2.2, turgor, a positive hydrostatic pressure, tends to increase the free energy of water and hence *raises* Ψ. The equation describing the water potential of a plant cell is, therefore:

$$\Psi = P - \Pi \tag{5.1}$$

This is an important, basic equation which you should remember. It can be applied to water outside a cell but P must then be generalized to hydrostatic pressure, the pressure above or below atmospheric pressure (which is taken as zero). Pressures below atmospheric (which are commonly described as *tensions*) occur, for example in xylem vessels.

◇ What is the water potential of a cup of salt solution of osmotic pressure 1 MPa?

◆ -1 MPa. P here is zero (atmospheric pressure) so $\Psi = 0 - 1 = -1$.

◇ What is the gradient of water potential between the salt solution just described and a plant cell placed in it with an initial turgor pressure of 0.2 MPa and osmotic pressure 0.6 MPa? Will water flow into or out of the cell?

◆ The water potential of the salt solution is -1 MPa and that of the plant cell is:

$$P - \Pi = 0.2 - 0.6 = -0.4 \text{ MPa};$$

so the gradient of water potential, $\Delta\Psi$, is:

$$-0.4 - (-1) = -0.4 + 1 = 0.6 \text{ MPa}.$$

Because $\Psi_{\text{salt solution}}$ is lower than Ψ_{cell}, water will move *out* of the cell, down the gradient of water potential.

Water movement into, out of or between living plant cells will be considered further in the next section but before this we need to refine and modify Equation 5.1 so that it is more accurate and can also be applied to systems that are *not* separated by semipermeable membranes. First, the Π term needs qualification. Cell membranes are never truly semipermeable (i.e. allowing no solute molecules at all to diffuse across). In the cell in salt solution example above, some salt molecules will diffuse into the cell, raising Π_{cell} and reducing $\Delta\Psi$ and, therefore, the rate of water diffusion out of the cell. **Differentially permeable** is a better description of such membranes and the extent to which solute molecules can diffuse across — membrane 'leakiness' — is expressed by multiplying Π by a term σ (Greek letter sigma), the

reflection coefficient. This has values between 0 and 1; for a truly semipermeable membrane, $\sigma = 1$ and for a membrane that is equally permeable to water and solutes (or if no membrane is present) $\sigma = 0$. Equation 5.1 then becomes:

$$\Psi = P - \sigma\Pi \tag{5.2}$$

One other factor that can affect Ψ must also be considered; this is the *presence of colloids and hydrophilic ('water-attracting') surfaces*. Recall that water molecules tend to stick to surfaces, which lowers the free energy of the water and, therefore, also lowers Ψ. Within living, fully hydrated cells this surface or *matric* effect is regarded as negligible but in the soil, in cell walls or in dry seeds it is of major significance. The degree to which surfaces adsorb water molecules at atmospheric pressure is termed the **matric pressure, *m***.

◇ Will *m* be added to or subtracted from the right-hand side of Equation 5.2?

◈ Subtracted: this follows because, like Π, *m* tends to decrease Ψ.

The final form of the equation then becomes:

$$\Psi = P - \sigma\Pi - m \tag{5.3}$$

Table 5.1 summarizes the factors that affect the components of water potential.

Table 5.1 Factors that affect water potential, Ψ.

Factor	Symbol	Measured in terms of:	Effect when increased
Hydrostatic pressure	P	pressure above (e.g. turgor) or below (tension) atmospheric	raises Ψ
Solutes	Π	osmotic pressure	lowers Ψ
Interfaces or colloids	m	matric pressure	lowers Ψ

Summary of Sections 5.1 and 5.2

1 Water may circulate in land plants from roots via xylem to leaves (where much is lost in transpiration) and back again in phloem.

2 The polar nature of water molecules results in hydrogen bonding between them. This cohesion explains the high tensile strength of water columns, the marked cooling effect of evaporation and the expansion of water when it freezes.

3 Water moves passively in plants, either by mass flow (in xylem and phloem) or by diffusion (into roots and from cell to cell).

4 The potential or freedom of water molecules in a particular system to move is defined as the water potential, Ψ, which is measured in units of pressure. It is a measure of the free energy of water in the system.

5 Water potential is lowered by solute molecules, pressures below atmospheric, and colloids or hydrophilic surfaces. It is raised by pressures above atmospheric. The water potential of pure water is defined as zero.

6 Any difference in water potential, $\Delta\Psi$, is a driving force for water movement and water moves down gradients of water potential.

7 Water potential inside living plant cells is determined by solute concentration, which is measured as osmotic pressure, Π, and by turgor pressure, P. Π lowers and P raises Ψ_{cell} so that $\Psi_{cell} = P - \Pi$.

8 If the cell membrane is differentially permeable rather than truly semi-permeable, then the equation for water potential is modified to $\Psi_{cell} = P - \sigma\Pi$, where σ is the reflection coefficient of the membrane, with values between 0 and 1.

9 If colloids or hydrophilic surfaces affect water potential substantially in a system, then water potential is defined by $\Psi = P - \sigma\Pi - m$, where m is the matric pressure.

Question 1 (*Objectives 5.1, 5.2 and 5.3*) Which of the following statements are true and which are false? Give reasons for your answers.

(a) Most of the water entering plant roots during daytime circulates round the plant in xylem and phloem.

(b) Along most of its route through a plant, water moves by diffusion either along cell walls or across protoplasts.

(c) If a detergent solution (which lowers the surface tension of water) were introduced into the xylem of a tree trunk, this could prevent water from reaching the upper branches.

(d) Gradients of water potential cause water to move into plant cells but not into animal cells.

Question 2 (*Objectives 5.1, and 5.5*) Each of (a) to (d) describes a situation involving two compartments, A and B. From the information given, decide whether the water potential in A, Ψ_A, is equal to, greater than or less than that in B, Ψ_B.

(a) Water in a glass, A, is being sucked into the barrel of a syringe, B.

(b) A lump of dry clay, A, was placed in a bowl of water, B: after 4 hours the clay had swollen and its weight had increased by 50%.

(c) A pig's bladder, A, (which behaves like a semipermeable membrane) was half-filled with seawater, sealed and suspended in a bowl of distilled water, B: after 1 hour the weight of the bladder had increased by 30%.

(d) A pig's bladder, A, was half-filled with seawater, sealed and suspended in a bowl of sugar solution, B: after 1 hour the weight of the bladder had not changed.

Question 3 (*Objectives 5.1, and 5.5*) For each of (a) to (d) in Question 2, what was the nature of the driving force, that is the chief component(s) of $\Delta\Psi$, that caused water movement in the observed direction?

Question 4 (*Objectives 5.4, 5.5 and 5.16*) Figure 5.4 shows a system in which two dilute solutions, X and Y, with osmotic pressures of Π_X and Π_Y, are separated by a differentially permeable membrane whose reflection coefficient, σ, is close to 1. The compartment containing Y is open to the atmosphere but the pressure exerted on X, P_X, can be varied by moving the plunger.

(a) If $P_X = 0$ (i.e. atmospheric pressure), $\Pi_X = 0.3\,\text{MPa}$ and $\Pi_Y = 0.6\,\text{MPa}$, what is the water potential gradient between solutions X and Y, and in what direction is the water moving?

(b) If the plunger is pushed in so that $P_X = 0.6\,\text{MPa}$, and Π_X and Π_Y are both $0.6\,\text{MPa}$, what is $\Delta\Psi$ and in what direction is the water moving?

(c) If the plunger is raised to create a suction so that $P_X = -0.3\,\text{MPa}$, and Π values remain as for (a), what is $\Delta\Psi$ and in what direction is the water moving?

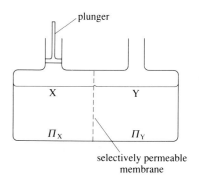

Figure 5.4 For Question 4.

5.3 WATER MOVEMENT IN PLANTS AND SOIL

Knowing the path of water movement in plants (Section 5.1) and that gradients of water potential cause movement (Section 5.2), the question addressed now is how do these gradients arise. Figure 5.5 shows typical values of Ψ in the soil, different parts of a plant and in the atmosphere.

◇ Is the water potential gradient steepest between soil and roots or between leaves and atmosphere?

◆ Between leaves and atmosphere. $\Delta\Psi$ here, from leaf xylem to atmosphere, is 27 MPa $(-3 - (-30))$ whereas that between soil and root xylem is only 0.27 MPa $(-0.03 - (-0.3))$.

The four main types of water movement that will be considered in this section and can be identified from Figure 5.5 are:

1 Into and out of living cells aross a cell membrane;
2 Through the soil;
3 Along the xylem;
4 From leaves to atmosphere (as water vapour).

Movement of water in phloem was considered earlier, in Chapter 4, and will not be discussed further here but it, too, is down a gradient of water potential, as shown in Figure 5.5b.

Figure 5.5 (a) The passage of water through a plant as in Figure 5.1 showing approximate water potentials at different points (scale on left). (b) Alternative routes for water in plants; the major route is indicated by the thicker line. Transport from leaves to roots in phloem is also shown.

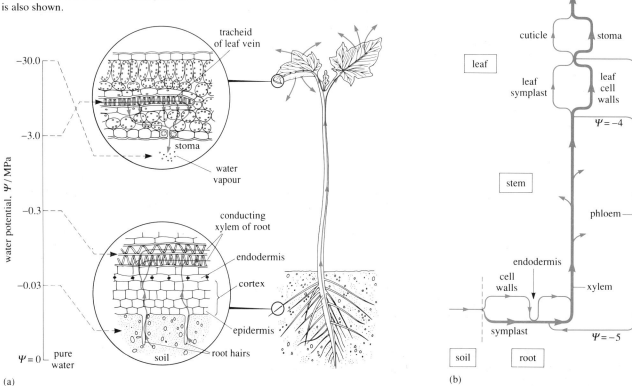

(a)

(b)

5.3.1 Water movement between living cells

Most of the information from earlier sections about water movement into living plant cells is summarized in Figure 5.6. Studying this carefully will not only revise and clarify the main points but also reveal some new ones.

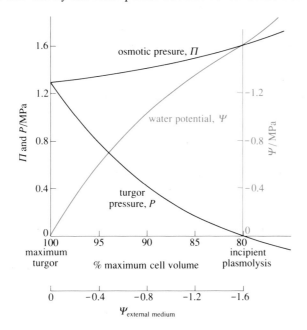

Figure 5.6 The relationship between the volume, water potential, osmotic pressure and turgor pressure of a cell with very elastic cell walls. Note that values for Ψ are negative.

The x-axis shows changes in cell volume as the cell swells or shrinks during water movement and the external water potential that causes movement. The volume limits shown are between maximum turgor, when volume is at a maximum and the cell wall can stretch no further, and **incipient plasmolysis**, when volume is the minimum at which the protoplast still touches the cell walls; when cells are **plasmolysed** the protoplast shrinks away from the walls, as shown in Figure 5.7.

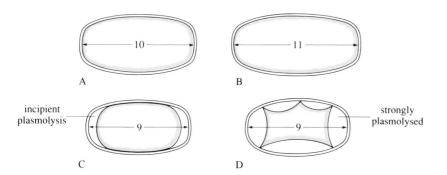

Figure 5.7 The behaviour of a living cell, originally 10 units long, in solutions of different concentrations (A–D). A is 9% sugar (slightly hypotonic); B is 2% sugar (hypotonic); C is 10% sugar (just hypertonic); D is 20% sugar (hypertonic).

At maximum turgor no more water can enter the cell and this state can be achieved only if the cell is in pure water; $\Delta\Psi$ is then zero and water potentials both outside and inside the cell are zero, as shown in Figure 5.6

◇ From this information and Equation 5.3 or from Figure 5.6, deduce the relative values of P and Π at full turgor, assuming that the reflection coefficient, σ, is 1.

◐ Since $\Psi = 0 = P - \sigma\Pi - m = P - \Pi$ (m can be neglected for living cells and σ is 1), then $P = \Pi$. Turgor pressure and osmotic pressure are equal at full turgor, which is what Figure 5.6 shows.

Now consider incipient plasmolysis. At this point so much water has moved out of the cell that the protoplast exerts no pressure on the walls and P is zero. From the equation $\Psi = P - \Pi$ it then follows that $\Psi = -\Pi$, i.e. water potential is numerically equal to osmotic pressure at incipient plasmolysis, which is again shown in Figure 5.6. We show later that these relationships can be used in conjunction with simple experimental methods for measuring Ψ and Π of tissues.

When water moves between living cells, exactly the same principles apply. As an example take the movement of water across a leaf from the xylem in the veins to the mesophyll cells bordering an air space close to a stomatal pore (as illustrated in Figure 5.1). Water evaporates from the latter cells, lowering turgor and increasing the concentration of solutes, i.e. increasing Π.

◇ What will be the effect on Ψ in these cells?

◐ From Equation 5.1, it will be lowered, becoming *more negative*.

This increases the gradient of water potential between cells adjacent to the air space and their neighbours, thus driving water from the latter into the former. The process is repeated until water moves out of the nearest vein. Figure 5.8 illustrates how water flow down a gradient of water potential might occur between living cells. Notice that there is not necessarily a gradient of either osmotic or turgor pressure.

Figure 5.8 Diagram to illustrate how water could flow between living cells along a water potential gradient $\Delta\Psi$, of 0.3 MPa which does not necessarily coincide with Π or P gradients.

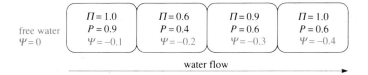

Question 6 at the end of this section is about a similar situation and you might like to try it now to check that you understand the arguments.

When water moves into phloem in a leaf vein it does so because solutes are actively pumped into sieve tubes or adjacent transfer cells (as described in Chapter 4), which raises osmotic pressure and, therefore, lowers water potential. The resulting influx of water generates high turgor pressure in sieve tubes of the leaf (a source organ) and this, according to the pressure flow theory, is what sets up the pressure gradient that drives water and solutes from source to sink by mass flow.

Measurement of Ψ, P and Π in plants

The water relationships discussed above give some clues about how Ψ and Π can be measured in living plant tissues. Many of the methods used to measure Ψ depend on the principle that if living tissue is in equilibrium with a medium of equal water potential, i.e. $\Delta\Psi = 0$, then there will be no net movement of water into the cells and tissue volume or weight or length (for a piece of stem) will remain constant. The results of a simple experiment using the constant-length principle are shown in Figure 5.9.

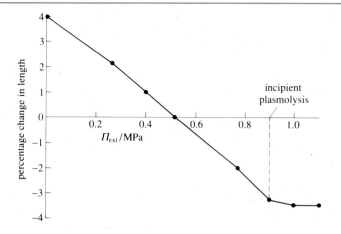

Figure 5.9 The results of an experiment for determining the water potential of stem tissue, Ψ_{tissue}. Changes in the average length of the segments were recorded (fresh weight may be measured instead).

Batches of uniform stem segments, all of identical length, were placed in a graded series of solutions ranging from distilled water to $0.5\,mol\,l^{-1}$. The solute used was mannitol, a sugar alcohol, which is not usually taken up by the tissue and for which the reflection coefficient is, therefore, 1. The osmotic pressure of these mannitol solutions was calculated from

$$\Pi = CRT$$

Where C is the molar concentration, R the gas constant, and T the absolute temperature. The tissue was allowed to equilibrate in these solutions for one to two hours until all cells were in equilibrium with the outside medium and then the segments were measured.

◇ From Figure 5.9, what is the *initial* water potential of cells in this stem tissue? (Remember that for the external solution, $\Psi = -\Pi$ at atmospheric pressure.)

◈ Approximately $-0.5\,MPa$. This is the point at which the graph crosses the horizontal axis, i.e. when no change in length had occurred, the net flux of water into or out of the tissue was zero and, therefore, $\Delta\Psi$ was 0. So at this point the external water potential, which was *minus* the osmotic pressure of the solution, equalled the water potential of the cells in the tissue.

This method can be used with any uniform plant tissue, for example pieces of potato tuber, which could be weighed instead of measured. It is not, however, very precise and for research purposes more elaborate set-ups are employed — although the same principles apply.

The same method also enables an estimate of the initial osmotic pressure of the stem segments to be made based on the fact that $\Psi = -\Pi$ at incipient plasmolysis. In Figure 5.9, incipient plasmolysis is the point at which tissue shrinkage became markedly slower (a point which is rather difficult to determine and hence the crudeness of the estimate). Figure 5.9 shows that stem segments equilibrated in a solution of osmotic pressure $0.9\,MPa$ were approximately at incipient plasmolysis, so this is their average initial osmotic pressure.

◇ What was the initial turgor pressure of the stem segments?

◈ About $0.4\,MPa$. This is determined from the equation $\Psi = P - \Pi$ (the reflection coefficient being 1 for the external solute), i.e. $-0.5 = P - 0.9$, therefore, $P = 0.9 - 0.5 = 0.4\,MPa$.

There are many other ways of measuring Π, however. Some are more sophisticated methods for determining the point of incipient plasmolysis but others measure the concentration of solutes in 'sap' obtained from a tissue sample. This sap is a mixture of cytoplasm, the solution from the cell vacuole (probably 80–90% of the total), and fluid from the cell walls and intercellular spaces. It can be obtained by squeezing, grinding or freezing and thawing the tissue to disrupt cell membranes. Solute concentration in the sap is then determined by the same methods as are used for body fluids in animals, for example by measuring the reduction in freezing point relative to distilled water. Osmotic pressures in plant cells are commonly in the range 0.7–2.5 MPa but, for reasons that will be discussed in Section 5.7, values as high as 5–10 MPa may occur in plants from saline or desert habitats.

Until the mid-1980s turgor pressure of higher plant tissues was usually determined by difference (from $\Psi = P - \Pi$) after measuring Ψ and Π. Another method was then developed which measured the degree to which tissue vibrated, the tissue becoming stiffer and vibrating less as turgor increases. Probably the most useful method, however, and one which has the advantage of measuring turgor in single cells rather than the average value of a tissue, is the pressure probe. Figure 5.10 shows how an early version of the probe worked when used to measure turgor of very large algal cells. Much smaller probes have now been developed which, when combined with a micro-manipulator, can measure turgor in most types of full-grown cells in higher plants.

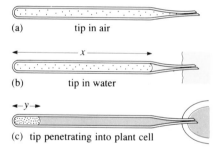

Figure 5.10 A method for determining turgor pressure in giant algal cells. A small capillary pipette closed at one end is used (a). The tip of the pipette is first placed in water (b) and then inserted into the vacuole of the test cell (c). Turgor pressure, P, compresses air in the capillary pipette and the amount of compression $(x - y)$ is proportional to P.

(a)   tip in air

x

(b)   tip in water

y

(c)   tip penetrating into plant cell

5.3.2 Water movement through soil

Soil consists of fine particles that are surrounded by a film of water with minute, water-filled pores between. The size and chemical nature of these particles, particularly their surface electrical charge, and the way they are arranged determine how much water can be held in the soil against the force of gravity and how 'tightly' the water is held. This water-holding capacity is one way of describing the water potential of soil and, as mentioned earlier, it depends mainly on the interface forces measured as the matric pressure. Since there is little variation in hydrostatic pressure in the surface layers of soil (so P = atmospheric pressure = 0) and few solutes are present (except in saline soils), Equation 5.3 simplifies to

$$\Psi = -m \tag{5.4}$$

When soil is fully wetted after draining under gravity, a state described as **field capacity**, its water potential is essentially zero. As plant roots take up water or the soil dries by evaporation at the surface, however, water potential falls (becoming more negative) as matric pressure rises. This is illustrated in Figure 5.11, which also shows the marked difference in water-holding capacity between a sandy and a clay-rich soil — a difference reflected in their

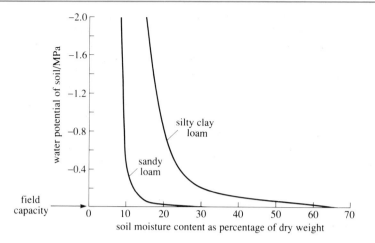

Figure 5.11 The water potential $(= -m)$ of a sandy and a silty clay loam soil plotted as a function of the moisture content of the soil.

description by gardeners as 'light' and 'heavy', respectively. At field capacity, sand contains only one-third as much water as clay and, when they have the same water content, clay soils also have a much lower water potential than sandy soils. Local drying of soil, therefore, is what creates a gradient of water potential, causing diffusion of water to that site. For plant roots to take up water at a particular rate, the same gradient of water potential must be maintained between roots and soil and, to achieve this, roots must either grow continuously to wetter areas or must decrease their own water potential (for example by increasing osmotic pressure).

5.3.3 Movement in xylem

You know from Chapter 3 that water and ions move from roots to shoots by mass flow in the xylem and, when plants are transpiring, this flow is called the **transpiration stream**. That such movement can occur, sometimes for over 100 metres to the tops of tall trees, is truly remarkable. How is it possible? The properties of water and the structure of plants provide the answer, which was first suggested around 1914 and is referred to as the **cohesion theory**. The main principles are illustrated in Figure 5.12 and the arguments are as follows:

1 Evaporation of water from cell walls into the air spaces of a leaf sets up gradients of water potential not only between living cells as described in Section 5.3.1 but also along the cell walls. The high surface tension of water means that the walls do not dry out but remain permanently wetted, so that water flows along the walls — ultimately from xylem in the leaf veins — to replace that which is lost. Removal of water from the xylem creates a negative pressure or tension — a pulling force — and this is the driving force for the transpiration stream.

◇ Apply Equation 5.3 to water movement along xylem cells and derive a simplified equation, as done for water movement in soils.

◈ Starting with $\Psi = P - \sigma\Pi - m$, m can be eliminated for a dilute solution when relatively few surfaces are present. Similarly, Π can be eliminated because no membranes are present in xylem and hence $\sigma = 0$. Thus $\Psi = P$ for movement of water *along* xylem and the driving force, $\Delta\Psi$, equals the pressure gradient, ΔP.

An instrument called the *pressure bomb* can be used to measure water potential in xylem, utilizing the fact that $\Psi = P$. A piece of tissue is cut, whereupon water in the xylem contracts away from the cut surface because of

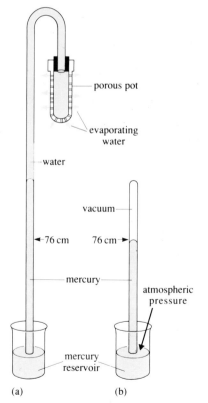

Figure 5.12 (a) A model demonstrating the principle of water movement up plant stems in the xylem. (b) A demonstration that atmospheric pressure supports a column of mercury 76 cm high in a vacuum.

the low xylem pressure. The tissue is clamped inside the chamber of a pressure bomb which can be pressurized and, as the pressure is raised, it forces water in xylem towards the cut surface. The pressure at which water just reaches this surface exactly balances the original tension and so measures the pressure and hence the initial water potential in the xylem.

2 The pulling force, a pressure gradient, is steepest in the leaves and small branches but is transmitted all the way down to the roots *provided* that continuous columns of water are maintained in the xylem. This is a very important proviso and at the heart of the cohesion theory. Continuous water columns under tension are possible because (a) the conducting cells of xylem have strong, rigid cell walls that do not collapse under tension; (b) water molecules adhere strongly to the hydrophilic walls of the xylem cells, which have a relatively small diameter, so the water column does not pull away from the walls; and (c), most important of all, the strong cohesive forces between water molecules hold the column together and prevent water from vaporizing at the low pressures experienced. This is why the cohesion theory is so named.

Cavitation and embolism

Despite cohesive forces, however, water columns under tension in the xylem are somewhat unstable and if the pressure falls too low, air enters at a wound or some untoward event, such as sudden jarring, occurs, then water will vaporize locally. This process is called **cavitation** and it results in xylem conduits filling with water vapour. As air comes out of solution an air bubble or **embolism** forms, blocking water movement. The water vapour or embolism is usually confined to a single vessel or tracheid because the pits which link up adjacent cells are lined by a 'pit membrane' — the middle lamella plus the remains of the primary cell wall (Chapter 1, Section 1.3) — and, through surface tension effects, this prevents the bubble from spreading. Bordered pits (Chapter 1, Section 1.5) act as more sophisticated valves that close when the pressure on one side rises (Figure 5.13).

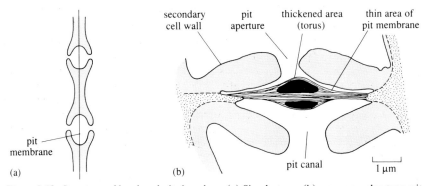

Figure 5.13 Structure of bordered pits in xylem. (a) Simple type; (b) more complex type with a thickened central torus found in conifers. The pit membrane is the remains of the middle lamella plus primary wall; when pressed sideways against the pit aperture, it prevents air bubbles passing from one xylem element to another.

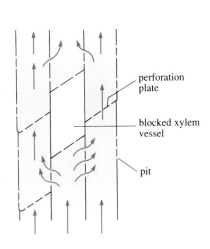

Figure 5.14 Diagram to illustrate how sideways flow of water through pits in xylem elements can circumvent a blocked vessel or tracheid.

Another feature of xylem that minimizes the effects of embolism is the ease with which water can move sideways through the pits, thus bypassing the blocked conduit (Figure 5.14).

Until the mid-1970s it was generally assumed that cavitation was relatively rare and did not seriously disrupt water flow in xylem. Ideas began to change in 1974, however, when John Milburn in Glasgow applied to whole plants a method of detecting cavitation events, which was developed earlier in his laboratory. This *acoustic detection method* depends on the fact that sudden release of pressure during cavitation causes the cell walls to vibrate, producing an audible click (Figure 5.15). With suitable amplifying equipment, the clicks can be counted and, from this and later work, it emerged that cavitation is a common event. During one day, embolism reduced water flow through the xylem of a corn plant (*Zea mays*) by about one half and in sugar maple saplings (*Acer saccharum*) flow reduction in the main trunk was 31% by the end of one summer and 60% by the end of the following winter, with many small twigs completely blocked (Tyree and Sperry, 1989). Damage on this scale is potentially a serious matter and two questions must be asked: what causes it and how is the damage repaired or circumvented?

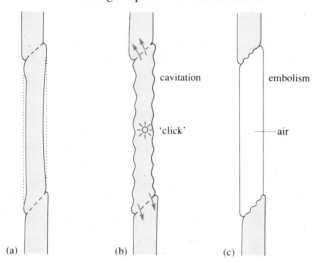

Figure 5.15 Representation of cavitation and embolism formation in a xylem conduit. (a) Under extreme tension (low pressure), the walls are strained inwards from the dotted line. (b) A cavitation 'bubble', filled with water vapour, forms and the strained walls vibrate, detectable as an audible 'click'. Water exits to adjoining conduits (arrows). (c) The evacuated conduit gradually becomes air-filled (embolism) due to gas coming out of solution from sap in the walls, etc.

Causes of cavitation and embolism

Chance mechanical damage from wind or grazing, for example, means that a low level of embolism will always occur but this does not explain the high frequencies described above. Two main factors appear to be responsible.

1 *Water stress* associated with high rates of transpiration and low xylem pressures, particularly in the leaves and small branches. What seems to happen is that tiny bubbles of air squeeze through pores in the thin pit membranes separating an accidentally embolized conduit from its neighbour. The bubbles act as seeds or nuclei for further cavitation and the bigger the pores the more likely is bubble formation. Pore size correlates roughly with vessel or tracheid diameter so that, although water flow is easier and faster through wide conduits, this must be balanced against an increased risk of cavitation.

2 *Freezing* of xylem in winter leads to extensive formation of air bubbles when thawing occurs and widespread embolism.

There is also a third cause, recently discovered but possibly of some importance.

3 *Pathogen-induced* embolism has been identified as the first major damage caused to xylem by the Dutch Elm disease fungus that has devastated British elm trees. This could well be true of other pathogens that cause rapid wilting (e.g. in *Clematis* wilt disease). Why the pathogens cause embolism is still not clear but an interesting suggestion is that they release into xylem sap compounds such as oxalic acid that lower surface tension. This greatly facilitates air-seeding at pit membranes.

Water stress during the day is the main cause of embolism in herbaceous plants and in woody plants during the growing season. Freezing causes the increase in embolized xylem during winter, as described for sugar maple saplings. How, if at all, do plants recover from such damage?

Recovery from embolism

The air in an embolism goes back into solution if xylem pressure rises to atmospheric pressure or very slightly below. Often this occurs nightly in herbaceous plants due to a phenomenon called **root pressure**. This arises because stomatal pores close and transpiration largely stops at night (Chapter 2) but active ion uptake and osmotic influx of water continue in the roots. Ions accumulated in xylem parenchyma diffuse into the dilute xylem sap (which is no longer moving rapidly upwards) and water continues to flow in, gradually raising the hydrostatic pressure until it becomes positive. Table 5.2 illustrates the steps involved, the root cortex being treated as a single unit separated from the soil and the xylem by semipermeable membranes; you can see that, at both stages, water movement is down a gradient of water potential.

Table 5.2 Data for a slowly transpiring plant showing stages in the development of root pressure (positive hydrostatic pressure in the xylem, at Stage 2). Units are MPa and note that the osmotic pressure of xylem sap is taken into account when considering water movement *into* xylem. Note also that values of m and Π are preceded by $(-)$ in the table because $\Psi = P - \Pi - m$.

	Root		Wet soil adjacent to root
	Xylem	Cortex	
Stage 1			
m	negligible	negligible	$(-)0.03$
Π	$(-)0.1$	$(-)0.5$	negligible
P	-0.3	$+0.4$	—
Ψ	-0.4	-0.1	-0.03
	← gradient of Ψ along which water flows		
Stage 2			
m	negligible	negligible	$(-)0.03$
Π	$(-)0.3$	$(-)0.5$	negligible
P	$+0.05$	$+0.4$	—
Ψ	-0.25	-0.1	-0.03
	← gradient of Ψ along which water flows		

Root pressures are usually small (0.1–0.2 MPa) but they are sufficient to induce re-filling of embolized conduits in herbaceous plants. They cause a slow movement of water through xylem, this time through a pushing force from below rather than a pull from above. One interesting consequence of root pressure, however, is the phenomenon known as **guttation**. You may have noticed drops of liquid looking rather like dew hanging from the tips of leaves at night or in humid conditions when transpiration is low. In fact this is guttation fluid forced out of the vein endings by root pressure and through stomata or, in some cases, specialized hairs called *hydathodes* onto the surface.

In trees positive root pressures do not develop during the growing season so it is vital that large-scale embolism does not occur. This seems to be achieved by closing stomata and reducing the rate of transpiration but at the obvious cost of cutting off the CO_2 supply for photosynthesis. Nevertheless, in the small, outermost twigs, which experience the most negative xylem pressures, massive embolism is not uncommon and such twigs die as a result.

Recovery from winter embolism caused by freezing is at least partly due to the growth of new xylem cells in spring. In some woody plants, however, notably birch, *Betula*, and grape vine, *Vitis*, large root pressures develop in spring, forcing air out of embolized xylem or causing it to enter solution. This 'rising of sap' is so pronounced in vines that fluid drips for several days from the cut ends of pruned branches and frothing can be seen as air is expelled.

Susceptibility to cavitation, tolerance of embolism and ability to recover from it have undoubtedly been major factors influencing the design and distribution of land plants, particularly their tolerance of drought and freezing. Only in the 1980s has the importance of embolism been recognized but it should be a key area of research for the future.

5.3.4 Movement of water from leaf to air

Because water moves from leaves to air as a vapour, Equation 5.3 cannot be used to describe the components of Ψ because the potential of water vapour, Ψ^{wv}, is affected by factors quite different from those affecting liquid water. If you have ever tried to dry wet clothes outside, common sense will probably tell you what these factors are: temperature and the 'wetness' of the air — its relative humidity (r.h.) or water content (C). This is described by the equation:

$$\Psi^{wv} \propto T \log_{10} C_o/C_{sat} \tag{5.5}$$

where C_o is the water vapour concentration in the system (below saturation) and C_{sat} is the maximum or saturation water vapour concentration at temperature T. C_o/C_{sat} defines the relative water vapour content of the system and is equivalent to r.h./100 when r.h. is expressed as a percentage.

The gradient $\Delta\Psi^{wv}$ between the intercellular spaces in a leaf and the external atmosphere is the driving force for transpiration. This gradient is often steep because air inside leaves is usually close to saturation (100% r.h.) whereas the relative humidity of the atmosphere rarely exceeds 90% and in temperate climates is more usually around 50–70%.

Table 5.3 shows how even small differences in r.h. greatly affect water vapour potential and gives typical values for leaf and atmosphere at 20 °C, taking into account the still layer of moister air (the *boundary layer*) that lies close to leaf surfaces.

Since leaves tend to heat up above the temperature of the surrounding air when exposed to the sun, this tends to increase Ψ^{wv} (making it less negative) and, therefore, steepens still further the gradient of water potential between leaf and atmosphere. Short-term fluctuations in Ψ^{wv} are usually due to variations in temperature but the primary factor that determines the magnitude of Ψ^{wv} is the gradient of relative humidity.

Table 5.3 Representative values of relative humidity and water potential for leaves and atmosphere at 20 °C.

	Relative humidity (%)	Water potential/MPa
Liquid water in leaf cell walls	—	−1.36
Air spaces in leaf (in equilibrium with the cell walls)	99	−1.36
Boundary layer of air just outside stomata	95	−6.9
Atmosphere	50	−94.1

Summary of Section 5.3

1 Water moves into or out of living plant cells down gradients of water potential.

2 If cells are in pure water ($\Psi = 0$), water moves in until cells are fully turgid and turgor is at a maximum. $\Delta\Psi$ between cells and outside medium is zero at this point so $\Psi_{cell} = 0$ and turgor is equal to osmotic pressure (from $\Psi = P - \Pi$).

3 Water moves out of cells if they are surrounded by a concentrated solution; cells shrink and turgor falls. At incipient plasmolysis, $P = 0$ and, therefore, $\Psi = -\Pi$.

4 A simple way of determining the water potential of a tissue is by identifying the external solution in which the tissue shows no change in water content (i.e. no change in volume or fresh weight).

5 Osmotic pressure can be estimated from the concentration of the external solution that causes incipient plasmolysis or by measuring the concentration of solutes in cell 'sap'.

6 Turgor pressure can be calculated by difference (knowing Ψ and Π) or measured using a vibration method (for whole tissues) or pressure probe (for single cells).

7 After draining under gravity, water movement in soils is driven by water potential gradients that depend only on matric pressure ($\Psi = -m$, so $\Delta\Psi = \Delta m$). Soil matric pressure depends on the chemical nature of soil particles and on water content: it provides a measure of the water-holding capacity of a soil. As soil dries out, m rises more rapidly in heavy clay soils than in light sandy soils.

8 Water flow along xylem is by mass flow and driven by pressure gradients ($\Psi = P$, so $\Delta\Psi = \Delta P$). The cohesion theory explains how negative pressure (tension) gradients arise in the xylem of transpiring plants and pull water columns to the tops of trees — provided that cohesion between water molecules is sufficient to prevent the columns from breaking.

9 Columns of water in xylem become increasingly unstable as pressure falls (tension rises). Breakage through cavitation and subsequent embolism is common under conditions of water stress (high rates of transpiration), winter freezing, or infection with certain pathogens. Acoustic methods can be used to detect cavitation events.

10 Air bubbles (embolisms) can be removed by positive pressure in xylem. In herbaceous plants, this may arise each night through root pressure, which

occurs in non-transpiring plants because of active ion uptake by roots and osmotic influx of water into xylem. Root pressure also causes guttation in herbaceous plants. It does not usually occur in woody plants except in spring in a few species, where the 'rising of sap' is pronounced. Apart from these species, woody plants recover from winter embolisms by developing new xylem tissue and avoid summer embolisms by closing stomata and reducing the rate of transpiration.

11 Evaporation of water vapour from leaves depends on leaf-to-air gradients of temperature and relative humidity (or relative water content). Δr.h. is usually the main component of the driving force, $\Delta\Psi^{wv}$ (the gradient of water vapour potential), but short-term fluctuations in $\Delta\Psi^{wv}$ result chiefly from changes in leaf temperature.

Question 5 (*Objectives 5.1, 5.6 and 5.8*) Which of the following statements are true and which are false?

(a) The water potential of vacuolated cells is greatest when they are plasmolysed.

(b) In a source tissue (e.g. a mature leaf), the water potential of phloem sieve tubes is lower than that of surrounding parenchyma cells.

(c) Plants will wilt if their cells are plasmolysed.

(d) The matric pressure and water potential of a dry soil are greater than those of a wet soil.

(e) The water potential in xylem can be raised through root pressure and this facilitates recovery from embolism.

(f) In air of 100% relative humidity, water vapour cannot move from leaf to air.

Question 6 (*Objectives 5.5, 5.6 and 5.16*) Table 5.4 gives data from which you can calculate Ψ for different parts of a potato plant during the daytime and at night. Potato tubers are borne on underground stems that branch from the main stem below ground level. Calculate values of Ψ and fill in Table 5.4. Then describe the pathways of water movement (a) at noon and (b) at midnight through potato plants (assuming that $\sigma = 1$ for all membranes). From your answers to (a) and (b), (c) explain the observation that potato tubers often swell up at night and shrink during the day.

Table 5.4 Average values (in MPa) for the components of water potential in different tissues of a potato plant.

Tissue	Noon (plant transpiring)			Midnight (not transpiring)		
	P	Π	Ψ	P	Π	Ψ
Root cortex	0.4	0.6		0.45	0.65	
Potato tuber	0.45	0.75		0.50	0.85	
Xylem at the root–stem junction	−0.5	0.1		−0.01	0.25	
Stem xylem 30 cm above ground	−0.6	0.1		0	0.2	
Leaf cells 30 cm above ground	0.2	1.0		0.55	0.75	

Question 7 (*Objective 5.7*) For each of tissues (a)–(c) select one or more techniques from (i)–(vi) that, singly or in combination, would allow measurement of the water potential.

(a) potato tuber tissue

(b) xylem in a woody twig

(c) phloem in the stem of a bean plant.

(i) pressure bomb

(ii) pressure probe

(iii) plasmolytic technique (tissue placed in graded series of solutions of differing osmotic pressure)

(iv) freezing point depression of sap (to measure solute content)

(v) vibration method (to measure turgor).

(vi) Aphid stylet technique both coupled to a manometer and used to sample cell contents.

Question 8 (*Objectives 5.6 and 5.8*) Explain the following:

(a) Freeze-dried pea seeds swell up when placed in water.

(b) Pots of unrooted plant cuttings are often placed inside a polythene bag and shaded from sunlight in order to prevent wilting.

(c) Rough handling of cut flowers may result in premature wilting in the vase.

5.4 RESISTANCE, CONDUCTANCE AND WATER FLUX

The last section described the various ways in which a gradient of water potential — the driving force for water movement — can arise within the soil–plant–atmosphere system. This tells you the direction in which water will move but not how much or how fast, i.e. the water or **hydraulic flux**, J_v. J_v of a pathway is the volume of water (m^3) moving across unit cross-sectional area (m^2) per unit time (s), so has the units $m^3\,m^{-2}\,s^{-1}$ which simplifies to $m\,s^{-1}$ (cancelling out the metres) — the units of velocity. Hydraulic flux depends on two factors: the driving force (water potential gradient) and the ease with which water moves along the pathway — its conductance, L_p or, put the other way round, the resistance to water movement, R, which is 1/conductance. If you are thinking about water movement across a boundary, the cell membrane, for example — a situation where $\Delta\Psi$ operates between two compartments separated by a path of negligible thickness — then the equation that describes this relationship is:

$$J_v = L_p.\Delta\Psi \text{ (or } \Delta\Psi/R) \qquad (5.6)$$

Hydraulic conductance, L_p, has units of distance moved per unit pressure per unit time ($m\,Pa^{-1}\,s^{-1}$), those for **hydraulic resistance**, R being the reciprocal ($s\,Pa\,m^{-1}$), and for $\Delta\Psi$ simply pressure, Pa.

Of all the boundaries to water movement in a plant — cell membranes, plasmodesmata, the cell wall, pits between xylem elements — that with the highest resistance (lowest conductance) is probably the cell membrane. This has significant consequences for water movement through living tissues. Consider, for example, the pathway from the epidermis across a root to the central xylem. Figure 5.16 shows two types of path, one via the apoplast (cell walls) and the other via the symplast (protoplasts and plasmodesmata).

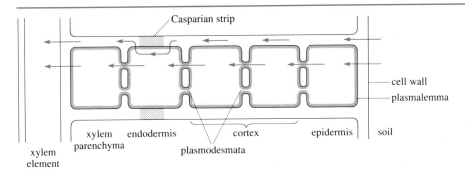

Figure 5.16 Pathways of radial water movement across a root from epidermis to xylem via the apoplast (cell walls) and symplast (protoplast) routes.

◇ For each path, what is the minimum number of times that water must move across a plasmalemma?

◈ Two: because of the waterproof Casparian strip in the endodermis, water moving via the apoplast must enter and then leave endodermal cells; on the symplast route, water crosses the plasmalemma of the epidermis and of the xylem parenchyma.

Because water must cross a plasmalemma barrier *at least* twice during movement across a root, the overall hydraulic resistance of roots is rather high and, if transpiration rates are high, can be rate-limiting for the whole plant. This characteristic of roots will be discussed further when we consider water movement through whole plants (Section 5.6).

5.4.1 | Resistance of soils to water movement

You already know that, as soils dry out, their water potential falls (becomes more negative) but at the same time the resistance to water movement increases dramatically. Table 5.5 shows some comparative data and emphasizes that, if soils are wet and have equal water potentials, water moves much more slowly through clay than through sand (by a factor of at least 10 000) because of the greater hydraulic resistance of clay. If soils are very dry, clay and sand have the same water potential and hydraulic resistance but the 'difficulty' factor goes up about 100-fold in clay and at least a million-fold in sand. Dry surface soil is obviously a very inhospitable environment for plant roots with both a very low water potential and very high resistance to water movement.

Table 5.5 Relative hydraulic resistance of different soil types with different moisture contents expressed in terms of approximate water potential.

Soil type	Soil water potential/MPa	Relative hydraulic resistance	Relative water flux
Wet sand	−0.01	1–10	10^{-2} to 10^{-3}
Wet clay	−0.01	10^5	10^{-7}
Dry sand or clay	−1.5	10^9	1.5×10^{-9}
Dry surface soil	−1.5 to −100	10^8 to 10^{11}	1.5×10^{-11} to 10^{-6}

This means that, if the soil is even slightly dried out, it would take days for plant roots to extract water from a slab only a few centimetres thick. The way round this problem is to have many fine roots ramifying through the soil —

which is what most plants do. Sending roots deep into the soil also avoids the problem of dry surface soil; surface rooters or plants on shallow soils usually have special characteristics, some of which will be described in later sections, that allow them to survive drought. Because the surface soil is usually much richer in mineral nutrients, many plants have both deep and shallow roots.

One other factor that can result in the development of large resistances to water movement from soil to root is root *shrinkage* in dry conditions. Root diameter may fall by 50% or more leaving an air gap between root and soil. It is thought that root hairs minimize this problem because their intimate contact with soil particles and the sticky mucilage they release drag soil along with the shrinking roots: you may have noticed how difficult it is to shake all the soil off fine roots when pulling up plants.

5.4.2 Water flow in xylem

Water movement between living cells and in the soil is by diffusion but in xylem, as in blood vessels, it is by mass flow along pipe-like conduits (vessels or tracheids). The factors that determine how fast and how much water flows in xylem (velocity, J_v, and volume flow (volume per unit time = $J_v \times$ area)) are the resistance of individual conduits, their diameter and their number (i.e. cross-sectional area).

Overall, resistance to flow in xylem, R xylem, is low. It is directly proportional to the number of pits or perforated plates that must be traversed between tracheids or vessels respectively, and inversely proportional to conduit 'bore' (cross-sectional area), that is, to the square of the radius. Volume flow, for individual conduits, which depends on both resistance and conduit area, is proportional to the *fourth* power of the conduit radius (see Table 5.6). And the *total* volume flow (through all conduits in an organ) depends further on the number of conduits.

Table 5.6 Relative resistances to flow and relative volume flow in pipes of different diameter.

Diameter/mm	Relative resistance	Relative volume flow
1.0	1	10 000
0.2	25	625
0.1	100	1

◇ Conifers have xylem tracheids but no vessels (Chapter 1). For equal pressure gradients and cross-sectional areas of xylem tissue, would you expect the volume flow to be greater for a pine tree or an oak?

◆ For an oak because xylem vessels usually have a greater diameter than tracheids and are also longer, with fewer end walls per unit length.

In fact so dominant is the influence of conduit radius that a few large vessels often carry most of the water flow when there is a mixture of different sized conduits. Other factors contributing to faster flow in vessels compared with tracheids are: (i) their greater length: xylem vessels are composed of vessel elements fused end-to-end and are commonly between one and 18 metres long; and (ii) the lower resistance of the perforated plates connecting one vessel to another compared with the simple pits that connect tracheids.

Vines, which supply large amounts of foliage via long, narrow stems, can maintain huge rates of sap flow because of the exceptionally large diameter and length of their vessels.

◇ Recall from Section 5.3.3, an explanation as to why all plants do not have very wide, long vessels.

◆ The risk of cavitation and embolism is higher in wide conduits and there is a balance between speed and safety.

This does not necessarily mean that plants with small-bore conduits always deliver less water to leaves per unit time than do plants with large-bore conduits (for equal values of $\Delta\Psi$). The compensating factor is the total area of xylem tissue: a large number of narrow conduits are equivalent to a small number of wide ones — and a lot safer when cavitation is likely.

5.4.3 Transpiration: water movement out of leaves

Transpiration — the evaporation of water from leaves — provides most of the energy for water movement through land plants because it sets up the gradient of water potential. What happens when water reaches the smallest veins in a leaf is that much of it evaporates from the walls of cells close to the vein ends. Movement to the leaf surface is then largely as water vapour via the interconnected air spaces and not as liquid water moving between mesophyll cells.

There is usually a very steep potential gradient, $\Delta\Psi^{wv}$, from leaf to atmosphere, caused by differences in leaf temperature and water vapour concentration or r.h. (Section 5.3.4), so the driving force for transpiration is large and often fairly constant. What determines the variable flux of water vapour in transpiration, J_{wv}, therefore, is the resistance to outward vapour diffusion or **diffusive resistance**. This is given the same symbol, R, as hydraulic resistance described earlier, but it is not strictly equivalent, and has units of time per unit length of pathway (e.g. $s\,m^{-1}$). At constant leaf temperature, the equation which describes this relationship is:

$$J_{wv} = \Delta C/R \qquad (5.7)$$

where ΔC is the difference in water vapour concentration inside and outside a leaf. If temperature is not constant, then ΔC must be replaced by $\Delta\Psi^{wv}$, which incorporates both T and C terms (see Section 5.3.4):

$$J_{wv} = \Delta\Psi^{wv}/R \qquad (5.8)$$

Diffusive resistance, R, is the largest resistance to water movement in the plant and, because it determines the rate of transpiration, normally determines the rate of water movement through the whole plant. We need to consider R more closely and find out what affects it.

Figure 5.17 shows that there are two pathways for the diffusion of water vapour out of leaves: via stomatal pores and via the cuticle. Each path offers a resistance to movement, R_{st} being *stomatal resistance* and R_c the *cuticular resistance*.

◇ Using common sense and Figure 5.17 as a guide, which resistance will usually be the larger, R_{st} or R_c?

◆ R_c; the cuticle has a waxy layer so that movement of water across it is very difficult and R_c is typically 10–100 times greater than R_{st}. Figure 5.17 shows that much more transpiration occurs through the open stomatal pore.

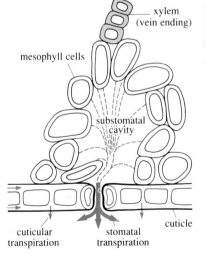

Figure 5.17 A section through a leaf showing the outward diffusion pathways for water vapour (green dashed lines). The thick and thin green arrows show, respectively, stomatal and cuticular transpiration. Note that most of the water vapour evaporates from walls of cells close to the vein ending.

'Open' is the operative word here, however, because you should recall from Chapters 1 and 2 that stomata can vary their aperture, so R_{st} is a *variable resistance* and we shall discuss how stomatal aperture is controlled in the next section.

In addition to these two resistances, all water vapour diffusing out of a leaf encounters a third resistance — that of the boundary layer of still, almost water-saturated air that covers the leaf surface. *Boundary layer resistance*, R_{air}, is also variable and depends on the thickness of the boundary layer, l, which is proportional to leaf width (w) and wind speed (v):

$$R_{air} \propto l \propto \sqrt{\frac{w}{v}} \tag{5.9}$$

◇ From Equation 5.9, (a) decide which leaf shape would give the greater boundary layer resistance, the flat leaf of a sycamore or the needle leaves of a pine tree. (b) Would R_{air} be greater in a high wind or still air conditions?

◆ (a) The flat sycamore leaf (R_{air} is directly proportional to leaf width), (b) In still air (R_{air} is inversely proportional to wind speed — the higher the speed the thinner the boundary layer and the smaller R_{air}).

Hairs on the leaf surface increase boundary layer thickness and hence R_{air}. Some plants, for example pine trees and marram grass on sand dunes, increase R_{air} because they roll up their leaves (as described in Chapter 1), enclosing the surface bearing stomata and a thick boundary layer. In general, however, plants have little control over the magnitude of R_{air} and *wind speed is the main factor determining boundary layer resistance*. With wind speeds of 10, 1.0 and 0.1 m s^{-1}, R_{air} has approximate values of 3, 10 and 30 s mm^{-1}, respectively, the corresponding thicknesses of the boundary layer being 0.4, 1 and 4 mm.

This brings us to another consideration. Since three resistances contribute to the total resistance, R_{total}, of a leaf to the loss of water vapour — R_{st}, R_c and R_{air} — what are their relative contributions to R_{total}, i.e. what is the main factor determining the amount of water transpired? The cuticle and the stomatal pores represent alternative pathways for vapour diffusion out of a leaf and, in this situation, we talk about *resistances operating in parallel*. They contribute in a fractional way to the overall resistance to diffusion from the leaf, R, given by:

$$\frac{1}{R} = \frac{1}{R_c} + \frac{1}{R_{st}} \tag{5.10}$$

So the combined resistance is:

$$R = \frac{R_c R_{st}}{R_c + R_{st}} \tag{5.11}$$

When different resistances occur at different places along a linear pathway, we talk about *resistances in series*, and the total resistance of the pathway is the sum of all resistances in series. So for boundary layer resistance, R_{air}, and the combined resistance of cuticle and stomata, R, the total resistance, R_{total}, is:

$$R_{total} = R_{air} + R = R_{air} + \frac{R_c R_{st}}{R_c + R_{st}} \tag{5.12}$$

These methods of summing resistances apply to any pathway and to hydraulic as well as diffusive resistances. Figure 5.18 represents diffusive resistances in parallel and in series, with a convention used for electrical circuits. For completeness we add a third resistance in series, R_{leaf}, which is the resistance of the air spaces inside the leaf: this varies with factors such as leaf thickness and leaf water content but, for a particular leaf, it does not exert a controlling influence on transpiration.

◇ From Figure 5.18, (a) write out a complete formula expressing the diffusive resistance to water vapour from the surface of leaf mesophyll cells to the outside atmosphere; (b) ignoring R_{leaf}, calculate R from the values given for the other three resistances.

◆ (a) $R_{total} = R_{leaf} + R_{air} + \dfrac{R_c R_{st}}{R_c + R_{st}}$,   i.e. you use Equation 5.12 but add in

one more resistance in series.

(b) The combined resistance of cuticle and stomata is

$\dfrac{1 \times 70}{70 + 1} = 70/71 = 0.99 \text{ s cm}^{-1}$,

so $R = 0.35 + 0.99 = 1.34 \text{ s cm}^{-1}$, applying Equation 5.12.

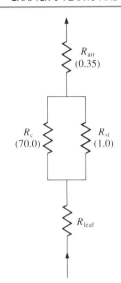

Figure 5.18 Resistances to the movement of water vapour out of a leaf. R_{leaf}, resistance of air spaces in the leaf; R_c, cuticular resistance; R_{st}, stomatal resistance; R_{air}, boundary layer resistance. Figures in brackets are representative values of resistance to the movement of water vapour, in units of s cm^{-1}, for a windy day when the stomata are fully open.

This calculation illustrates that, for the conditions described in Figure 5.17, stomatal resistance is the major component of the total diffusive resistance. For many plants, for much of the time, stomatal resistance does indeed limit the rate of transpiration. But because both R_{air} and R_{st} are variable resistances, we need to consider how variations in both affect the control of transpiration — always assuming that the driving force for water movement, ΔC (or Δr.h.) remains constant. Figure 5.19 illustrates the results of an experiment to determine the effect of the size of stomatal pores (degree of opening) on the rate of transpiration in still and moving air, which determines the value of R_{air}.

In moving air, the rate of transpiration increases almost linearly with size of stomatal aperture, so R_{st} clearly exerts the stronger control. In still air, however, it is only when stomata have a very small aperture and are almost closed that they influence transpiration rate significantly; boundary layer resistance otherwise exerts the main control. In fact 'in nature' air is hardly ever really still and R_{air} rarely rises above 1.7 s cm^{-1} so, in general, *stomatal aperture is the dominant factor controlling the rate of water loss from leaves*: exceptional circumstances where this is not true are described in Section 5.6. In the next section we describe how stomatal aperture can change in a controlled way.

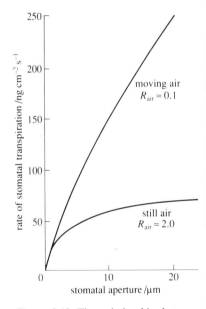

Figure 5.19 The relationship between the rate of transpiration and the diameter of the stomatal pores in still and in moving air. (The units of R_{air} are s cm^{-1}.)

Summary of Section 5.4

1 The amount of water moving along a pathway (hydraulic flux, J_v) depends on the gradient of water potential ($\Delta \Psi$) and the hydraulic conductance or resistance of the pathway, L_p or R.

2 Resistance is high for membranes and, therefore, for water flow across roots.

3 Soil resistance depends on the soil type (e.g. clay or sand) and, in particular, on soil moisture level. The drier the soil the greater the value of R.

4 Resistance to flow along xylem is low and depends on the width ('bore') and number of conducting cells. R_{xylem} is inversely proportional to conduit radius: the wider the conduit the greater the flux of water for a given gradient of water potential.

5 Diffusive resistance to water vapour movement from leaf to air is often the largest resistance in a plant. It has three major components: cuticular and stomatal resistance operating in parallel and boundary layer resistance operating in series.

6 When resistances operate in parallel, Equation 5.10 applies ($1/R = 1/R_1 + 1/R_2$) and water or water vapour moves chiefly via the pathway of least resistance. When resistances operate in series, total resistance is obtained by summing ($R = R_1 + R_2$).

7 Cuticular resistance to water vapour, R_c, is usually very high and most vapour therefore moves through stomata. Stomatal resistance, R_{st}, and boundary layer resistance, R_{air}, vary, respectively, with the size of stomatal pores and boundary layer thickness (which depends mainly on wind speed).

8 In theory, R_{air} will be the controlling resistance for transpiration if stomata are wide open and the air is still. In practice, stomata are rarely wide open and air is hardly ever 'still', so stomatal resistance is typically the major factor controlling the rate of transpiration at a fixed gradient of water vapour potential.

Question 9 (*Objective 5.9*) At root tips, the Casparian strip is not developed (cf. Figure 5.16). (a) Explain why radial flow of water is mainly along the apoplast pathway at root tips. (b) Write out an equation that defines the total resistance to radial flow.

Question 10 (*Objectives 5.5 and 5.9*) Select from (a) to (h), factors likely to *increase* the rate of transpiration. For each factor selected, indicate how it acts to promote transpiration.

(a) strong wind

(b) low wind speed

(c) dry air

(d) humid air (high relative humidity)

(e) leaf temperature above ambient air temperature

(f) leaf temperature below ambient air temperature

(g) stomata closed

(h) stomata wide open.

Question 11 (*Objectives 5.5 and 5.9*) Consider two hypothetical plants, X and Y. The stem of X has a single, central xylem vessel with a radius of 1 mm; that of Y has 100 xylem vessels each with a radius of 0.1 mm. The frequency of perforated cross-plates is equal in the xylem of X and Y.

(a) For equal gradients of water potential, in which of X and Y will the flux of water (J_v) be highest and how much higher will it be?

(b) By what factor does volume flow ($J_v \times$ area) differ for individual conduits in X and Y, with equal gradients of water potential?

(c) Without changing the diameter of xylem vessels, what two kinds of change could result in equal volumes of water being transported through the stems of X and Y per unit time?

5.5 STOMATAL MOVEMENTS

5.5.1 The role of stomata

The aerial parts of plants and animals on land are nearly always 'waterproof' in some degree. Human skin (epidermis), insect cuticle and the waxy cuticle of plants are examples of waterproof and gas-proof layers that allow little diffusion into or out of the body. It may seem paradoxical, therefore, that 2–5% of the surface of leaves is punctured by holes — stomatal pores — through which up to 98% of the water absorbed evaporates (Figure 5.20). The point, of course, is that stomata are required, not to let water *out* of leaves but to let carbon dioxide *in*. Entry pores for the CO_2 used in photosynthesis are essential because the cuticle presents just as great a barrier to the diffusion of CO_2 as it does to water.

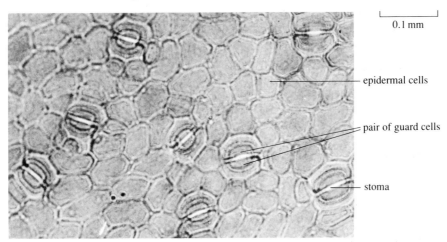

0.1 mm

epidermal cells

pair of guard cells

stoma

Figure 5.20 Surface view of an epidermal peel of a Box (*Buxus* sp.) leaf showing stomata and guard cells.

Stomatal transpiration has a useful side too. First, it creates the transpiration stream that carries mineral nutrients rapidly from roots to growing shoots and, second, it cools leaves in hot weather or strong light, having a similar effect to sweating in animals. When water is plentiful, therefore, stomata may play a role in temperature regulation. On balance, however, transpiration is more of a necessary evil than a blessing and the need to obtain CO_2 conflicts directly with the need to conserve water (Chapter 1). Many plants, particularly in dry habitats, maintain a precarious balance between starvation and desiccation and you have already seen (Section 5.3.3) that the risk of runaway embolism in trees often requires a drastic slowing down of transpiration. This is why the ability to open and *close* stomatal pores is absolutely essential and why they are regarded as turgor-operated valves that regulate the exchange of both CO_2 and water vapour. Stomata play a vital role in maintaining plant homeostasis and how the pores open and close and the factors that control these processes are considered next.

5.5.2 The mechanism of stomatal movement

Stomata (singular stoma, which means 'mouth') are found on most aerial organs but are most frequent on leaves and young stems.

◇ Recall from Chapter 1 where stomata occur on leaves.

◆ They may occur on both surfaces of a leaf but are often more common and sometimes confined to the lower surface only.

To understand how stomata function, you need to know about their structure and we describe this first, before considering the processes that cause stomatal pores to change size.

The structure of stomata

A stoma consists of a pore surrounded by two kidney-shaped or (in grasses) dumb-bell-shaped **guard cells** (Figure 5.21). Adjacent epidermal cells are intimately involved in guard cell functioning and referred to as *subsidiary cells*; the whole — guard cells plus subsidiary cells — is called the *stomatal complex*.

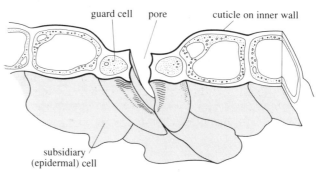

Figure 5.21 A section through a stoma showing the undersurface of the leaf. Note the cuticle extends round the guard cells and some way along the inner surface of other epidermal cells.

The diameter of a pore depends on the shape of the guard cells which, in turn, depends on properties of their cell walls and on their turgor (Section 5.2.2) relative to the turgor of neighbouring cells. Changes in size of the pores come about because of certain peculiar properties of guard cells:

1 They can rapidly and reversibly alter their turgor (i.e. their water content) and, as this happens, the volume of the cell changes, sometimes almost doubling as turgor increases.

2 Guard cell shape at high and low turgor depends on wall properties. The arrangement of cellulose microfibrils (Chapter 1) in guard cell walls is such that the inner wall (next to the pore) is less elastic in a longitudinal direction than the outer wall. This is mainly because of microfibril orientation but partly also because the inner wall is often thickened. The significance of the diagonal arrangement of microfibrils (shown in Figure 5.22), is that it makes kidney-shaped guard cells behave like radial tyres:

> 'the incorporation of inelastic reinforcements into the wall allows a change in shape of the cross section of the tyre or cell but not an expansion in the circumference of that cross section' (Raschke, 1979).

So when turgor and cell volume increase, the outer wall elongates more than the inner wall and the guard cells become bow-shaped (pores 'open'); when turgor decreases the cells are more or less straight (pores 'closed'), as shown in Figure 5.22.

3 The bowing of guard cells is possible only because they attain a higher solute content and, therefore, a greater turgor than neighbouring epidermal cells: a turgid guard cell literally pushes its neighbours out of the way.

4 Guard cells usually (but not always) differ from other epidermal cells in two further ways: they are not linked by plasmodesmata to adjacent cells; and they contain chloroplasts. If you look at a strip of leaf epidermis under the microscope, guard cells usually stand out as the only green cells present.

From this brief description you can see that guard cells are very specialized epidermal cells and the problem of how stomata alter their size boils down to a question of how guard cells change their relative turgor.

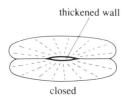

thickened wall

closed

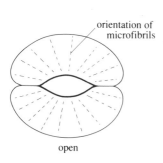

orientation of microfibrils

open

Figure 5.22 Closed and open stomata viewed from above. Note the radial arrangement of cellulose microfibrils (dashed lines) and the thickened walls of the guard cells adjacent to the pore (not present in all species).

Changes in turgor in guard cells

(i) Stomatal opening

As a closed stoma opens, there is a massive increase in the solute content of its guard cells, for example from about 1.8×10^{-12} to 7.2×10^{-12} osmoles (or 1.8 to 7.2 pico-osmoles) in broad bean leaves.

◇ How will this affect the osmotic pressure, the water potential and the turgor pressure of guard cells?

◈ Π will increase and so, from Equation 5.1 ($\Psi = P - \Pi$), Ψ will become more negative; water will move into the cells down a gradient of water potential and turgor pressure, P, will rise until the system is once more in equilibrium.

The increase in osmotic pressure involves a rapid migration of specific ions into guard cells from the surrounding tissue and cell walls (apoplast). Without exception, the dominant cation is K^+, although Na^+ ions may also be involved in halophytes: potassium ions move into guard cells when stomata open and move out when they close (Figure 5.23).

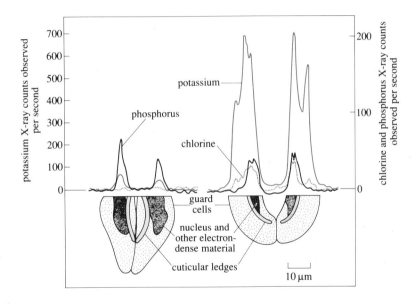

Figure 5.23 Profiles of the relative amounts of potassium (continuous green line), chlorine (broken green line) and phosphorus (black line) in guard cells of a closed and an open stoma. Data were obtained by electron-probe micro-analysis of freeze-dried cells. This technique involves scanning the material with a fine beam of electrons: atoms of different elements emit X-rays of characteristic wavelengths and these can be measured. Note the different scales for potassium on the one hand and for chlorine and phosphorus on the other.

The key event leading to K^+ uptake is the active pumping of protons out of the guard cells by a membrane-bound H^+-ATPase — in other words this is a classic chemiosmotic mechanism obeying the same principles as cation uptake by roots (Chapter 3). The electrochemical gradient which builds up across guard cell membranes as a result of proton export allows passive inward diffusion of potassium ions down a gradient of electrical charge. So K^+ uptake is a secondary active process, i.e. a passive process that depends directly on an active process, and it occurs through highly selective potassium channels in the plasmalemma.

So far so good: protons, derived from water, move out of guard cells leaving behind hydroxyl ions (OH^-) so that, as external pH falls, intracellular pH rises and the membrane potential rises (inside of the cell more negative) leading to K^+ influx (Figure 5.24a). However, this cannot go on for long: OH^- does not serve as the balancing anion for K^+ in the long term because intracellular pH is strictly controlled, so there must be other balancing anions.

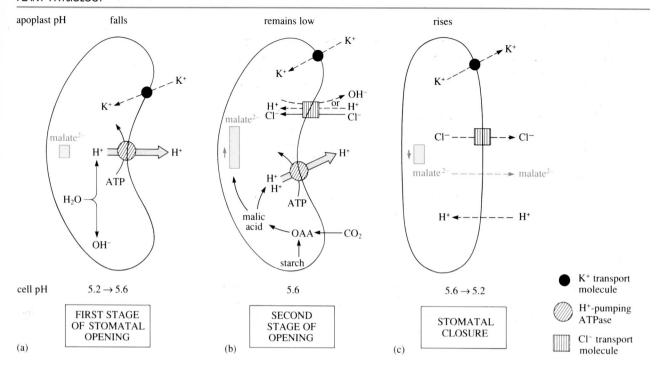

apoplast pH   falls   remains low   rises

cell pH $5.2 \rightarrow 5.6$ 5.6 $5.6 \rightarrow 5.2$

FIRST STAGE OF STOMATAL OPENING	SECOND STAGE OF OPENING	STOMATAL CLOSURE

(a)   (b)   (c)

● K^+ transport molecule

◐ H^+-pumping ATPase

▦ Cl^- transport molecule

Figure 5.24 A summary of ion movements and metabolism in guard cells during stomatal opening and closure. The broken arrows indicate secondary active transport inwards and passive movement outwards; the broad arrows indicate primary active pumping. The green rectangles indicate the level of malate, rising in (b) and falling in (c), with malate derived from oxaloacetate acid, OAA.

◇ From Figure 5.23 suggest one possible anion that might be used.

◆ The chloride ion, Cl^-: Figure 5.23 shows that chloride levels rise considerably when these stomata open, although not as much as potassium levels.

Chloride uptake by guard cells is also through specific channels but is driven not by the electrical gradient (which favours Cl^- efflux) but by the pH gradient. The precise mechanism is not yet known but two possibilities are illustrated in Figure 5.24b: Cl^- may be exchanged for an OH^- or entry may be linked to proton uptake (co-transport: see Chapter 3, Figure 3.7). However, Cl^- is rarely the *major* balancing anion for K^+. This is commonly the organic anion, *malate*, derived from the dicarboxylic malic acid (Figure 5.24b), whose synthesis is coupled to stomatal opening in a neat way. As intracellular pH rises, this activates the enzyme PEP carboxylase (where PEP stands for phosphoenolpyruvate). This enzyme fixes CO_2 to produce oxaloacetic acid, which can be reduced to malic acid — in the same way as described for C_4 plants in Chapter 2. Malic acid ionizes at the prevailing pH, providing not only balancing anions for K^+ but also hydrogen ions for the proton pump. The ultimate precursor for malic acid synthesis is stored starch and if, as in the onion, *Allium cepa*, guard cells contain no starch, then chloride is the sole balancing anion.

(ii) Stomatal closure

According to this chemiosmotic model, closure occurs when the proton pump is switched off. Provided that their specific channels remain open, K^+ and Cl^- move passively out of guard cells along electrochemical gradients and, indeed, increases in the levels of these ions in the epidermal cells surrounding stomata have been measured following stomatal closure. Some malate may be metabolized within guard cells but there is also evidence that malate efflux occurs during closure (Figure 5.24c).

(iii) Energy sources

An important and much debated question concerns the energy source for stomatal opening, i.e. where does the ATP that fuels the proton pump come from? Current thinking (in 1990) is that there are two and possibly three separate sources of ATP which are used to different extents under different conditions. One is photophosphorylation by guard cell chloroplasts, operating mainly at medium-to-high light levels; a second is oxidative phosphorylation (respiration), which can operate in darkness; and a possible third (whose existence is still uncertain) is a photosystem powered by blue light and quite separate from photosynthesis, which can operate at low light levels, such as occur in shade or at dawn.

The crucial question now is what kinds of signals trigger active stomatal movements and control the size of stomatal pores.

5.5.3 The control of stomatal aperture

A bewildering array of signals influence stomatal aperture, some originating within the plant (e.g. levels of certain growth regulators) and others from the outside environment (e.g. air humidity, quality and quantity of light). The system becomes easier to understand if you bear in mind, firstly, the functions that stomata perform (Section 5.5.1):

1 They facilitate the entry of CO_2 in amounts sufficient to meet the 'demands' of photosynthesis.

2 They control water loss in such a way that tissue damage from desiccation or embolism is avoided and (if water is plentiful) they promote transpiration when leaf cooling is necessary.

The second point to bear in mind is the stomatal mechanism itself (Section 5.5.2): rapid changes in aperture must involve changes in ion fluxes across the plasmalemma.

◇ Suggest two ways in which this might be induced.

◆ The obvious way is by changing the activity of the proton-pumping ATPase and thus the electrochemical gradient across the plasmalemma; another way is by changing plasmalemma permeability to relevant ions (e.g. the number or availability of K^+ or Cl^- channels).

More gradual changes in aperture could involve factors such as the supply of water for maintenance of guard cell turgor, the supply of substrates for respiration or malate synthesis, or a shift in the steady-state supply of ATP for the proton pump. Against this background, consider some commonly observed patterns of stomatal movement and the associated environmental signals.

Patterns of stomatal movement and environmental signals

1 A commonly observed pattern for plants that are well supplied with water and not exposed to extremely high irradiance or temperature is shown in Figure 5.25. Stomata open at dawn and close at dusk; the extent of opening is lower in cloudy than in sunny conditions and there is a gradual reduction of aperture from about midday onwards. This pattern can be related to the plant's *requirement for CO_2* and involves two main control factors: light and intercellular CO_2 levels, with the afternoon decline probably related to air humidity, which is discussed later.

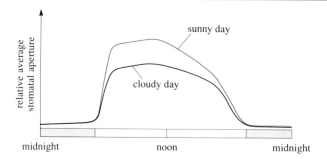

Figure 5.25 Typical patterns of stomatal opening and closure on a sunny and a cloudy day.

Intercellular CO_2

Low CO_2 promotes stomatal opening and high CO_2 causes stomatal closure which, since intercellular CO_2 serves as a marker for photosynthetic activity, nicely ties in photosynthetic demand for CO_2 with degree of stomatal opening. That CO_2 concentration is sensed within guard cells, is shown by the ability of isolated pairs of guard cells to respond to the gas but how CO_2 causes stomatal movements is still unclear. One hypothesis suggests that because CO_2 is fixed (via PEP carboxylase) into malic acid in guard cells, the levels of one or both of the dissociation products — malate and H^+ ions (i.e. cytoplasmic pH) — reflects CO_2 level and influences membrane permeability or the steepness of the proton gradient. Another hypothesis suggests that, in the light, CO_2 affects the rate of photophosphorylation and the supply of ATP for proton pumping.

Light

Stomatal opening is promoted by light but there are two distinct ways in which this comes about. First, photosynthetically active radiation (PAR) initiates photosynthesis, which lowers intercellular CO_2 and, therefore, promotes stomatal opening.

◇ Would this effect operate (a) with an isolated leaf and (b) with isolated epidermal strips?

◆ Bearing in mind that the main site of CO_2 fixation is the leaf mesophyll, the answers are (a) yes and (b) probably no (since no mesophyll is present, normal epidermal cells contain no chloroplasts and guard cell fixation into malate would have very little effect on CO_2 levels).

This indirect effect of light is thought to be important for C_4 plants but not, usually, for C_3 plants. The latter are influenced strongly by the second effect of light, which operates directly on guard cells, can be detected with isolated guard cells and is independent of CO_2 concentration. Again there are disagreements about how light operates but the simplest hypothesis is that it stimulates proton secretion by providing ATP for the proton pumps. To complicate matters further, however, two separate light-absorbing systems may be involved, one being the usual chlorophyll system in chloroplasts (PAR photosystem) and the other a system absorbing blue light and based on a membrane-bound pigment outside chloroplasts. The PAR photosystem seems to dominate at medium-to-high irradiances and supplies ATP through photophosphorylation; it seems to be essential for very wide stomatal opening in the light. The blue-light photosystem does not promote such wide stomatal opening as the PAR system and saturates at very low irradiances. It is present in grasses (with dumb-bell-shaped guard cells) but not, usually, in species with kidney-shaped guard cells, and it has been suggested that this system is responsible for the rapid opening of stomata at dawn. In the lady's slipper orchids, *Paphiopedilum* spp., which grow in very shady habitats, the kidney-shaped guard cells lack chloroplasts and their response to light is apparently mediated largely through the blue-light photosystem.

◇ Explain how this feature of *Paphiopedilum* makes physiological sense.

◆ In deep shade, photosynthesis will be limited by the availability of light and not CO_2, so wide stomatal opening may be unnecessary for obtaining sufficient CO_2.

Fast stomatal responses to light anticipate the photosynthetic demand for CO_2 and may, therefore, be described as *feedforward* responses.

2 Other patterns of stomatal movement can be related to *water conservation* by the plant and the two patterns illustrated in Figure 5.26 occur commonly in dry conditions or when transpiration rates begin to exceed the rate of water supply through roots and stems.

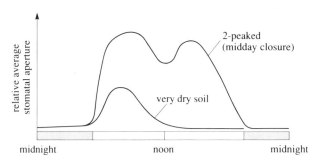

Figure 5.26 Patterns of stomatal opening and closure when there is moderate water stress (midday closure, upper curve) and severe water stress due to very dry soil (lower curve).

The upper, two-peaked pattern in Figure 5.26, with partial or complete stomatal closure around midday, is quite common in trees. Leafy tree canopies may transpire an enormous quantity of water, especially if the air is dry and temperatures are high, and midday stomatal closure could be an important factor preventing embolism and cavitation (Section 5.3.3). Many herbaceous plants growing in seasonally dry habitats also show this two-peaked pattern at the start of the dry season. Later in the season, however, when the soil is much drier, stomata may open only in the morning and a single-peak pattern (lower curve in Figure 5.26) is found: the extent of stomatal opening is often less during this late-season morning peak than earlier in the season. The question is: what signal(s) override the low $[CO_2]$ and high light signals at midday — both promoters of stomatal opening — and cause stomata to close? And why do stomata usually open less widely if the soil is dry?

There are no simple answers to these questions because several 'override' signals linked to water relations have been found and scientists disagree about their relative importance. In general:

1 Stomatal closure at midday seems to be controlled by the outside environment, principally the relative humidity of the air and, to some extent, leaf temperature.

2 Sensitivity to these signals appears to be determined by the water status (water potential or turgor) of the plant, and that of the roots seems to be more important than that of the leaf; this probably explains the seasonal shift in opening pattern described above.

3 In a real emergency, when leaves are about to wilt and turgor is effectively zero, a hormonal signal is released from mesophyll cells and causes rapid closure of stomata.

Look at the left-hand side of Figure 5.27, which relates to the semi-desert plant *Artemisia herba-alba* early in the dry season.

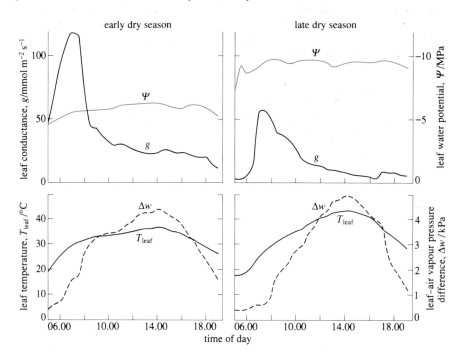

Figure 5.27 Diurnal changes in leaf water potential, Ψ, leaf conductance, g, leaf temperature, T_{leaf}, and the leaf–air vapour pressure difference, Δw, for leaves of sagebrush, *Artemisia herba-alba*, in Israel, early and late in the dry season.

◇ Which factor correlates most closely with the degree of stomatal opening (measured here as leaf conductance, g, which is the reciprocal of leaf resistance): leaf water potential, air humidity (assessed by the difference in water vapour concentration between leaf and air, Δw), or leaf temperature?

◆ Air humidity, which correlates inversely (although not very closely): both leaf water potential and temperature change very little over the period of rapid stomatal closure.

How air humidity exerts this effect is still not clear but one hypothesis is that it affects the rate of water loss across the cuticle and thus the turgor of the epidermis. Reduced epidermal turgor is then supposed to trigger stomatal closure. Because the response to humidity anticipates any change in water status of the leaf as a whole, it is described, like the response to light, as a *feedforward* response.

If you now compare the right side of Figure 5.27 (late dry season, soil very dry) with the left, decreasing air humidity (increasing Δw) again shows the best correlation with stomatal closure but it does not explain why leaf conductance is much lower (i.e. stomata open less widely) later in the dry season. There is, indeed, a much lower (more negative) leaf water potential but, although this may be a contributing factor, other evidence (which will not be discussed) suggests that it is not the primary cause. Rather, it is water stress (more negative water potential) in the *roots* that seems to be more important, although what signal passes from roots to guard cells and how it acts on guard cells is unclear.

The signal from the leaf that causes emergency stomatal closure *is* known, however; it is a plant growth inhibitor, **abscisic acid** or **ABA**. When this ABA effect was first discovered in the early 1970s it caused much excitement because here, it seemed, was a method of controlling transpiration in crop plants: spray on a harmless natural compound and prevent wilting in very dry conditions. So far, this idea has not worked in practice because the control system is more subtle than first thought. Leaves and even the upper and lower epidermis on the same leaf may vary in their sensitivity to ABA, for example, but Figure 5.28 illustrates what is thought to happen.

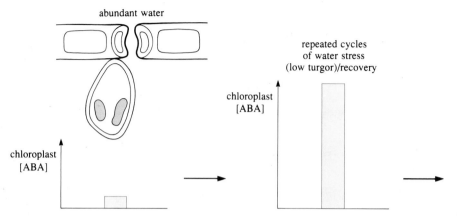
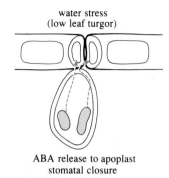

Figure 5.28 The accumulation of ABA in the chloroplasts of leaf cells during repeated cycles of water stress (low turgor) and recovery. Eventually, accumulated ABA is released during water stress and passes via the apoplast to guard cells, where it triggers stomatal closure.

When leaves wilt (zero turgor), this triggers the synthesis of ABA in the chloroplasts of mesophyll cells. At first this has no effect on stomata but with repeated cycles of wilting and recovery, ABA levels build up and eventually a fall in leaf turgor (not necessarily to wilting point) causes release of stored ABA into the cell walls (apoplast). From here ABA can reach the plasmalemma of guard cells where, even in minute amounts, it affects ion pumping and causes stomatal closure. As the leaf rehydrates, apoplastic ABA is broken down. So ABA in the mesophyll acts initially as a barometer of leaf water stress and then as a potent trigger for emergency stomatal closure.

3 The pattern of stomatal movements for *CAM plants* is, as you might expect from Chapter 2, radically different from that of other plants. Stomata open at night and close during the day (Figure 5.29).

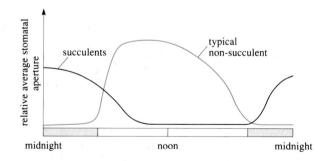

Figure 5.29 Patterns of stomatal opening and closure in succulents (CAM plants). The green line shows the pattern typical of non-succulents.

The normal responses to light are overridden completely here but CO_2 and humidity remain important stomatal regulators.

◇ How does internal $[CO_2]$ vary in CAM plants and how will this influence stomatal opening?

◆ CO_2 fixation and storage (as vacuolar malic acid) is switched on at night, so internal $[CO_2]$ will be low and promote stomatal opening. During the day, when malic acid is released from vacuoles and decarboxylated to give CO_2, the high levels of internal CO_2 promote stomatal closure.

Higher air humidity at night also tends to promote stomatal opening and this probably acts as a fine tuning mechanism for stomatal movements. However, if there is a prolonged drought during which soil dries out completely and roots wither, then stomata remain closed the whole time — the ultimate override control!

To *summarize* the main factors that control stomatal aperture (Figure 5.30):

☐ Two signals light and intercellular $[CO_2]$, control stomatal aperture in relation to the plant's demand for CO_2 for photosynthesis. Low intercellular $[CO_2]$ promotes stomatal opening, thus raising $[CO_2]$ and leading to stomatal closure (negative feedback loop (b) in Figure 5.30) unless photosynthesis and assimilation maintain low $[CO_2]$ (feedback loop (a)). Conditions of light and temperature that promote assimilation act via loop (a) but, in addition, the degree of stomatal opening is affected by the supply of energy and assimilates (positive feedback loop (d)). There is also a direct, feedforward effect of blue light, (e), in some species.

☐ At least three signals — air humidity, ABA levels in the leaf apoplast and some unknown signal from the roots — control stomatal aperture in relation to water supply. Humidity (or the water vapour pressure difference between leaf and air) acts via cuticular evaporation in a feedforward manner (path (f)). ABA levels, which serve as a monitoring system for leaf turgor, cause stomatal closure through feedback loop (c): when stomata open, evaporation increases, leaf turgor falls and ABA is first synthesized and accumulated and then released. When the soil is very dry and water supply to roots does not keep pace with demand by shoots, stomata open less widely in response to some signal from the roots (feedback loop (g)).

A poorly understood response is the tendency of stomata to open more widely when leaf temperature becomes very high. This occurs only if water is plentiful and leaf turgor does not fall but it may override a decline in net photosynthesis and associated rise in intercellular $[CO_2]$. It can be of considerable importance in keeping leaves cool through **transpirational cooling**, but whether temperature is acting directly on stomata (tentative path (h)) or in some more general way on leaf metabolism is not known.

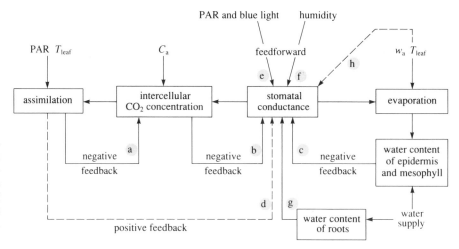

Figure 5.30 Summary diagram to illustrate the factors that control stomatal conductance and the feedback and feedforward control paths. PAR, photosynthetically active radiation; T_{leaf} leaf temperature, C_a and w_a, atmospheric concentrations of CO_2 and water vapour, respectively.

Summary of Section 5.5

1 Stomata are pores on the leaf surface that allow CO_2 to enter and water vapour to diffuse out. They have a variable diameter and act like turgor-operated valves.

2 Stomatal diameter depends on the turgor of the two guard cells that surround the pore. Their turgor and volume can change rapidly and reversibly and, when turgor is high, the inelastic inner walls (next to the pore) cause guard cells to become bow-shaped (pore 'open'). Adjacent epidermal cells have lower turgor than turgid guard cells and are pushed aside.

3 Stomatal opening occurs because solute levels rise in guard cells, causing an influx of water and thus a rise in turgor. Outward proton pumping leads to an influx of potassium ions. The balancing anion is commonly malate, synthesized from starch within guard cells, but chloride influx may also occur. ATP for this process may derive from photophosphorylation, respiration or, at low light levels, possibly by a blue light-driven photosystem.

4 Stomatal closure occurs when the proton pump is switched off and efflux of K^+, Cl^- and malate occurs. Malate may also be metabolized within guard cells.

5 Stomata commonly open during the day and close at night. If water is in short supply or xylem tension falls too low, midday or afternoon closure occurs. CAM plants show night opening and daytime closure. Wider opening at high temperatures may occur when water is plentiful and results in transpirational cooling.

6 These patterns of stomatal movement reflect chiefly the balance of conflicting demands between the need to conserve water and the need to obtain CO_2 for photosynthesis.

7 Light and low $[CO_2]$ promote stomatal opening; darkness and high $[CO_2]$ promote closure. These signals reflect photosynthetic demand for CO_2.

8 Related to water conservation, low air humidity promotes stomatal closure and is often responsible for midday closure. Emergency closure when leaves are in danger of wilting is triggered by abscisic acid (ABA); this accumulates in mesophyll cells during earlier episodes of low turgor and is subsequently released into the apoplast of guard cells when turgor falls. A signal from the roots that reflects their water status lowers sensitivity to light and CO_2 and probably explains small stomatal apertures and prolonged daytime closure late in the season.

9 All these signals operate through a complex web of feedback and feedforward loops.

Question 12 (*Objective 5.10*) Which of the properties or features (i)–(vi) are (a) commonly found in guard cells and (b) essential if stomata are to function properly?

(i) an inner cell wall (facing the pore) that is less elastic than the outer cell wall

(ii) a K^+ pump in the cell membrane

(iii) the capacity to synthesize malate

(iv) the possession of chloroplasts

(v) a proton pump in the cell membrane

(vi) the capacity to generate higher turgor than adjacent cells

Question 13 (*Objective 5.11*) Suggest an explanation for the following:
Certain fungi release a toxin that induces or stimulates active pumping of protons out of cells; when infected with these fungi, plants tend to wilt severely and isolated leaves wilt if treated with the toxin.

Question 14 (*Objectives 5.11 and 5.16*) Figure 5.31 shows the results of an experiment with leaf discs of broad bean (*Vicia faba*). Discs floating on water

Figure 5.31 Changes in stomatal aperture and guard cell K^+ in leaf tissue of broad bean (*Vicia faba*) in response to water stress. Discs cut from leaves were floated on water. At time 0, discs were removed from the water, dried for 10 minutes in moving air and then transferred to saturated (100% r.h.) moving air. (Above) Changes in fresh weight of discs as percentage of fully turgid weight when floating on water. (Below) Stomatal aperture and change in guard cell K^+, measured on disc samples from which epidermal strips were removed. Stomatal aperture was measured microscopically and K^+ determined from the percentage area of guard cells occupied by a K^+-specific stain.

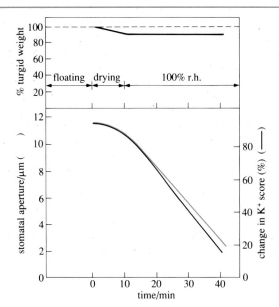

were removed and dried in moving air for 10 min; they were then transferred to moving, water-saturated air. Fresh weight, stomatal aperture and the K^+ content of guard cells were monitored from the start of the drying period.

(a) Do these data support the hypothesis that stomatal aperture is controlled directly by leaf turgor? What other control factor(s) may be involved?

(b) Do these data prove that the turgor of guard cells depends solely on their K^+ content?

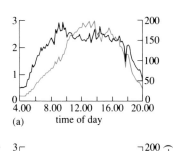

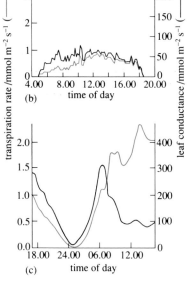

Figure 5.32 Diurnal changes in transpiration and leaf conductance (a measure of the degree of stomatal opening) for (a) sun leaves of beech (*Fagus sylvatica*) on a sunny day; (b) shade leaves of beech on the same day; (c) leaves of the arctic cushion plant *Dupontia fisheri* in Alaska on a day in mid-summer.

5.6 MOVEMENT OF WATER THROUGH WHOLE PLANTS

After the digression into stomatal physiology in the last section, this section draws together information about the driving force for water movement $\Delta\Psi$ (Section 5.2.3) and resistances to movement (Section 5.4). The aim is to give an overall picture of how transpiration, root physiology and soil conditions interact to determine the water flux and water status of plants.

5.6.1 Overall control of transpiration rate

For most of the time, the water flux through a plant depends on the rate of transpiration, J_{wv}, so we need to consider first what determines J_{wv}. Is it always just stomatal resistance or conductance? Compare the three parts of Figure 5.32, which show how transpiration and leaf conductance (a measure of stomatal opening: Figure 5.27) vary for sun and shade leaves of beech (*Fagus sylvatica*) and for an arctic cushion plant, *Dupontia fisheri*, over one day.

◇ For which type of leaf does stomatal resistance appear to be of overriding importance in controlling transpiration?

◆ For the shade leaf of beech (Figure 5.32b); the correlation between leaf conductance and transpiration is closer here than for the other two leaves.

For *Dupontia* in particular, the correlation is much less close and despite partial stomatal closure after 06.30 transpiration remains high. To understand what is happening here remember first that transpiration depends on the driving force, $\Delta\Psi^{wv}$, and the resistance to vapour diffusion, R: $J_{wv} = \Delta\Psi^{wv}/R$. $\Delta\Psi^{wv}$ depends on both the difference in water vapour concentration between leaf and air, ΔC, and on temperature T (Equation 5.5), so if transpiration does not correlate closely with stomatal resistance it must be because one or both of these factors has changed and perhaps also because boundary layer resistance — another component of the total resistance — has decreased. This is exactly what you might expect to happen for leaves exposed to strong sunlight in the middle of the day, when the air becomes drier and leaves heat up.

The full complexity of the situation is indicated in Figure 5.33. The factors that affect the driving force for transpiration — air humidity (affecting ΔC), light (affecting leaf temperature) and wind (which cools leaves and reduces humidity) — also affect resistance to evaporation.

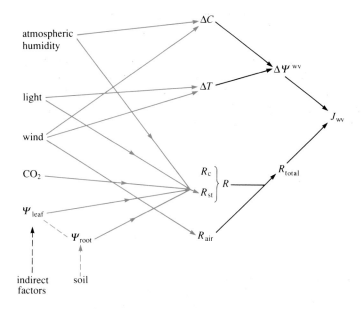

Figure 5.33 Factors affecting transpiration, J_{wv}. Black lines indicate components that determine J_{wv}; green lines indicate factors affecting components.

A complex web of interacting factors determines the rate of transpiration and the controlling factor commonly varies from minute to minute. The question that must be asked now is: what happens within the plant when transpiration rate varies?

5.6.2 Roots and absorption lag

When the rate of transpiration increases, the rate at which water is supplied to leaves must also increase and there is only one way in which this can be achieved: the gradient of water potential along the pathway from roots to leaves must increase. This adjustment takes time, because there are resistances along the main pathway of water movement, and the result is a delay between root absorption and leaf transpiration. Figure 5.34 illustrates this **absorption lag**.

The extent of the absorption lag depends on the magnitude of the largest resistance along the pathway of water flow and, in moist soil, this rate-limiting resistance is located in the *roots*.

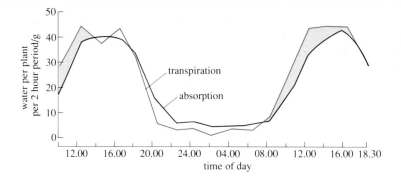

Figure 5.34 Diurnal variation in rates of transpiration and absorption of water by sunflower leaves. The absorption lag is shaded.

◇ Recall the pathway of water in roots: why is there a high resistance element in this pathway?

◆ Because the root endodermis has waterproof walls, water must cross a cell membrane when moving from soil to root xylem, and membranes have a very high resistance to water flow (Section 5.4).

So it is *the high resistance of cell membranes in roots* that accounts for most of the absorption lag in moist soils and you can see from Figure 5.35 that removing the roots of a plant greatly reduces the absorption lag.

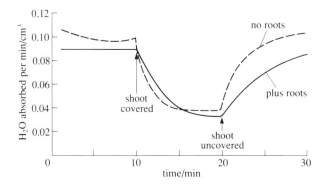

Figure 5.35 The effect of removing roots on the lag between absorption and transpiration. Transpiration was reduced by covering the shoots for 10 minutes. Note the slower increase in absorption by rooted plants compared with rootless plants when the shoots are uncovered and transpiration increases.

One consequence of the lag between absorption and transpiration is that during the day, when transpiration rates are high, water tends to move out of cells that are situated beside the main water pathway (e.g. living xylem parenchyma and partly blocked xylem elements in the heart wood of trees) and into the functioning xylem. This results in a reduction in 'donor' cell volume and tree trunks may shrink during the day and swell up at night as cells rehydrate (Figure 5.36).

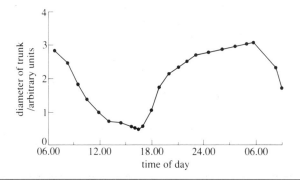

Figure 5.36 Shrinkage in the trunk of an avocado tree at midday caused by a water deficit resulting from rapid transpiration.

There are also certain environmental conditions under which the absorption lag becomes so severe that plants actually wilt and may die even when the soil is wet. One of these is *low soil temperature*. The most important effect of this is to increase the hydraulic resistance of cell membranes in roots, reducing the rate of water movement across them. Cold roots combined with high leaf temperatures, strong winds and dry air — all of which promote transpiration (and cuticular transpiration continues even when stomata are closed) — often kill plants through desiccation. The danger is particularly great in alpine environments but it is plants from warm, tropical habitats that are usually most susceptible to this form of damage.

Poor aeration of the soil, especially when combined with conditions favouring rapid transpiration, can also cause injury through desiccation. Lack of oxygen necessitates anaerobic respiration (glycolytic fermentation) in roots and greatly reduces the supply of ATP. The end-products of fermentation tend to injure root cells and increase their hydraulic resistance and, in addition, shortage of ATP restricts the roots' ability to absorb ions actively and maintain the low water potential necessary to draw in water from the soil. Flooding of soils, therefore, often results in permanent injury to crop plants such as tobacco or cabbages. Marsh or water plants (hydrophytes; Chapter 1, Section 1.4.2) that are naturally tolerant of wet soils avoid the problem by having a built-in aerating system. You may have noticed the hollow stems and spongy texture of such plants (illustrated in Chapter 1): oxygen can diffuse from shoots to roots through these giant, interconnected air spaces so that roots remain oxygenated even in totally anaerobic soils.

Thus root structure, the presence of an endodermis with waterproof cell walls, and any factors that affect water movement across root cell membranes or the capacity of roots to accumulate ions all influence the size of the absorption lag. This, in turn, affects the steepness of the water potential gradient necessary to drive water through the plant at a rate sufficient to match transpiration losses. What happens when the going gets even tougher and the soil begins to dry out is discussed next.

5.6.3 Osmoregulation and the response to drying soil

As soil dries out water potential decreases (becomes more negative) and hydraulic resistance increases. Both these factors reduce the flux of water into roots and there is only one way that plants can compensate for this: the water potential of the root, Ψ_{root}, must decrease. From the basic equation for cell water potential, $\Psi = P - \Pi$, there are two possible ways in which this can occur: osmotic pressure could increase or turgor decrease. In drying soil, the main way is by an increase in osmotic pressure, partly by enhanced accumulation of inorganic ions and partly by increasing the levels of organic solutes. Such **osmotic adjustment** or **osmoregulation** can occur without any fall in root turgor, which is important because turgor is the driving force for cell expansion and, therefore, root growth. When high rates of transpiration steepen the gradient of water potential through the plant, Ψ_{root} may fall because water moves into the xylem and then, of course, root turgor does fall. Root shrinkage may occur in this situation (described in Section 5.4.1) causing not only a reduction in root growth but also reducing contact between the root and the soil particles.

Figure 5.37 shows an idealized version of the changes in leaf and root water potential as the soil dries out, and experimental work generally supports this model.

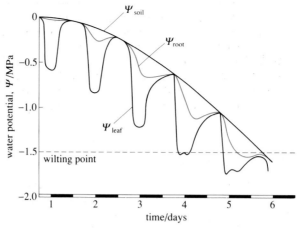

Figure 5.37 Predicted changes in leaf water potential, Ψ_{leaf}, water potential at the root surface, Ψ_{root}, and the water potential of the soil, Ψ_{soil}, as a plant rooted initially in wet soil transpires and the soil dries out. The same evaporative conditions are considered to prevail each day. The alternating white and black areas on the horizontal axis indicate day and night, respectively.

There is a daily cycle of daytime water stress (shortage of water), caused by transpiration, high resistance in the roots and damage from embolism and cavitation in the xylem. This is followed by recovery at night and is superimposed on long-term changes related to the water potential of the soil. Once leaves have reached the point of wilting, water potential can decrease no further without causing permanent injury through desiccation; so if the water potential of the soil continues to decrease then the plant will die.

The effect on a plant of these daily cycles of low water potential depends strongly on the extent of xylem recovery from cavitation (Section 5.3.3) and also on whether or not osmoregulation occurs in the leaves. If it does not, then the fall in Ψ_{leaf} inevitably causes a fall in leaf turgor, which can trigger stomatal closure and, for reasons described in the next chapter, reduce cell expansion in young leaves. This situation is illustrated by the middle, dotted line in Figure 5.38, the distance between this line and that for Ψ_{leaf} indicating leaf turgor. The lower, dashed line in Figure 5.38 shows what happens when there is osmoregulation: leaf turgor is held constant until day 4 or 5 of soil drying and wilting does not occur until day 7.

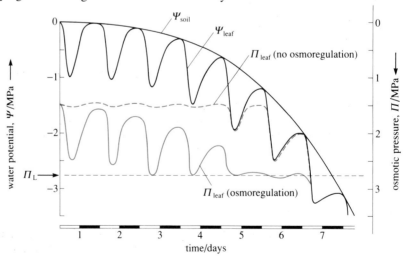

Figure 5.38 Predicted changes in leaf and soil water potential and leaf osmotic pressure for a leaf where there is no osmoregulation (green dashed curve) and where there is osmoregulation (continuous green curve). Leaf turgor is the difference between Ψ_{leaf} and Π_{leaf}. Π_L is the osmotic pressure at the maximum possible level of solute accumulation — higher solute concentrations cause cell damage.

◇ When does wilting occur in the leaf *without* osmoregulation?

◆ Briefly on day 4, for a longer period on day 5 and constantly on days 6 and 7.

The capacity to osmoregulate is clearly useful when water is in short supply and plant breeders have attempted to select strains of crop plants, such as rice, wheat and barley, with good osmoregulatory powers. These powers involve two distinct processes. By far the most important for lowering cell water potential is the accumulation of ions and electrically charged organic solutes in the *vacuole*. However, vacuole and cytoplasm must be in osmotic equilibrium and raising the concentration of ions in the cytoplasm is not feasible in plants because this would interfere with the conformation and functioning of macromolecules, especially proteins. Another process is used and we describe this briefly.

Osmotic adjustment of the cytoplasm

Instead of accumulating ionic solutes, specific uncharged organic molecules that do not seem to affect protein functions are synthesized and accumulated in the cytoplasm. These are known as **compatible solutes** and, among angiosperms, two of the most commonly occurring are the amino (or, strictly, imino) acid, *proline*, and quaternary ammonium compounds (substances in which the four H atoms of NH_4^+ are substituted by organic groups) such as *betaine*. In *Hordeum vulgare* (barley), for example, proline accumulates in the cytoplasm in proportion to the degree of soil dryness and, in very dry conditions, may even appear in vacuoles. Proline also accumulates in the halophyte *Aster tripolium* (sea aster), in proportion to external salinity, whilst the salt-marsh grass *Spartina townsendii* (cord grass) accumulates betaine.

Overall adjustment of osmotic pressure in higher plant cells is thus by the accumulation of ions (and sometimes organic solutes in the vacuole) and by the accumulation of compatible organic solutes in the cytoplasm. But an interesting question is: how are these adjustments controlled? What is the 'sensing' mechanism that signals when osmotic adjustment is needed? A currently favoured hypothesis is that pressure receptors on the plasmalemma monitor cell turgor and, if turgor falls sufficiently as cells lose water, this triggers the synthesis or active accumulation of solutes.

◇ How will this affect water movement?

◈ It will raise osmotic pressure and thus lower water potential causing water to move back into cells.

You should recognize this as another elegant example of regulation by negative feedback.

A marked ability to osmoregulate is characteristic of many plants that grow on dry or saline soils and is one of several features that permit survival in such difficult conditions. The next section considers these features and the general question of plant survival when water is in short supply.

Summary of Section 5.6

1 Water flux through a plant commonly depends on the rate of transpiration. This, in turn, depends on the gradient of water vapour potential between leaf and air (a function of humidity and temperature gradients) and the total resistance (stomatal, cuticular and boundary layer) to water loss.

2 Internal and environmental factors interact to control these variables and the factor controlling transpiration may change from minute to minute.

3 There is an absorption lag between an increase in transpiration rate and the increase in the supply of water from roots. This is caused mainly by the high resistance to water movement of membranes in root cells.

4 The absorption lag causes stem or trunk shrinkage as water is drawn out of cells adjacent to functioning xylem. Low soil temperatures and poor aeration of the soil exacerbate absorption lag.

5 When soil is relatively dry, an increase in transpiration rate may cause root shrinkage because the uptake of water from the soil does not keep pace with water movement into xylem. This reduces root turgor and growth rate.

6 In the longer term, the fall in soil water potential as soil slowly dries out is balanced by a fall in root water potential through osmotic adjustment or osmoregulation. This maintains the water potential gradient from soil to root.

7 Osmotic adjustment involves an increase in osmotic pressure (and hence a reduction in Ψ) and is brought about by the accumulation of inorganic and organic ions in cell vacuoles and of compatible organic solutes such as proline or betaine in the cytoplasm.

Question 15 (*Objectives 5.12 and 5.13*) Suggest explanations for statements (a)–(c):

(a) Pot plants in a sunny position may wilt if watered with very cold water at midday.

(b) Garden or pot plants that grow naturally in shady habitats often wilt in full sun even if the soil is moist.

(c) Many garden plants are more susceptible to drought and grow poorly on a windy, exposed site compared with a sheltered site.

Question 16 (*Objectives 5.12, 5.14 and 5.16*) Figure 5.39 shows the diurnal courses of transpiration and leaf conductance ($= 1/\text{resistance}$) for a shrub, growing in desert conditions during the early part of the season (soil moist). Use Figure 5.39 and information from elsewhere in this chapter to explain:

(a) the different shapes of the transpiration and leaf conductance curves during the day.

(b) the likely changes (and their causes) in the water potential of leaf, stem and root cells over the 24 hour period (starting at 06.30).

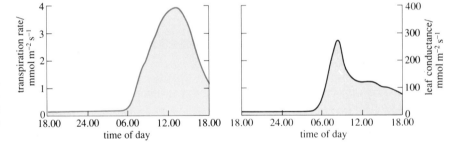

Figure 5.39 Transpiration and leaf conductance (a measure of stomatal aperture) during one day for leaves of *Prunus armeniaca* growing in a desert environment (Negev, Israel).

5.7 SURVIVAL DURING WATER SHORTAGE

Whenever the water potential of the soil falls to a low value it becomes more difficult to absorb water and plants are in danger of desiccation. The problem arises with saline soils (where high salt levels lower soil water potential), frozen soils and, of course, the dry soils of deserts. In all these situations special features have evolved that allow plants to survive.

For **halophytes** there are additional problems due to high levels of ions (discussed in Chapter 3) and we described in the last section their main adaptation for obtaining water from saline soil: they osmoregulate very efficiently and also develop extremely high osmotic pressures. Values as high as 10 MPa are not uncommon compared with typical values of 1–2 MPa for most crop plants. Going back to Figure 5.38, this is equivalent to lowering the dashed line indicating the limit of solute accumulation (Π_L).

We describe below some of the features that allow plants to survive drought or frozen conditions. Broadly, there are two main strategies: tolerance of extreme dehydration is one strategy whilst the other covers a range of structural and physiological features that help plants to obtain, store or eke out limited supplies of water.

5.7.1 Tolerance of desiccation

Resurrection plants

For most plant cells, severe desiccation — loss of water, with water potential falling to a very low value — causes either death or permanent injury. However, there are many lower plants (including fungi, lichens, algae and bryophytes) and a small number of vascular plants (about 70 species of ferns and fern allies and 60 species of angiosperms) where vegetative tissues show a remarkable ability to tolerate drying out. The popular name for these desiccation-tolerant species is **resurrection plants**: they can dry out to a crisp and then recover within a few hours of being re-wetted. Figure 5.40 shows a fern with these properties, the rustyback or scaly spleenwort (*Ceterarch officinarum*), which grows on rocks and walls in the UK.

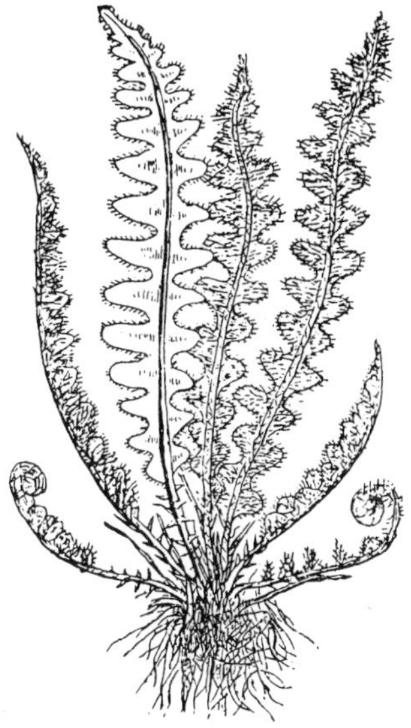

Resurrection plants typically have no special mechanisms for preventing loss of water or increasing its uptake: their main adaptation to water shortage is tolerance of desiccation and they have a specialized cell biochemistry which permits this. At present we have virtually no idea what this specialization involves. In some species, enzymes, cell membranes and photosynthetic pigments are all retained undamaged in the desiccated state. In other species, some pigments and enzymes are lost and membranes may be disorganized but they are rapidly repaired or re-synthesized after hydration.

Figure 5.40 The rustyback fern, *Ceterarch officinarum*, a resurrection plant.

However, desiccation tolerance is also a feature of the majority of reproductive propagules — seeds and spores — and some information about how this is achieved is beginning to emerge.

Seeds

Late in their development, seeds usually dry out to a moisture content of 10% or less, which is a high level of desiccation. Work in the late 1980s has shown that drying out is accompanied by a rise in the level of ABA, the growth regulator that accumulates in leaves when turgor falls (Section 5.5.3) and this brings about changes that are essential for 'safe', i.e. reversible, desiccation. The most important change seems to be activation of a gene that brings about the synthesis of a specific 'dehydration protein', whose chief characteristic is that it is almost entirely hydrophilic (water-loving), with no hydrophobic regions (domains). Maize mutants that cannot synthesize this protein have seeds that do not dry out but germinate on the cob, a phenomenon called *vivipary* that occurs naturally in many alpine and tundra plants. Currently it is not known what this protein does but it is synthesized also in the roots and leaves of young seedlings subjected to slow, partial dehydration and presumably is important for drought survival in these organs too.

5.7.2 Drought survival without desiccation

Among the plants that cannot tolerate extreme desiccation but grow in very dry places, three main survival strategies can be distinguished: water tappers, water savers and water storers.

Water tappers

Some plants grow remarkably long roots and are able to tap water supplies deep underground. Figure 5.41 shows one example which grows on sand dunes. Such 'water tappers' are also characteristic of dried-up river beds in deserts, where the roots of acacia trees may go straight down for 30 m before branching in moister soil. Often, these plants have no special adaptations that reduce water loss by shoots but the part of the root that passes through very dry soil is usually covered by waterproof, corky layers that restrict water loss.

Some epiphytic bromeliads, particularly species of *Tillandsia*, actually tap dew or water vapour from the *air*. The species grow mostly in coastal deserts where there is very little rainfall but frequent fog and dew, and they have special water-absorbing hairs on the leaf surfaces.

Water savers

All vascular land plants are to some extent water savers simply by virtue of having a waterproof cuticle, but the species that grow in very dry habitats, a group mentioned in Chapter 1 and known as **xerophytes**, go a lot further than this. Some of their characteristics are discussed below but first a word of warning. There is a strong temptation to describe every strange feature of xerophytes as an 'adaptation to water shortage', even if the usefulness of the 'adaptation' is obscure or unproven. This kind of thinking has been increasingly challenged and it is now apparent that several classic xerophyte characters, such as tough, hairy leaves or spiny surfaces, may have more or as much to do with deterring grazing animals as with reducing water loss.

Succulent xerophytes are a special group discussed in the next section but three characteristics shared by all non-succulent xerophytes are:

1 *The ability to close stomata rapidly and completely* before cells are damaged by desiccation. Non-xerophytes are quite often incapable of complete stomatal closure and respond sluggishly to a fall in leaf water potential.

2 *The ability to osmoregulate and possession of high osmotic pressures.*

◇ What is the significance of these features in desert conditions?

◆ A high value of Π means that roots can absorb water from dry soils of low water potential and osmoregulation helps to maintain turgor during daily cycles of low water potential (Section 5.6.3).

Despite osmoregulation, however, xerophytes often have rather low turgor pressures, teetering on the edge of wilting. It has been suggested that the high degree of woodiness in many xerophytic leaves and stems, caused by large numbers of sclerenchyma fibres (Chapter 1, Section 1.5), helps to counteract this problem by holding the leaf rigid even when close to wilting. As mentioned above, however, woody leaves may also be an anti-grazer strategy.

3 *Possession of a thick and highly waterproof cuticle*, often covered by waxy or resinous layers. Cuticular resistance to vapour loss, R_c, may be $120 \, \text{s cm}^{-1}$

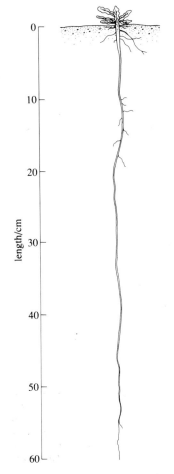

Figure 5.41 The deeply penetrating root system of a one-year-old seedling of cat's ear (*Hypochoeris radicata*) growing on a sand-dune.

length/cm

or more in xerophytes compared with values in the range 10–30 s cm^{-1} among non-xerophytes. This means that, once stomata have closed, xerophytes lose very little water.

A critical factor for xerophytes is the proportion of time that they can maintain open stomata for CO_2 uptake because, if they are to grow and survive, they must maintain a positive carbon balance. Two strategies are related to this problem: opportunistic leaf production and mechanisms that reduce transpiration when stomata are open.

4 *Opportunistic leaf production* is well illustrated by the ocotillo (*Fouquieria splendens*), which grows in the Sonoran desert. This plant produces leaves after rain but sheds them when the water supply runs out, leaving the thickened midribs and petioles as spines (Figure 5.42). It may produce several crops of leaves a year and, during leafless periods, green cells in the stem maintain a low rate of photosynthesis.

(a)

(b)

Figure 5.42 Ocotillo (*Fouquieria splendens*). (a) Young branch which is sprouting new leaves after losing its previous crop. (b) A branch in full leaf after adequate watering. Note spines, which are the midribs and petioles of old primary leaves.

5 *Mechanisms that reduce transpiration losses* are probably more common than (4), however. From Equation 5.8 (Section 5.4.3):

$$J_{wv} = \frac{\Delta \Psi^{wv}}{R_{st} + R_{air}}$$

(in which stomata are assumed to be open and cuticular resistance is, therefore, ignored), you can see that two possibilities exist for reducing J_{wv} apart from increasing R_{st} (i.e. closing stomata): (a) a decrease in $\Delta \Psi^{wv}$ or (b) a substantial increase in boundary layer resistance, R_{air}.

Reductions in $\Delta \Psi^{wv}$ are achieved mainly through modifications that reduce the leaf-to-air temperature gradient (bearing in mind that temperature is one component of $\Delta \Psi^{wv}$ — Section 5.3.4). Small leaves dissipate heat more readily and are more easily cooled by convection currents than large leaves, and are commonly seen among xerophytes. Leaves aligned parallel to the sun's rays (seen in some species of eucalyptus, for example) absorb less radiation and so do leaves with pale or shiny surfaces. Another mechanism that effectively reduces $\Delta \Psi^{wv}$ is the release of volatile oils, producing the aromatic smell typical of eucalyptus and many Mediterranean species. The oils increase the average density of gas in the boundary layer and this slows down the rate at which water vapour diffuses across, just as if air humidity had increased.

Figure 5.43 illustrates one structural modification, very common in xerophytes, that increases boundary layer resistance above stomatal pores — which is where it really matters. Stomata are deeply sunk in pits below the leaf surface, creating 'still-air' conditions above (see Figure 5.19). Many xerophytes have stomata confined to one surface and, when leaf turgor falls sufficiently, the leaf rolls up enclosing stomata in a protected, humid chamber (Chapter 1, Section 1.4.2): Figure 5.44 illustrates this for marram grass, which grows on sand dunes.

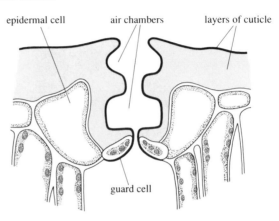

Figure 5.43 A deeply sunken stoma of the desert plant *Dasylirion* sp.

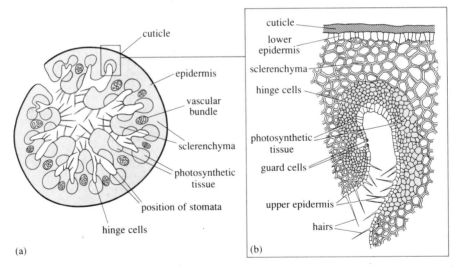

Figure 5.44 (a) Diagrammatic transverse section of a rolled-up leaf of marram grass (*Ammophila arenaria*) showing protected stomata. Turgor loss by the hinge cells cause the leaf to roll up. (b) Part of (a) enlarged to show stomata on the sides of the ridges.

Water storers

In this group are the fleshy plants so typical of both Old and New World deserts — cacti and euphorbias, for example — but it includes also many species of orchids and bromeliads that live as epiphytes on the upper branches of rainforest trees, a kind of aerial desert habitat. The desert species can often survive droughts that last for years, their most obvious characteristic being highly vacuolated water-storage tissue in leaves or stems (Figure 5.45). This is linked with a range of adaptations for conserving water very efficiently indeed and succulents are probably more successful than any other group at surviving water shortage.

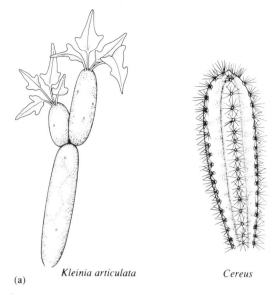

(a) *Kleinia articulata*   *Cereus*

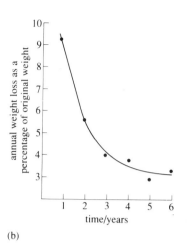

(b)

Figure 5.45 (a) Two typical succulents with water storage tissues and few leaves (*Kleinia articulata*) or leaves that are reduced in size (*Cereus*). (b) The rate of water loss from a stem succulent kept in dry conditions.

The succulent strategy can be summarized as:

1 Extensive, shallow roots absorb surface water efficiently, even heavy dew being utilized, and most of the water taken up is stored.

2 This storage is possible because losses through transpiration are very low: the cuticle is extremely thick and stomata are few in number, often deeply sunk and, above all, open only at night.

◇ Recall the photosynthetic strategy that allows this nocturnal stomatal opening.

◆ Crassulacean acid metabolism (CAM), which was described in Chapter 2, Section 2.5.4.

3 As external water becomes scarcer, roots wither and stomata remain open for shorter and shorter periods, eventually remaining closed altogether. Succulents then survive by recycling respiratory CO_2 and their thick cuticles prevent virtually all loss of water. During one year in dry air, a piece of *Opuntia* cactus lost only 8% of its water and 3% of its dry weight. Similar data are shown in Figure 5.45b. The water-storing strategy of succulents is thus linked to a frugal life-style and, not surprisingly, their growth rates are typically low. This is not usually the case for xerophytes. They do not store water when it is plentiful but expand new leaves, keep stomata open and may transpire quite rapidly. Only when the water supply begins to run out do they restrict stomatal opening and increase osmotic pressures. Desert succulents do not develop high osmotic pressures. The xerophyte strategy can perhaps best be described as 'spend when there is plenty and slowly starve whilst saving water' and it is effective for relatively short (e.g. summer) droughts. For prolonged periods of water shortage lasting more than one season, the succulent strategy of 'store and eke out resources' or the desiccation strategy of resurrection plants appears a better recipe for survival.

5.7.3 Frost drought and freezing tolerance

After a sharp frost you may have noticed the leaves of evergreens, such as rhododendrons, drooping rather dismally. They are experiencing **frost drought**, a situation when frozen soil or blocked, frozen xylem prevents water

221

reaching leaves. Even with stomata closed, leaves may lose sufficient water across the cuticle to reduce turgor but there is another reason why low turgor nearly always accompanies freezing temperatures. Withdrawal of water from living cells into intercellular spaces, a kind of controlled dehydration, is usually essential to survive freezing. It prevents the formation of ice crystals inside cells, which is lethal because the crystals disrupt membranes and cell organization.

Evergreens in cold regions, therefore, often have xerophytic features such as thick cuticles and the ability to close stomata completely, combined with a capacity to initiate and tolerate partial desiccation. Huge ice crystals may form in intercellular spaces and, when they melt, water moves back into the cells. Much the same happens in the twigs of deciduous trees. The ability to tolerate freezing in this way involves changes in cell biochemistry (which are not understood yet) that develop slowly in autumn, a process described as **cold hardening** or **acclimation**. Plants vary greatly in their degree of hardiness and not all can develop freezing resistance (hence all those 'tender' garden plants that die in cold winters). Acclimation is also inhibited by factors, such as high-nitrogen fertilizers or pruning, that promote plant growth. It is of considerable importance in agriculture too because winter cereals that are sown in autumn yield up to 40% more than spring-sown varieties but, because of limited cold hardiness, cannot be sown in large areas of the USA and USSR. If cold hardiness could be increased in winter wheat by chemical treatment or breeding new varieties, it could have a significant impact on world food production.

Summary of Section 5.7

1 Water becomes more difficult to obtain in three situations: when it is saline, in very short supply (drought), or frozen.

2 Halophytes tolerate saline soils because they develop very low water potentials (and high osmotic pressures) through osmoregulation.

3 Plants that survive extreme drought fall into two broad groups: those that tolerate desiccation (resurrection plants), where tissues dry out and then revive; and those that do not tolerate desiccation.

4 Among desiccation-intolerant plants there are three main strategies for drought survival: water tapping, water saving and water storage.

5 Water tappers typically have deeply penetrating root systems that reach water far underground.

6 Water savers (xerophytes) have structural and physiological features that minimize water loss and prevent damage from desiccation when water is in short supply. They also osmoregulate and can absorb water from very dry soils (i.e. soils of low water potential).

7 Water storers (succulents) store water when it is available and then conserve it very efficiently. Because they have Crassulacean acid metabolism (CAM), stomata open only at night when transpiration losses are much lower than in the day.

8 Frozen soil or xylem create frost drought and leafy plants can survive these conditions only if they have become cold hardened. This involves a controlled withdrawal of water from cells into intercellular spaces and is coupled with xerophytic features (such as thick cuticles) that reduce water loss.

Question 17 (*Objectives 5.1 and 5.15*) Which of the following statements are true and which are false? Qualify your answers where necessary.

(a) For both halophytes and xerophytes, the ability to generate high osmotic pressure, tolerate low turgor and restrict water loss is important for survival when water is difficult to obtain.

(b) Increased hairiness in leaves might be expected to correlate with increased drought resistance because of the influence of hairs on boundary layer resistance.

(c) Desert plants usually have deeply penetrating root systems.

(d) Under identical conditions of water shortage (e.g. unwatered pot-plants), a xerophyte is likely to close its stomata sooner than a non-xerophyte.

Question 18 (*Objectives 5.1 and 5.15*) Describe at least two simple experiments or observations that you could carry out to distinguish between a potted halophytic succulent and a potted desert succulent if you did not know which was which.

Question 19 (*Objective 5.15*) List two similarities between xerophytes and freezing-resistant evergreens and two differences that relate to their behaviour or state in conditions of water shortage.

OBJECTIVES FOR CHAPTER 5

Now that you have completed this chapter, you should be able to:

5.1 Define and use, or recognize definitions and applications of, each of the terms printed in **bold** in the text. (*Questions 1, 2, 3, 5, 17 and 18*)

5.2 Describe the pathway of water movement from soil to leaves through a transpiring vascular plant and identify where water movement is by mass flow or by diffusion. (*Question 1*)

5.3 Relate the properties of water to its behaviour in plants and identify the main types of water movement and their location within the plant. (*Question 1*)

5.4 Explain the origin of turgor pressure in plant cells and its significance for water movement into cells. (*Question 4*)

5.5 Explain or calculate the direction and flux of water into, out of, or within plants or other systems in terms of gradients of water potential, plant structure and resistances to water movement. (*Questions 2, 3, 4, 6, 10 and 11*)

5.6 (a) Describe the major factors that affect water potential in a soil–plant–atmosphere system or other system and write equations for Ψ and $\Delta\Psi$ in terms of these factors.

(b) Perform simple calculations using these equations to determine the value of Ψ or $\Delta\Psi$ and predict the direction of water movement in a given situation. (*Questions 4, 5, 6 and 8*)

5.7 Describe simple methods by which Ψ, P and Π can be measured in plants. (*Question 7*)

5.8 Explain how water can move to the tops of tall trees and describe the main causes of, significance of and means of recovery from cavitation and embolism in xylem. (*Questions 5 and 8*)

5.9 Describe (a) the main types of hydraulic resistance and diffusive resistance (to water vapour) in a soil–plant–atmosphere system and (b) the factors that may affect the size of these resistances. (*Questions 9, 10 and 11*)

5.10 Relate the structure and properties of stomata to their function. (*Question 12*)

5.11 Explain how stomatal aperture may be varied and describe the factors that control the size of the aperture. (*Questions 13 and 14*)

5.12 List or recognize the main factors that influence the rate of transpiration and explain how they operate. (*Questions 15 and 16*)

5.13 List and explain the consequences of high hydraulic resistance in roots. (*Question 15*)

5.14 Explain the significance of osmoregulation for plants in dry or saline soils. (*Question 16*)

5.15 Describe or recognize the mechanisms by which plants may survive periods of water shortage and explain the principles on which they operate. (*Questions 17, 18 and 19*)

5.16 Interpret data about plant water relations in ways that are consistent with principles given in this chapter. (*Questions 4, 6, 14 and 16*)

PLANT CELLS AND GROWTH

6.1 INTRODUCTION

In Chapters 2–5, we have considered how plants obtain materials and energy from their environment, and how these materials are distributed to different parts of the plant.

◇ What are the three major items required by plants?

◈ (a) energy, as light, which is used for synthesizing organic compounds
(b) mineral nutrients
(c) water.

Some of the materials absorbed by plants are used to maintain and repair existing tissues. The remainder is available for the construction of new tissues: in other words, for growth. If there is a shortage of any of the above requirements, growth can be severely limited.

◇ How would the growth of a plant be affected by light levels that were always below the light compensation point of the plant?

◈ At the compensation point, gross photosynthesis G and respiration R balance and net photosynthesis $NP = 0$. Below the compensation point, R would be greater, so there would be no net increase in organic material and no growth could occur (see Chapter 2, Section 2.2).

Other examples where environmental conditions can restrict growth include plants growing in soils which are deficient in certain minerals (or where the mineral is unavailable to the plant due to the soil pH). Shortage of a particular mineral nutrient may produce characteristic deficiency symptoms as well as severely restricting the growth of the plant (see Chapter 3, Section 3.1). Water forms over 90% by weight of plant tissues, so plants experiencing drought conditions are unable to grow efficiently. New cells produced by cell division expand by taking up water, and so if water is in short supply this step in the growth of cells will be restricted.

When there is no shortage of any of these items, healthy growth can occur; however, the precise form of this growth is not rigidly fixed and will depend both on internal factors, such as the age of the plant or its stage of development, and on the environment in which it is growing. Under different conditions, growth may be concentrated in different parts of the plant. Stems and roots may increase in length or girth or both, lateral buds may develop, flowers may be formed, or nutrients may simply be transported to storage organs to be used for growth at a later time.

This variation can be seen clearly in regular seasonal changes in growth patterns. A hawthorn bush, for instance, produces new leaves in April, flowers in May and loses its leaves in autumn. However, in beech, only mature trees and bushes actually shed their leaves in autumn. Those in the juvenile stage retain brown dead leaves because the ability to produce a new layer of cells at the base of leaf petioles (the abscission layer) is not yet developed.

The characteristic seasonal patterns shown by many plants can be abolished by keeping them in greenhouses which are uniformly lit and heated throughout the year. There are many commercial uses for manipulation of conditions to produce particular growth patterns, such as obtaining flowering chrysanthemums in midwinter simply by increasing the number of hours of continuous light each day in the greenhouse. This suggests that in plants, as in animals, processes such as the initiation of reproduction and the length of winter dormancy are often controlled by environmental cues.

Another characteristic of plant development is that plants respond to external stimuli by means of growth; examples are the resumption of upward growth by stems that have been fixed in a horizontal position (roots treated similarly grow downwards) and the bending of seedlings towards a source of light. These responses are produced by unequal growth of two opposite sides of the structure resulting in growth in a particular direction and are examples of **tropisms**. Animals respond to such stimuli by simple movement.

Plants also have the capability of regenerating structures when damaged or cut even when they are at a mature stage of their life cycle. In animals, regeneration of amputated parts is usually only possible at the embryo stage and is rare in adult organisms (except in mythological creatures such as the Lernean Hydra, which supposedly sprouted three heads where one had been cut off). Taking cuttings from plants is, however, a common method of propagation, and pinching out the growing point at the apex of a plant causes an increase in the growth of lateral shoots that is remarkably like the response of the mythical Hydra.

This flexibility of plant development is largely due to the presence of immature cells in specialized regions (**meristems**) throughout the plant. Under appropriate conditions, these meristematic cells can undergo repeated cell division and the products of these divisions can give rise to all types of differentiated plant cells.

Figure 6.1 shows the structure of a typical cell from the meristematic tissue at the tip of a growing shoot. The average size of such a cell is about $6 \mu m \times 6 \mu m \times 12 \mu m$, although in other parts of the plant such as the vascular cambium as described in Chapter 1, meristematic cells may be up to $20 \mu m$ in diameter and up to $150 \mu m$ long.

◇ How does the structure of the meristematic cell differ from that of a typical parenchyma cell?

◆ The average size of a meristematic cell is much smaller than that of a typical parenchyma cell and the meristematic cell has neither a large vacuole nor chloroplasts (Chapter 1, Section 1.7).

Figure 6.2 shows meristematic cells drawn to the same scale as a selection of differentiated cell types.

All the cells shown in Figure 6.2 b–e are considerably larger than the meristematic cell from which they have developed, but they vary in how much the cell structure is modified. The structure of some cells and how they are adapted to their function are described in more detail in Chapter 1. Figure 6.2 shows that in addition to cell division two other processes — expansion and structural differentiation — are involved in the development of plant cells.

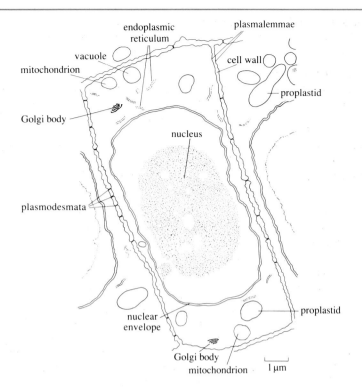

Figure 6.1 A simple diagram of a shoot tip cell.

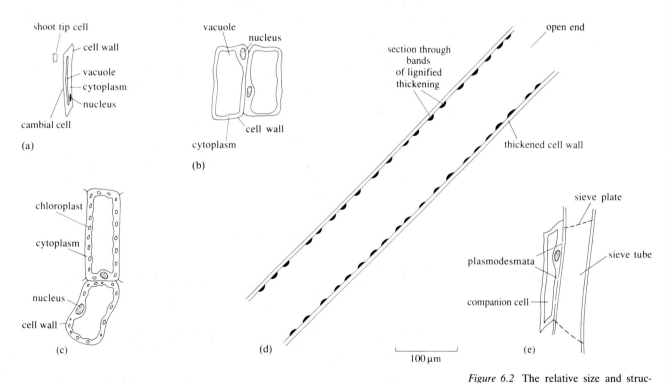

Figure 6.2 The relative size and structure of different types of plant cell. (a) Meristematic cells. (b) Parenchyma. (c) Leaf mesophyll. (d) Xylem. (e) Phloem sieve tube and companion cell.

Plant cells usually develop in three distinct stages: *division* followed by *expansion* followed by *differentiation*. These phases can often be identified in different parts of the developing organ, giving rise to the zonation of the root as shown in Figure 6.3.

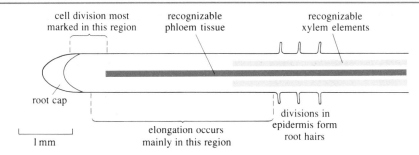

Figure 6.3 The location of the three phases of plant cell development in root tissue. Such maps are constructed from large numbers of sections from which the percentage of cells in mitosis, the cell size and the degree of differentiation are estimated.

There is usually some overlap between these three phases but it is still convenient to discuss them independently before considering how they are coordinated. Section 6.2 will deal with the way in which cell division occurs in meristematic zones. Sections 6.3–6.5 will explain how cell expansion is controlled and Section 6.6 will be concerned with the control of cell differentiation.

The division, expansion and differentiation of plant cells are not only affected by environmental factors but are also subject to control by internal factors, particularly the **plant growth regulators (PGRs)**. These are substances produced in small quantities by plant cells which influence their development or that of other cells in the plant. As well as naturally occurring PGRs, there is a range of artificial and synthetic chemicals which exert various effects on plant tissues, sometimes mimicking the effect of natural PGRs.

PGRs have some similarities in their mode of action to animal hormones, and are sometimes referred to as 'plant hormones'. However, for reasons that will become clear later in this chapter, the use of the term 'hormone' is not always justified. These PGRs can be divided into five groups, each of which appears to be able to influence all three phases of development of a plant cell. However, by studying some of the less complex examples of their action, it is possible to deduce how PGRs may act to coordinate and control the development in plants.

This chapter will consider the role of PGRs in the growth process. However, a word of caution before we start: experiments that investigate the influence of PGRs on plant cells are usually carried out on plants growing under ideal conditions — that is, where no environmental limitations are imposed. While this is necessary to allow investigation of a PGR and its role in coordinating *normal* growth, it does make it difficult to apply the findings from such experiments to plants growing in conditions of environmental stress. Internal factors such as PGRs may provide the fine control and coordination of development under ideal conditions, but the growth pattern may be determined more by environmental factors when these are less than ideal.

6.2 THE LOCATION AND ACTIVITY OF DIVIDING CELLS

Some simple algae consist of filaments or sheets made up of cells that are all similar to one another. There is little division of labour between different cells, and cell division takes place throughout the structure. Growth occurs because the daughter cells enlarge before entering further rounds of division followed by expansion. In the larger algae such as *Laminaria* (oar-weed), as in all the more complex plants, cell division tends to be concentrated into particular meristematic areas.

In higher plants, cells that retain the ability to divide are restricted to certain zones within the plant.

◇ Where are the two main regions of cell division in a young seedling such as the pea shown in Figure 6.4a?

◆ These are the apical meristems at the tip of the root and the shoot (Chapter 1, Section 1.7)

Older plants contain many more than two apical meristems, having a growing point at the tip of each branch as shown for the frond of the alga *Fucus* in Figure 6.4b or the bushy angiosperm in Figure 6.4c.

◇ How does the activity of an apical meristem lead to an increase in the length of a stem or root?

◆ After each division of a meristem cell, the daughter cells enlarge until they attain the same size of the parent cell.

Either one, or sometimes both, of the daughter cells retain the ability to divide (i.e. remain meristematic) so the process can continue resulting in an increase in length of the stem or root. Increase in the width of the growing structure occurs if the meristematic cell divides in a plane at right angles to that which produces increase in length (see Chapter 1, Section 1.7)

Apical meristems are therefore the source of all the cells in the tissues behind them. Indeed whole organs can be generated from isolated apices. When root tips are cut off and placed in a suitable medium, they continue to grow and may form long branched roots. Isolated shoot tips behave differently, because not only is a complete new shoot formed but it may develop its own roots; in effect a complete plant is formed from a small piece of shoot apical tissue. This technique, called **meristem culture**, a special type of tissue culture, is extremely important in horticulture because it can be used to obtain virus-free plants. (Tissue culture is described in more detail later, in Section 6.6.1.) The viruses, which usually multiply within differentiated cells, cannot attack the types of cells found within apices. We examine these and other micropropagation techniques in Section 6.6.

In higher plants there are other zones of cells, apart from those at apical meristems, which retain the ability to divide and can generate new growth.

◇ Recall one example of a meristematic zone within a mature plant.

◆ Cambium cells, which are found between xylem and phloem tissue in stems and roots (Chapter 1, Section 1.8).

Other examples are: cork cambium cells, which produce secondary (suberized) cortex cells in old stems (forming the protective bark), cells that initiate lateral root growth and cells which are stimulated to divide actively when tissue is damaged, so healing wounds.

6.2.1 Investigating cell division in apical meristems

The apices of a variety of lower plants, including certain larger algae, mosses and ferns, have one feature in common: a single, large, pyramid-shaped cell is found at their centre. This is shown diagrammatically in Figure 6.5, and it is thought that division of this one cell gives rise to all other cell types, even in comparatively complex ferns. Single apical cells are not found in gymno-

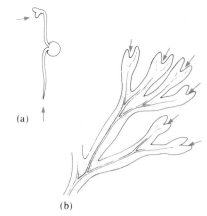

(a)

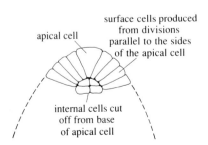

(b)

(c)

Figure 6.4 The position of apical meristems. (a) In a pea seedling. (b) In a *Fucus* frond. (c) In a bushy plant with a well developed root system.

surface cells produced from divisions parallel to the sides of the apical cell

apical cell

internal cells cut off from base of apical cell

Figure 6.5 The structure of the apices of lower plants, showing the characteristically large apical cell.

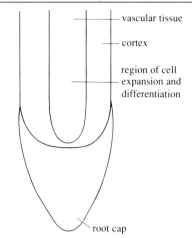

vascular tissue

cortex

region of cell
expansion and
differentiation

root cap

Figure 6.6 Zones of tissue recognizable
in longitudinal sections of root apices.

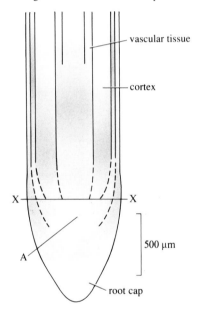

vascular tissue

cortex

X   X

500 μm

A

root cap

Figure 6.7 The distributions of cell divisions in the root tip of an onion. Several
roots were sectioned and the number of
cells in mitosis counted. The extent of
division is indicated by the intensity of
the colour.

sperms and angiosperms, and it appears that more complex structures are
needed to produce the levels of organization seen in these higher plants.

Dividing cells can be seen clearly in apices studied under the light microscope.

◇ Recall how dividing cells can be recognized.

◈ The individual chromosomes can be stained and are then visible as strands.
Also, cells undergoing division (mitosis) have no nuclear envelope.

When the site of cell division is studied, either by counting the number of cells
in mitosis or by examining autoradiographs of sections of roots incubated with
tritiated (^{3}H) thymidine (a radioactive precursor of DNA) the pattern shown
in Figure 6.7 is obtained. (Figure 6.6 shows a low-power map of tissues of the
root apex.)

◇ Recall the name of the region marked 'A' at the centre of the meristem.

◈ This the quiescent centre where the cells divide very rarely (Chapter 1,
Section 1.7.1).

Above and below this region however, there are actively dividing cells. Those
nearer to the tip contribute new cells to the root cap, while those behind the
quiescent centre form parts of the growing root. Removal of the root cap, or
cutting the root at the line XX in Figure 6.7, or destroying the dividing cells
by X-ray irradiation, causes the rate of cell division in the quiescent centre to
increase.

◇ What does this suggest about the role of the quiescent centre?

◈ The quiescent centre appears to function as a 'reserve bank' of cells, which
can undergo mitosis if the normal dividing cells are damaged. The products
of these divisions are then responsible for regenerating the missing
portions of the root and for continuing root growth.

Other techniques can be used to build up an overall picture of the activity of
the meristem. It is possible to tag (label) particular cells so that they can be
recognized in later stages of development. One way of doing this is to use an
inhibitor of cell division called **colchicine**. Cells that are in mitosis when
colchicine is added replicate their DNA but do not divide. They can,
however, enlarge, so forming large cells with twice the normal amount of
DNA. When the inhibitor is removed, these cells can continue dividing by
mitosis (even though they have double the normal amount of DNA), and all
the daughters of tagged cells will have large nuclei like their parents. The fate
of the tagged cells can be followed by allowing growth for various times after
the inhibitor is removed. Figure 6.8 shows the shoot apex of a *Datura* plant
that developed from a seedling treated with colchicine.

Figure 6.8 Sections through the apex of
a lateral bud of a *Datura* seedling. (a)
Untreated apex. (b) Apex treated with
colchicine. The cells with the larger nuclei contain twice as much DNA as the
others.

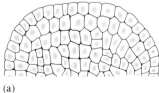

(a)

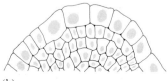

(b)

◇ What can be deduced from this experiment?

◆ The cells affected are only those in the outermost layer. It is impossible to decide whether all the cells were derived from a single tagged cell but it does show that the products of division of cells in the outer layer do not become mixed with cells from other layers.

A similar technique can allow us to draw cell lineage maps of developing roots, such as that shown in Figure 6.9. Each of the small areas enclosed by black lines (e.g. the shaded zone) represents a column of cells that presumably arose by repeated divisions in the same plane resulting in growth in length.

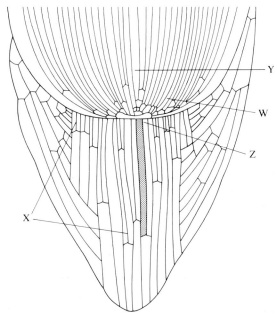

Figure 6.9 Cell lineage map constructed from a longitudinal section of a maize root.

◇ What has occurred at the cells at the positions marked X in Figure 6.9?

◆ A dividing wall has formed parallel to the length of the root. If each daughter cell then continues to divide normally (i.e. at right angles to their length) they will give rise to two files of cells where there was only one before. This will cause an increase in width of the developing root.

Two further points emerge from Figure 6.9. First, if a single file of cells is studied, you should be able to see all the stages in which that type of cell develops. The second point is that the fate of a cell depends very much on its position in the apical meristem.

◇ From your knowledge of root apex structure, predict the fates of the cells in the files marked W, Y and Z in Figure 6.9.

◆ Cells marked W will all develop into the parenchyma cells that are the major component of the cortex, while those in file Y will differentiate into the conducting tissue of the stele; cells marked Z become root cap cells. (See Chapter 1, Sections 1.6.2 and 1.7.1)

Figure 6.10 shows the structure of a typical shoot apical meristem. The shape is different from a root apex, because both leaves and buds are formed at this early stage. The cellular structure of the tip of a *Datura* meristem is shown in Figure 6.8, and the cells further back have similar structures.

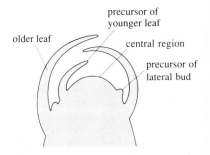

Figure 6.10 Shoot apex of a higher plant.

Files of cells are often less easy to pick out in shoot apex tissue than in root apices, but experiments like those used to obtain Figure 6.8b, confirm that there is zonation of tissues here too. Figure 6.11 shows a simplified arrangement with one layer of surface cells and one column of internal cells, indicating how their division pattern leads to growth in length.

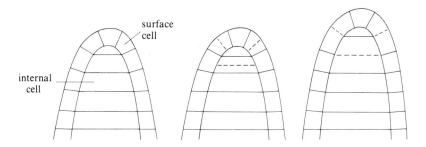

Figure 6.11 How division of surface and internal cells leads to growth in apex length.

◇ What other type of growth takes place in the shoot apex?

◆ Formation of the lateral tissues that develop into leaves and buds.

There are many ways in which leaves can be arranged on a stem, and each species has a characteristic pattern. Leaves may be single or in pairs, and arranged on opposite sides of a stem or in a spiral. When the apical dome above the youngest leaf (Figure 6.10) reaches a certain size, one or more of the cells in the second layer divides at right angles to the normal direction, producing the 'bump' (or primordium) that eventually develops into a leaf. This is shown schematically in Figure 6.12.

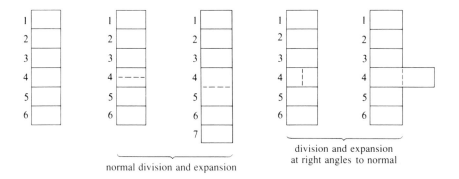

Figure 6.12 The pattern of cell division leading to the initiation of lateral tissues.

The precise arrangement of leaves that is produced in different plants have been explained by proposing the presence of 'anti-leaf initiation factors'

◇ How could inhibitory substances account for the formation of pairs of leaves in which each pair is at right angles to the pairs immediately above and below?

◆ Developing leaves could produce a diffusible inhibitor that prevents the formation of further leaves. Until the apical dome above the youngest leaves reaches a certain size, sufficient amounts of the factor will be present in all cells to inhibit leaf initiation. The inhibitor will be at its lowest level midway between two developing leaves, so that is where it will first fall below a critical level and allow another leaf to form.

Another type of explanation for spatial patterns of leaf primordia is based upon the distribution of mechanical forces. Studies of forces that are generated across the growing apex suggest that the bumps that form leaf primordia occur precisely in those positions which most effectively relieve the stress that builds up across the growing apical dome.

Similar hypotheses involving diffusible substances and/or mechanical forces have been proposed to explain pattern formation in animals*.

Bud formation is initiated in the same way as for leaves, and a bud develops at the base of each leaf soon after the formation of the next leaf above has been initiated. By the time a particular collection of cells ceases to be in the apical region, leaves have developed to the stage where immature vascular tissues can be seen and the apices that will give rise to side shoots are already present (in an active form) in the buds. This contrasts with the development of lateral roots, whose apices are formed by the resumption of activity of potentially meristematic cells well below the surface at a much later stage. (See Chapter 1, Section 1.8.1).

6.2.2 Growth of a leaf

The pattern of development of leaves is very variable, but the increase in size from leaf primordium to mature leaf is mainly due to cell expansion. In *Phaseolus* (bean) leaves, cell division stops when the leaf has attained about 20% of its eventual size, so the remaining 80% increase is due solely to the expansion of cells already present. Other leaf shapes are generated by a combination of meristematic activity at certain zones of the leaf followed by cell expansion. The meristematic zones can be at the margins or near the leaf base, but this meristematic activity typically ceases well before the leaf has gained its mature size, and the expansion of the individual cells of the leaf gives rise to the overall shape.

◇ Recall the three major zones of tissues that can be seen in a transverse section of a leaf.

◆ The upper and lower epidermis (outer cell layer) with a middle layer the mesophyll (consisting of palisade and spongy mesophyll) sandwiched between the epidermal layers. Vascular bundles (veins) run through the mesophyll (Chapter 1, Section 1.4.1).

◇ What would happen if the expansion rate of one tissue exceeded the expansion rate of another tissue in the developing leaf?

◆ If the mesophyll tissue expanded faster, the epidermal tissue would be stretched; if the epidermal tissue expanded faster, the arrangement of cells in the mesophyll would be disrupted.

In fact, the second of those possibilities is what occurs: epidermal cells continue to expand after the mesophyll cells have stopped growing, and this pulls the mesophyll cells apart (particularly those adjacent to the lower epidermis). This creates air spaces which are essential for the efficient diffusion of gases through the mesophyll zone of the leaf.

---

*Pattern formation in animals is discussed in Brian Goodwin (ed.) (1991) *Development*, Hodder & Stoughton Ltd in association with The Open University (S203 *Biology: Form and Function*, Book 5).

6.2.3 Growth in a coleoptile

The patterns of growth in apical meristems discussed earlier show a clearly identifiable zone of cell division (at the apex) and a zone of elongation due to cell expansion behind the apex. This pattern is not universal to structures that grow in length. In monocotyledons, such as grasses or wheat, the first structure to emerge above ground as the seed germinates is not the stem but a protective sheath (called a **coleoptile**) through which the stem will grow, eventually breaking out of the top of the coleoptile.

The following investigation compares the growth of coleoptiles from normal seeds and from seeds which had been given a short-term dose of γ (gamma) radiation when dry and before they germinated.

Figure 6.13 shows that there are essentially two types of cells in the growing coleoptile — the outermost epidermal cells and the inner cortical cells. (In microscope preparations one layer lies over the other so you would not see both in focus at the same time.)

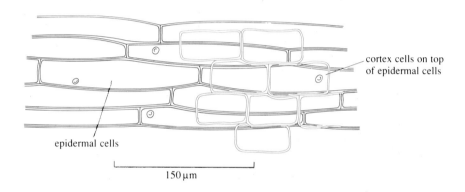

cortex cells on top of epidermal cells

epidermal cells

Figure 6.13 Epidermal and cortical cells of a wheat seedling coleoptile.

150 µm

By viewing strips of coleoptile tissue under the microscope, two measurements can be made: first, the average length of an epidermal cell and a cortical cell can be found by placing the equivalent of a microscopic ruler across the field of view and measuring the length of several cells directly. Dividing this average cell length into the coleoptile length gives a measure of the total number of each type of cell. This estimate is called the **cell index**. Cells in the coleoptile are arranged in files (rather like the arrangement in the root apex) and when they divide they do so in one plane only. This results in an increase in length of the coleoptile; there is no increase in width (i.e. there is no cell division at right angles to the normal plane). For this reason a simple estimate of cell index in coleoptiles provides a good indication of total cell numbers in one file of cells.

Consider the data in Table 1 and answer the questions that follow.

Table 6.1 Length, cell index and average cell length of 7-day wheat coleoptiles. (a) Growing from normal seeds. (b) Growing from seeds previously treated with γ-irradiation.

	Coleoptile length/mm	Cell index epidermis	cortex	Cell length/µm epidermis	cortex
(a) normal	27.0 ± 4.1	41 ± 7	228 ± 22	845 ± 226	129 ± 12
(b) irradiated	11.7 ± 1.6	39 ± 12	85 ± 17	351 ± 120	149 ± 17

Note: values are mean $\pm$ standard deviation which indicates the spread of measurements in a sample.

◇ What effect does γ-irradiation of wheat seeds have on the growth of the cortical cells in the coleoptile?

◆ The cell index of cortical cells where the seeds were previously γ-irradiated is about one third (37%) that for normal cortical cells; there is a similar difference between the lengths of normal and γ-irradiated coleoptiles. Irradiation seems to prevent cortical cells from dividing. However, the average length of each cortical cell is slightly greater in the irradiated coleoptiles.

◇ What effect does γ-irradiation have on the epidermal cells?

◆ Irradiation has no effect on the epidermal cell index, suggesting that epidermal cells don't normally divide. However, there is a marked reduction in the average cell length of epidermal cells of irradiated coleoptile. They are only about 40% of the average length of epidermal cells growing from untreated seeds.

The difference in length between normal and irradiated coleoptiles occurs because the cells of the cortex of irradiated coleoptiles cannot divide. In normal coleoptiles, the cortical cells divide and then expand to about 130 μm in length and then divide again and expand again. This increase in the number and size of cortical cells stretches the outer layer of epidermal cells (which do not divide) so the coleoptile grows in length.

Cortical cells in irradiated coleoptiles are able to expand, but when they reach the length that would prompt division in normal cortical cells (130 μm), they are unable to do so. Instead, they may continue to expand slightly. Consequently the epidermal cells are stretched only to the extent that the original cortical cells can expand. So the irradiated coleoptiles are shorter.

The 'driving force' behind the growth of coleoptiles is thus expansion and division (followed by expansion) of the cortical cells; the surrounding epidermal tissue is stretched passively by the actively growing cortical cells.

Note that this is the reverse of the mechanism seen in growing leaves; as a young leaf expands, the epidermal cells enlarge and in doing so disrupt the mesophyll cell layer below, which causes air spaces to form in the mesophyll tissue.

Summary of Section 6.2

1 Cell division in higher plants is located at specific zones (meristematic regions).

2 Growth in size of a plant is due to expansion of cells produced from meristems.

3 Zones of cell division, expansion and differentiation can be identified in some organs (e.g. root and shoot tips) but cannot always be completely separated.

4 The pattern of cell division in meristematic cells determines the direction of growth.

5 In some structures, the rate of growth of some tissues may be greater than that of adjacent tissues; this can cause the stretching of some tissues (e.g. epidermis of coleoptiles) or the disruption of underlying tissues (as in the mesophyll of leaves).

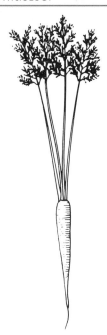

Figure 6.14 A young carrot plant (for Question 1).

Question 1 (*Objective 6.3*) Mark on Figure 6.14 at least three positions where you would expect to find actively dividing cells in the young carrot plant shown.

Question 2 (*Objectives 6.3 and 6.4*) The formation of lateral roots is initiated by cell division in a single layer of cells (pericycle) surrounding the vascular tissue. In which direction would these cells divide (a) as the primary root became thicker and (b) when lateral roots are initiated?

Question 3 (*Objectives 6.3 and 6.4*) Geranium (*Pelargonium*) plants can be propagated from sections of stem, each bearing a single leaf. Where will cell division be initiated when such cuttings are taken?

Question 4 (*Objectives 6.2 and 6.3*) What proportion of cells would you expect to have more than the normal amount of DNA if you examined the apex of a fern plant that had been briefly treated with the drug colchicine?

6.3 PLANT GROWTH REGULATORS AND THE CONTROL OF GROWTH

Questions that arise from the descriptions of the growth of the different plant organs described in Section 6.2 are: how are the processes of cell division and cell expansion controlled? To what extent are the growth processes controlled by internal rather than external factors? The rest of this chapter will be concerned with these issues. First, we will examine two examples where chemicals produced by cells within the plant are necessary to promote the observed growth pattern.

6.3.1 Normal versus dwarf mutants

Gregor Mendel, in his classic experiments on inheritance, crossed 'short' (**dwarf variety**) and 'tall' (normal) pea plants. However, it is only recently that the nature of the difference between the two varieties has been understood. Dwarf varieties exist in a wide range of plants including tomato, maize and wheat. Some types of dwarf plants have an inability to manufacture a particular growth-promoting compound called gibberellic acid, one of a class of PGRs called **gibberellins**. The structure of gibberellic acid or GA_3 is shown in Figure 6.15.

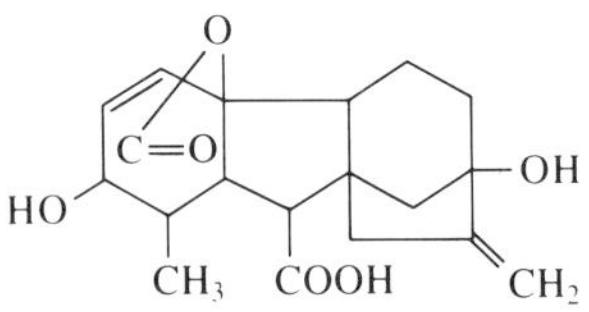

Figure 6.15 Gibberellic acid (=gibberellin A_3 = GA_3).

This group of dwarf mutants have specific blockages in the pathway for the manufacture of gibberellins. When a suitable gibberellin is applied externally to a dwarf plant it can grow to normal size. (Figure 6.16)

◇ Does this prove that this gibberellin is a growth promotor needed for plants to grow normally?

◆ It may be tempting to answer with an unequivocal 'yes' to such a question. But the evidence so far shows only that this gibberellin is required somewhere in the growth controlling system, and not that normal growth is 'caused' by the presence of this gibberellin acting on the cells. Other possibilities might be that this gibberellin is an intermediate in the

dwarf-2 | dwarf-2 +GA | NORMAL +GA | NORMAL

Figure 6.16 The effects of a gibberellin on the growth of a maize dwarf mutant.

biochemical pathway which has a growth promotor as its end product, or that this gibberellin may sensitize cells to another chemical which actually elicits the growth response. Even the demonstration that an externally applied gibberellin restores normal height of a dwarf plant does not necessarily tell us anything about the control of 'normal' growth, since there could be several mechanisms involved any of which could be overridden by the application of externally produced gibberellin

Whatever the complexities of the growth control system that involves gibberellins, there has been widespread commercial use of the knowledge we have of gibberellin synthesis and its effect on growing plants. Reducing the stem length of cereal crops increases the ratio of harvestable to non-harvestable material, by producing the same amount of 'grain' on much shorter stems. This can be achieved either by exploiting the use of dwarfing genes to produce new varieties (as has been successfully achieved for wheat) or by applying substances that block the synthesis of gibberellin in an otherwise normal variety. One such compound is CCC (2-chloroethyl-trimethylammonium chloride). The action of CCC is not species-specific, and it is used in a wide range of crops, to the extent that it is one of the most widely used artificial PGRs. So without knowing the details of the mechanism by which a PGR exerts its effect, it is still possible to exploit commercially what limited information we have.

6.3.2 Amphibious plants and the 'flood response'

Certain amphibious plants, such as *Nymphoides peltata*, the fringed water-lily, will grow surrounded by either air or water. However, the growth rate of the leaf petioles dramatically increases when leaf blades are submerged in water (Figure 6.17). The growth of the stem or petiole continues at a faster rate under water until the flower bud or the leaf lamina breaks the water surface.

A similar effect, called the **flood response**, can be seen in several grassland species. The creeping buttercup (*Ranunculus repens*) is common in meadows which experience occasional flooding. When surrounded by water, the newly produced leaves grow long petioles, resulting in the leaf being taken to the surface of the floodwater and this presumably enables the plant to survive if flooding is prolonged over a period of five days or so (the time it takes to generate new leaves with long petioles). However, there are disadvantages to this strategy: when the flood waters recede, the leaves with long petioles (no longer supported by the buoyancy of the water) collapse and lie flat on the ground. When walking through a recently flooded field, you may have observed such long, straggly stems and petioles.

◇ Suggest a mechanism by which this increase in petiole length could be achieved.

◈ Increase in length could be due to (i) an increase in cell division, followed by expansion, and then further division and expansion; or (ii) by increasing the length of existing cells; or (iii) a combination of the two. The first possibility would be indicated by a higher cell index (a measure of the number of cells present) over the time of the response.

In a study of the cells in such a flooded petiole, it was found that the epidermal cells divided and enlarged in the same way as other cells in the centre of the petiole, and so could be used as an indicator for what happens to the cells across the whole petiole while it grows.

Figure 6.18 shows the range of epidermal cell lengths in different zones along the length of a petiole grown in moist air and when submerged; there is clearly an increase in the average cell length, particularly in the cells that are near to the leaf lamina in the submerged petiole. The cell index shows that there are approximately double the number of cells present in the submerged petiole than in a petiole grown in air. Both petioles grow by cells expanding, then dividing and then repeating this cycle, but in submerged petioles there is more division and greater enlargement, especially near the apex. In different species the relative importance of these two factors can differ. Some rely almost totally on an increase in individual cell length, without increasing cell numbers. In other species there is an increase in cell number due to increased cell division, but all the cells end up the same size as those in petioles grown in air.

What causes cells to respond in this way and generate an increase in growth? Rather surprisingly, it has been shown that submerged stems and petioles accumulate the gas **ethylene** (ethene). All plant cells have an ethylene-producing capacity, manufacturing it from methionine.

◇ Suggest why submergence should lead to an increase of ethylene levels in the tissues of the plant.

◈ Either the ethylene production pathway is stimulated, or that production continues at the same rate but the gas can no longer diffuse out.

The rate of diffusion of ethylene is 10 000 times slower in water than in air, so the consequent 'entrapment' of the gas in submerged tissues would explain

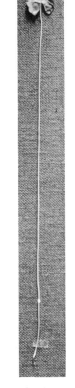

Figure 6.17 Extended petiole demonstrating flood response. Left, grown in air; right, grown under water.

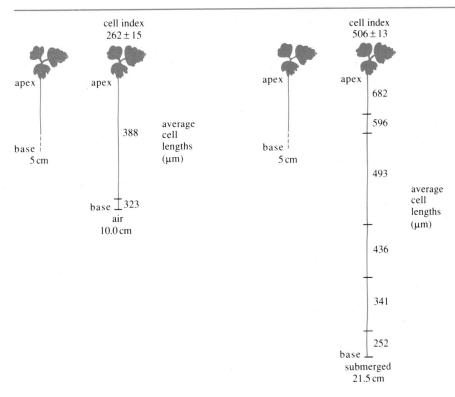

Figure 6.18 Total length, epidermal cell lengths and index of total epidermal cell number in petioles of *Ranunculus repens* which developed in air or submerged.

the rise. In some species (e.g. floating rice) the increase in ethylene level is due to a combination of increase in production and entrapment.

So there is a correlation between ethylene levels and the growth in length in the petiole; but does the increase in ethylene levels *cause* the increased growth rate or does it occur as a *consequence* of increased growth?

Figure 6.19 shows the result of an experiment in which the length of cells and the cell index in petioles of *Ranunculus repens* growing submerged and in ethylene-enriched air are compared.

◇ Compare the data in Figure 6.19 with those in Figure 6.18. Which set of data, petioles grown in air or submerged, more closely matches the data obtained from petioles grown in ethylene-enriched air?

◆ The cell indices of the submerged petioles and the ethylene-treated petioles are similar (about twice that of the petioles grown in air) and the pattern of increasing cell size nearer the lamina, seen in submerged petioles, is also observed in the ethylene-treated petioles.

The overwhelming evidence from this, and from many other studies on other species that show the flood response, is that retention of high ethylene levels causes an increase in cell numbers and/or cell elongation resulting in the faster growth rate in petioles, and sometimes in stems. However, care needs to be taken when interpreting the role of ethylene in the growth-promoting process. For instance, while high levels of ethylene do induce faster growth rates in these tissues, this can occur only if gibberellins or auxins (another group of PGRs) are also present. Ethylene may be acting to increase the sensitivity of the tissue to these other PGRs. More baffling still is that ethylene has quite different effects on the majority of terrestrial plants — a point that will be taken up in the review of the range of PGRs in the next section.

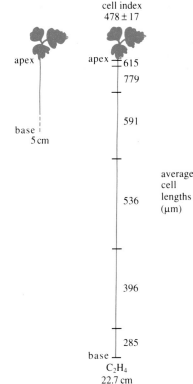

Figure 6.19 Total length, epidermal cell length and index of cell numbers in a petiole of *R. repens* grown in air enriched with ethylene.

6.3.3 The range of PGRs

The examples discussed above illustrated only two of a vast range of compounds produced by plant cells which have an effect on the growth and development of plants. One indication of this range is that over 70 compounds classed as gibberellins have now been isolated (sensibly designated GA_1 to GA_{70} rather than named individually). Fortunately, it has been shown recently that only a proportion of these are active forms, the remainder probably being intermediates in the pathways of the active ones.

An important feature of the role of PGRs on plant growth is that the same chemical (at identical concentrations) can have quite different effects on two different tissues; moreover, different concentrations of the same chemical can elicit different responses in the same tissue. Combinations of different PGRs may work synergistically or antagonistically or produce quite different effects than when applied separately on developing tissue. Given such a bewildering array of chemicals and their effects it is difficult to know where to begin any classification of PGRs.

It is usual to divide PGRs into five main groups according to their chemical nature rather than the responses they elicit. These are the **gibberellins**, **auxins**, **cytokinins** and two 'single member groups' — **abscisic acid** and **ethylene**. Two other groups of chemicals have recently been found to have important effects on plant development: these are the polyamines and brassins (steroid growth promoters). Knowledge of these two groups is only just emerging, but it is likely that future PGR classifications will include them alongside the five traditional groups.

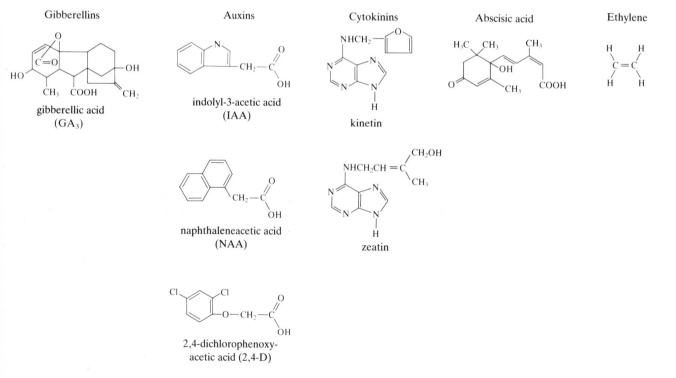

The extraction and identification of most PGRs is a tedious and time consuming business. Many of the problems result from the fact that plant tissues contain extremely small amounts of these chemicals. Typically, the concentration of one of the most widely studied PGRs, an auxin (IAA) is between 1 μg and 100 μg per kg of plant material. The lower figure represents a concentration of 1 part in 10^9, which indicates the difficulty of obtaining these chemicals in any quantity. In fact, representatives of the first three groups of PGRs to be discovered (auxins, gibberellins and cytokinins) were first purified from non-plant sources (respectively, from human urine, from a medium in which certain fungi had been grown and from a rather old preparation of yeast DNA).

The identification of the active ingredient in a complex mixture, whatever its source, is a chancy business. There may be enough hints during the procedure to support an informed guess about the nature of the unknown substance. Alternatively, as happened during the separation of the auxin from urine, much time may be wasted studying the wrong component.

Auxins

An auxin was the first PGR to be isolated, resulting from studies following observations by Charles and Francis Darwin, on the bending of coleoptiles towards light(phototropism). Some fifty years after the original observations by Darwin, it was shown that when the auxin indolyl-3-acetic acid (IAA) was applied to one side of the coleoptile using small blocks of a gel-like substance (agar) infused with IAA, a curved growth occurred, apparently mimicking the tropic response (see Figure 6.20). However, it was only recently that IAA was found to be present naturally in coleoptiles.

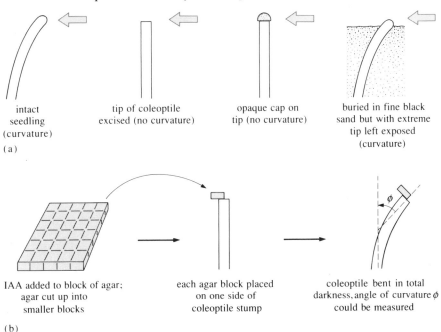

Figure 6.20 (a) The Darwins' observations which suggested that the tip of the coleoptile acts as the 'site of perception' for unidirectional light (arrow) and the growing region below the tip acts as the 'effector'. (b) 1926 experiment showing that an uneven supply of the auxin IAA to excised coleoptiles could produce curved growth apparently similar to phototropism.

We consider this example in more detail in a case study in Section 6.5. Two other naturally occurring auxins have been isolated from plant tissue: 4-chloro-indolyl-3- acetic acid (4-chloroIAA) and phenylacetic acid (PAA). The first of these is found in only a few plant tissues. The second, while present in a wide variety of tissues, is much less active than IAA. This has led many workers to consider that IAA is the only important naturally occurring auxin. It is implicated in many development processes in plants, including promoting growth in length of shoots and roots.

The increase in length of shoots and roots in the presence of auxin seems to occur mainly by increasing cell size (by facilitating cell expansion), although this may be accompanied by increased cell division. Other gross effects that were thought to be due to the presence of IAA include apical dominance, where the apical bud exerts an inhibitory effect on the development of lateral buds; the inhibitory effect can sometimes be mimicked by replacing the apical bud by agar blocks containing IAA.

◇ Does this observation show that IAA is the inhibitor responsible?

◢ No. Although it provides supporting evidence, other criteria have to be met: (i) naturally occurring levels of IAA within the plant must be capable of eliciting the response (and these levels are usually much lower than the levels applied via agar blocks in such experiments); (ii) IAA produced naturally within the plant must be shown to be distributed to the lateral buds and exert an inhibitory effect there.

Confusingly, some investigations show more IAA to be present in growing lateral buds than in non-growing ones, and application of IAA by agar blocks sometimes promotes their growth!

Auxins are synthesized mainly in the shoot apex, young leaves and buds, and are transported downwards in stems. They promote root initiation and promote the formation of lateral and adventitious roots from primordia (often located between nodes on woody stems such as willow). Growth from such root primordia is the basis of asexual reproduction in many species, and stem cuttings can be induced to generate roots by applying IAA.

An artificial auxin (naphthaleneacetic acid, NAA) has a commercial use as a 'hormone rooting powder', promoting the same sort of root initiation as natural IAA produced within the plant. Another artificial auxin 2,4-dichlorophenoxyacetic acid (2,4-D) is used as a selective weedkiller, causing little effect in monocot plants (such as grasses) but disrupting the growth pattern of the dicot weeds; whether this is due to differential effects of the same concentration of the PGR acting on different tissues, or whether the dicots take up a greater proportion of the substance through their larger leaf surface area, is open to conjecture. Figure 6.21 shows the chemical structure of IAA and these two artificial auxins.

◇ What are the common features of the structures of artificial auxins and IAA?

◢ Both NAA and 2,4-D have a short side chain with a carboxylic acid group (—COOH) joined to a six-carbon ring. IAA has a carboxyl group attached to an indole ring; however, the shapes of the molecules suggest that all three might interact with the same 'auxin receptor'.

Artificial auxins can be used in experiments where application of natural auxins is difficult (partly because natural auxins are so difficult to isolate). Their main experimental advantage is that they act as a continuing auxin supply because, unlike IAA, they are destroyed only very slowly by plant tissues.

◇ What disadvantage is there of using such artificial auxins in such experiments?

◢ However much their effects seem to mimic IAA, it is impossible to be sure that they are acting in an identical way to natural auxins produced within the plant.

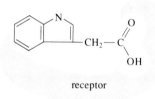

receptor

indolyl-3-acetic acid
(IAA)

(a)

naphthaleneacetic acid
(NAA)

2,4-dichlorophenoxy-
acetic acid (2,4-D)

(b)

Figure 6.21 (a) Naturally occurring auxin. (b) Artificial auxins.

A further rich source of IAA is crown galls of dicot plants. The gall is induced by an invading bacterium (called an agrobacterium); this bacterium can apparently donate a section of its DNA into the host cells of the plant. This section of DNA contains the genetic coding for IAA synthesis — hence the high production of IAA in the infected area. This exciting discovery opens up a new avenue of research, because these genes can now be inserted by genetic engineering techniques into specific tissues, and this makes it possible to investigate the effects of auxins produced by the plant tissue rather than relying on externally applied artificial auxins. However, most experiments to date have relied on the application of artificial auxins. We consider such experiments and their interpretation in Section 6.5.

Ethylene (ethene)

The fact that ethylene (a simple hydrocarbon gas, C_2H_4) could affect plant growth and development was recognized long before it was discovered that plants produced ethylene within their tissues. Town gas contained a proportion of ethylene, and gas leaks could often be detected by observing the odd growth pattern shown by some plants or trees affected by such leaks: leaf fall in trees, reduced growth of stems and sometimes a curious twisting of the leaf petiole (epinasty) caused by accelerated growth on one side of the petiole compared to the other.

ethylene

Ethylene is also implicated in the onset of fruit ripening — one ripe tomato added to a set of unripe ones will induce ripening even in the dark. One of the earliest recorded investigations by a botanist for a commercial company concerned the problem of bananas ripening too quickly while in transit from Jamaica to Europe: the botanist recommended that the bananas be kept in a separate hold to the cargo of oranges, since "some emanation from the oranges was inducing early ripening of the bananas". The 'emanation' eventually turned out to be ethylene, probably produced by fungal infection on the surface of the oranges since oranges do not produce large enough quantities of ethylene to induce this effect on their own. Other effects of ethylene include the inhibition of elongation of roots and stems (except for those amphibious plants discussed earlier) but enhanced growth in width.

An unexpected effect of ethylene is to promote female flower production in cucumber plants (an antagonist to ethylene action, silver nitrate, has the reverse effect, inducing male flowers). The advantage of being able to manipulate the proportion of female:male flowers was quickly seized on by commercial growers, and ethylene is widely used in cucumber greenhouses to increase the yield by inducing more female flowers, which are the only ones to develop the cucumber fruit. Using silver ions as an antagonist to ethylene helps to limit the damage to roots of commercially produced bulbs: damaged roots produce ethylene which promotes further breakdown of tissue and even more ethylene production (an autocatalytic effect) and this can be halted by treatment with silver ions.

Ethylene can also be used to induce synchronous flowering in some bromeliads, such as pineapple. The traditional way of encouraging synchronous flowering of the pineapple crop was to light fires around the crop fields — smoke contains ethylene. Ethylene also accelerates senescence of some flowers, hence 'no smoking' at flower shows!

As with auxin, the main sites of synthesis of ethylene are meristems and young growing tissue. Ethylene usually exerts its effect in the presence of auxin, and sometimes has similar effects on the growth of tissues to those shown by adding IAA. Artificially high levels of auxin can promote ethylene production, and this causes problems when trying to interpret results of experiments where auxin is applied externally. Ethylene may require the presence of auxin, or may be acting to sensitize the tissue to the auxin already present.

Gibberellins

As shown by the growth of dwarf and normal varieties (in Section 6.3.2), extremely small amounts of gibberellin can have marked effects on the growth pattern of whole plants. This is in contrast to the natural auxins which affect growth at the cell or tissue level.

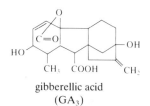

gibberellic acid
(GA$_3$)

The discovery of gibberellins followed investigations of a disease of rice seedlings, where the growth of the seedling is promoted to such an extent they keel over and die; the disease was aptly named 'bakanae' (literally, 'foolish seedling') and was found to be caused by infestation by the fungus *Gibberella fujikuroi*, which produced gibberellic acid (GA$_3$) in small quantities but sufficient to induce the dramatic increase in growth rate in the seedlings. Gibberellins are synthesized in seeds, young leaves and young apical tissues, particularly roots. In experiments where roots are progressively cut back, the amount of gibberellin in the shoots drops, suggesting that gibberellin synthesized in the roots is supplied to other organs. Gibberellin is probably transported upwards in the xylem.

The range of effects of gibberellins includes promotion of germination in seeds and of flowering (these topics are dealt with in more detail in Chapter 7). Gibberellins are used commercially to produce seedless fruit, since they promote parthenocarpic development (the fruit develops without prior fertilization of the ovule), and to alter the shape of bunches of grapes, so that the grapes are spaced further apart, which makes them less susceptible to fungal infections.

Cytokinins

These act primarily on cell division, although they may affect cell expansion as well in certain tissues. They are also implicated in the delay of senescence in leaves. As with auxins, the isolation of the first active compound was from a non-plant source and this compound has never been found in plant tissues (sources rich in this first cytokinin included yeast DNA and herring sperm DNA!).

zeatin

However, active cytokinins have subsequently been isolated from a range of plant tissues particularly 'nursery' tissues surrounding young growing structures. Zeatin is a cytokinin obtained from kernels of barley seeds. Several other cytokinins have been extracted from coconut milk which is a liquid storage tissue surrounding the embryo plant within the coconut seed. In growing plants, cytokinins are produced in growing root tips, and are transported upwards (in either the xylem or phloem or both) to the stem and leaves. Here they accumulate, although if the supply is cut off these young tissues can synthesize at least some of the cytokinins they require.

Cytokinins are usually derivatives of purines.

◇What is the role of purines in cell nuclei?

◆They are one of the two types of base in nucleic acids (the other type being pyrimidines).

This has led to speculation that cytokinins are derived from RNA, possibly fragments of transfer RNA. Radioactive labelling of possible precursors of cytokinins, however, suggests that the main source is from the metabolic pathway which produces adenine rather than from breakdown products of nucleic acids.

◇ Mutant strains of plants (e.g. dwarf plants) were used to investigate the role of gibberellins. Would it be possible to use a similar approach to investigate the role of cytokinins?

◆ No, not unless a mutant could be identified which could not convert adenine to cytokinin; any block in the production of adenine itself would have very serious consequences. It would be impossible for cells to produce nucleic acids or ATP. Also, mutants deficient in cytokinin, where cell division could not be stimulated, would be unlikely to survive at all.

Abscisic acid (ABA)

Abscisic acid is a growth inhibitor. Its discovery followed investigations by two independent research teams into dormancy and leaf fall (abscission). Only later was the active compound involved shown to be the same. ABA levels are high in dormant buds and in fruits just before they fall. Early experiments implicated ABA as the promoter for the formation of the abscission layer (the layer of cells that forms a 'weak' zone on a leaf petiole or fruit stalk resulting in eventual fall). However, more recent evidence questions this role (see Chapter 7), and so we are left with a PGR with a singularly inappropriate name!

Since its discovery, ABA has also been implicated in many other processes, including the control of stomata: ABA levels rise prior to active stomatal closure (see Chapter 5, Section 5.5.3). This role of ABA has led to the search for an artificial analogue which could be used as an 'anti-transpirant'. The difficulty here is that artificial compounds suffer the same high turnover rate as natural ABA, but when high levels are administered there is damage to the leaf tissue.

abscisic acid (ABA)

6.3.4 Mechanisms of PGR action

This brief summary of the range of PGRs indicates the extraordinary complexity of their effects on plant growth and development.

It would be useful to have a clear idea of the mechanism of action of these different PGRs, which would provide a common framework for explaining their varied effects. With animal hormones, we can make sense of the wide variety of effects shown by different hormones because we understand the mechanism of action at the cell level[*]. The concept of the hormone **receptor**, either membrane-bound or within the cytosol, provides a common basis for explaining the mechanism involved.

Similar attempts have been made to explain the mechanism of PGRs using the receptor concept. Indeed, this has been implied in some of the descriptions of PGRs given above. Artificially produced compounds seem to behave

[*] Animal hormones and receptors are discussed in Chapters 1–3 of Michael Stewart (ed.) (1991) *Animal Physiology*, Hodder & Stoughton Ltd in association with The Open University (S203 *Biology: Form and Function*, Book 3).

in a similar way to naturally produced PGRs because there is a similarity in the overall shape of the molecule, which would allow it to fit a receptor molecule to form the equivalent of a hormone–receptor complex.

The hunt for auxin receptors has been a most frustrating search. As mentioned above, auxin levels are extremely low (about 1 part in 10^9), but auxin binding sites seem to occur everywhere — at the membrane and within the cytosol. One frustrated worker, on finding that auxin binds to a component in bovine serum, wondered whether to suggest that he had discovered a putative auxin receptor there! This sort of finding illustrates a further problem: because the auxin binds to a particular substance, it doesn't necessarily mean that a hormone–receptor complex has been identified. The auxin may just be held in a bound, inactive form.

The structure of gibberellins shows some similarity to certain animal steroids or neurotransmitters; the search for gibberellin receptors has concentrated on soluble cytoplasmic receptors rather than membrane bound receptors. Unfortunately, unlike studies on the binding of animal steroids to cytoplasmic receptors, the identification of possible gibberellin receptors is proving extremely difficult, not least because binding experiments *in vitro* cannot yet be repeated *in vivo*.

Nevertheless, there is a general assumption in current (1990) research work that plant cells do have receptors for PGRs, and that the first step in a physiological response involving a PGR is the formation of a PGR–receptor complex.

Our review of PGRs indicates the very wide range of effects they have on a variety of tissues. Rather than trying to cover this range further, in the next two sections we will concentrate on one feature of growth: the expansion of individual cells and the role of a particular auxin (IAA) in facilitating and controlling this cell expansion.

Summary of Section 6.3

1 PGRs are implicated in the control of a wide range of growth processes. Two examples are gibberellin, which can restore normal growth in some dwarf varieties of plants, and ethylene which is responsible for causing rapid elongation of submerged petioles in some amphibious plants (the flood response).

2 There are five main classes of PGR — gibberellins, auxins, cytokinins, abscisic acid and ethylene. These PGRs can exert a wide range of effects. Some exert quite different effects in different plants, and sometimes the same PGR can elicit different effects on different tissues within the same plant. Effects of one PGR can be modified by the presence of other PGRs, providing an extremely complex set of different interactions in different circumstances.

3 Artificial PGRs can mimic the effect of natural PGRs produced by plant cells. Artificial PGRs typically have a similar structure to natural counterparts. This implies that receptor molecules are involved in the mechanism of PGR action. PGRs are known to bind to protein components in or on the surface of cells, but it is difficult to show that such binding is related to any physiological response of the cell.

4 Despite limited understanding of PGR action, there is widespread commercial application of PGRs and their antagonists, resulting in successful increase in yield or reduction of damage to crops and other plants.

Question 5 (*Objective 6.2*) When sections of pea stems are slit along half their length and placed in water, the split ends curve outwards as shown in Figure 6.22. When they are placed in water containing $30\,\mu\text{mol}\,l^{-1}$ IAA they curve inwards as shown. What can you deduce about the response of cells in different parts of the stem to the application of IAA?

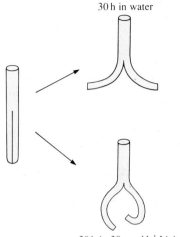

30 h in water

30 h in $30\,\mu\text{mol}\,l^{-1}$ IAA

Figure 6.22 The response of split pea stems to auxin. (For Question 5)

Question 6 (*Objective 6.2*) Discs removed from radish leaves turn yellow within three days if incubated in water, but remain green if incubated in a $30\,\mu\text{M}$ solution of kinetin, an artificial cytokinin. What further experiments would you carry out to test whether the yellowing of leaves is due to a fall in cytokinin levels?

Question 7 (*Objectives 6.1, 6.5 and 6.6*) Give one example of a commercial application of each of (a) an artificial auxin, (b) ethylene, (c) an antagonist to ethylene and (d) a suppressor of gibberellin production.

6.4 GROWTH IN A SINGLE CELL

In Section 6.2 it was shown that the growth of a structure such as coleoptile or leaf or leaf petiole is ultimately dependent upon certain cells which expand following cell division. But what are the forces that act upon these cells, and why should these forces result in cell elongation rather than, for instance, increase in diameter of the cell? What causes these cells eventually to stop elongating and so prevents growth rates similar to Jack's beanstalk? Here we look at the forces responsible for generating cell expansion and the way expansion is regulated.

6.4.1 Forces generating cell expansion

One way of analysing the events that occur during cell expansion is to consider the forces that cause an increase in the cell volume due to uptake of water.

◇ Recall the expression that relates water flux to the water status of the cell.

◆ $J_v = L_p \times \Delta\Psi$ (6.1)

where L_p is the hydraulic conductance of the cell membrane and $\Delta\Psi$ is the difference in water potential between the inside and outside of the cell, $\Delta\Psi = \Psi_{\text{ext}} - \Psi_{\text{int}}$ (Chapter 5, Section 5.4).

Recall that the value of Ψ for pure water is 0, so if a cell is surrounded by pure water then $\Psi_{ext} = 0$ and

$$J_v = L_p \times (-\Psi_{int}) \tag{6.2}$$

This relationship is shown graphically in Figure 6.23a.

An alternative approach when analysing the forces involved is to consider the balance of 'push' forces on the cell wall and the 'give' or **extensibility** of the wall. The push forces are generated by turgor pressure P. For cell expansion to occur, P has to be above a critical value P_{crit}, and the push force involved is represented by $(P - P_{crit})$. The rate of growth G is dependent upon this push force:

$$G \propto (P - P_{crit})$$

$$G = E_x(P - P_{crit}) \tag{6.3}$$

where the proportionality factor E_x represents the yielding property of the cell wall — the more extensible the wall the more the cell will enlarge under the hydrostatic pressure caused by $(P - P_{crit})$.

Note that the maximum P for a cell will be equal to the osmotic pressure Π and that $\Psi = P - \Pi$. So as P approaches its maximum value, Ψ_{int} approaches zero. We can represent Equation 6.3 graphically using the same axes as for Equation 6.2. This is shown in Figure 6.23b. Note that we have placed the horizontal axis in the same orientation as for Figure 6.23a; Ψ_{int} becomes more negative towards the right-hand side of the graph.

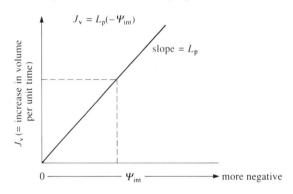

(a)

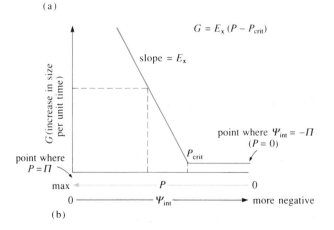

(b)

Figure 6.23 The relationship of factors that cause cell enlargement. (a) Using the water relations equation. (b) Using the growth equation.

For both Figure 6.23a and Figure 6.23b, the horizontal axis is the same, Ψ_{int}. In Figure 6.23a the vertical axis indicates the rate of uptake of water (the amount of water taken in per unit time); in Figure 6.23b the vertical axis indicates the increase in size of the cell per unit time (i.e. growth rate). These two scales are really measuring the same thing, because the increase in volume of the cell will match the volume of water taken up (a safe assumption providing the cell is not manufacturing or taking up large quantities of other materials). So for every cell, a particular value of Ψ_{int} will be associated with a corresponding rate of expansion whichever equation we use (as shown by the broken lines on each graph). We can thus combine the information of the two graphs. This is shown in Figure 6.24.

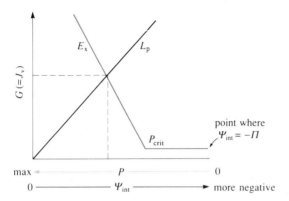

Figure 6.24 Combining the two graphs in Figure 6.23a and b.

Figure 6.24 shows that the rate of expansion of a cell is determined by a combination of the following factors: (i) turgor pressure above a critical value, $P - P_{crit}$; (ii) the hydraulic conductance L_p of the cell membrane; and (iii) the cell wall extensibility E_x. Altering the values of any of these factors could theoretically change the rate of expansion, as shown in Figure 6.25.

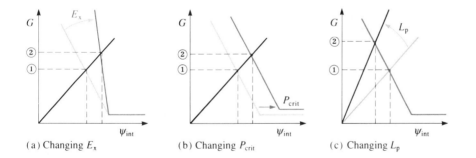

(a) Changing E_x (b) Changing P_{crit} (c) Changing L_p

Figure 6.25 The effects on the rate of cell enlargement of each of three factors.

Figure 6.25a shows that if E_x is increased, the rate of cell expansion would increase.

◇ What happens to Ψ_{int} when E_x rises?

◈ Ψ_{int} falls (becomes more negative). More water will be taken up by the cell as it enlarges.

Similarly, increasing the value of $(P - P_{crit})$ by reducing P_{crit} would also result in the cell expanding more; this is also accompanied a lowering of Ψ_{int} (Figure 6.25b).

◇ From Figure 6.25c, what change in hydraulic conductance L_p would result in increased cell expansion?

◈ L_p must rise (i.e. the slope of the line must become steeper) for the rate of expansion to increase.

Note, however, that in this last case Ψ_{int} increases (moves towards zero) and turgor P also increases. For cells already in a turgid state, when Ψ_{int} is already close to zero, increasing the hydraulic conductance will not actually make much difference to the flux of water entering the cell; P is already close to maximum, preventing further uptake of water.

So which of these factors is the most important 'control point' for determining the rate of growth of the cell?

To date, there is very little evidence that the hydraulic conductance of the cell membrane can be changed enough to alter water flux significantly and affect growth. Experiments where water is pumped into cells using probes (similar to the pressure probes shown in Chapter 5, Figure 5.10) to simulate more water uptake by the cell, as if the hydraulic conductance had altered, did not show the expected change in G or Ψ_{int}. So when water is in plentiful supply, hydraulic conductance does not seem to limit or control cell expansion. However, for cells growing in plants where water is in short supply, hydraulic conductance may play an important role.

Lowering P_{crit} (so that $P - P_{crit}$ is larger) would have the effect of generating a larger push force on the cell wall to start the process of cell enlargement; this would also lower the value of Ψ_{int} (Figure 6.25b). Growing cells do maintain a lower water potential than other cells in the plant; thus water moves into these cells and turgor is maintained even when there is a shortage of water elsewhere in the plant. Lowering the value of P_{crit} would be one way that cell expansion could be achieved, and there is evidence that this actually happens in the giant cells of the alga *Nitella*, but there is little convincing evidence for this in higher plants at present.

A fall in turgor, P, would decrease $P - P_{crit}$, and this is probably the main factor causing fluctuations in growth rate for plants growing in air, the real world for most. Even though turgor is maintained in growing tissues, it *does* fall under conditions of water stress, and this affects the growth rate, G. However, there is no evidence that PGRs affect P directly.

We do have considerable evidence that the remaining factor, cell wall extensibility E_x, does change during the development of a cell. E_x is high in cells while they enlarge and then falls when they stop growing, when the cell walls become thickened. A further important feature of cell wall extensibility is that some PGRs, particularly the auxin IAA, can increase its value.

◇ The main component of cell walls is cellulose. Can the microfibrils of cellulose be stretched?

◆ No. The tensile strength of a cellulose microfibril is very similar to that of steel (Chapter 1, Section 1.3.1.)

So how can the cell wall be extensible when the main component of it is unstretchable? How can the extensibility be altered at different times of the growth cycle of the cell and how can a PGR such as IAA effect such changes? We explore these questions in the next two sections.

6.4.2 The cell wall and cell expansion

Figure 6.26 shows the changes in length that occur when a piece of pea stem is immersed first in a concentrated solution of mannitol then in distilled water. (Remember that it is important to use a substance, such as mannitol, that is not taken up by plant cells when studying water relations — see Chapter 5, Section 5.3.1).

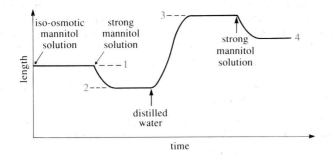

Figure 6.26 The changes in length observed when a piece of pea stem is immersed in distilled water and in mannitol solution.

◇ What will happen to the cells when they are placed in concentrated mannitol solution?

◈ They will lose water by osmosis and become plasmolysed.

When the length of the strip is measured again, it will have decreased (1→2 in Figure 6.26).

◇ What is the force whose removal has led to the contraction?

◈ Turgor pressure. Remember in plasmolysed cells $P = 0$ (Chapter 5, Section 5.3.1)

Now consider what happens when the plasmolysed pea stem is placed in distilled water. The water potential of pure water (zero) is greater than that of plant cells (about -6×10^{-5} Pa for pea stem cells). Water therefore flows into the cells, and the tissue expands. Water continues to flow into the cells down a gradient of water potential after they have regained their original size, and further expansion is observed. The total expansion is represented by 2→3 in Figure 6.26. If the stem is returned to strong mannitol solution, it again contracts by the same amount as during the previous immersion in strong solution (3→4 = 1→2). It does not return to its original length before it was placed in distilled water (at 2).

This demonstrates that there are two components of the expansion of the cell. The first is an **elastic expansion**, represented by the length change 2→1 (4→3 would indicate the same change); this increase in length is reversed when the turgor pressure within the cell is released, and the cell wall is behaving rather like a piece of elastic.

The second component is non-reversible, where the wall has been stretched and does not regain its original length when the force causing the stretching is removed. In this case the wall is behaving more like a piece of chewing gum which when deformed by stretching retains the new shape even when the force causing the stretching has been removed. This is termed **plastic expansion**.

◇ Which length change in Figure 6.26 represents the non-reversible increase in length?

◈ 1→3 (= 2→4)

An essential point to note about the cell wall structure is that the cellulose microfibrils themselves cannot be stretched, and therefore if the cell is to expand, the microfibrils must slip past each other as shown in Figure 6.27.

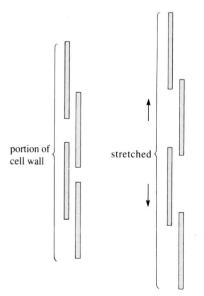

portion of cell wall

stretched

Figure 6.27 Relative movement of cellulose microfibrils as a portion of a cell wall is stretched.

Several models have been proposed to explain the nature of reversible and non-reversible expansion of cell walls. One rather 'mechanistic' model is illustrated in Figure 6.28. Imagine two rods (representing cellulose microfibrils) with hooks at intervals along their length. The rods are linked by rubber bands; as the rods are pulled apart and slide past each other, the rubber bands stretch until they can stretch no more (Figure 6.28b). However, if one end of each cross-linking band is detached and rejoined to a hook further along the rod (Figure 6.28c), the rods can be pulled apart again (Figure 6.28d).

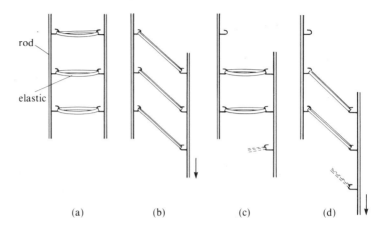

Figure 6.28 One proposed model of reversible and irreversible expansion of cell walls.

(a)   (b)   (c)   (d)

◇ Which parts of Figure 6.28 represent the elastic expansion shown by the change 2→1 in Figure 6.26?

◆ The change from (a) to (b) in Figure 6.28.

Plastic expansion (1→3 in Figure 6.26) must involve the breaking of bonds between components of the cell wall, represented by the change from (b) to (c). This explains why plastic expansion is irreversible.

Other models place more of an emphasis on the gel matrix that surrounds the cellulose microfibrils that make up the cell wall. The nature of this matrix (acting as a glue to the microfibril rods) may hold the key to explaining why the cell wall can become more flexible or less flexible at different times during the growth cycle of the cell; as the glue sets so the wall extensibility decreases. To consider the relative merits of these two explanations of the cell wall behaviour, we need to investigate the cell wall structure in more detail.

6.4.3 The molecular architecture of cell walls

Cell walls are relatively tough structures that can be separated from other subcellular components. As a first step in the investigation of their structure, cell walls are often broken down by hydrolysis. The major products are a variety of sugars together with smaller amounts of amino acids.

◇ What classes of macromolecule are present in primary cell walls?

◆ Cellulose is usually the main component. Microfibrils of cellulose are laid down in a matrix containing pectins, hemicelluloses (non-cellulose polysaccharides) and proteins. Polysaccharides and protein may sometimes be covalently bound forming glycoprotein. (Chapter 1, Section 1.3.1.)

There is much controversy about the function of cell wall protein, and we shall therefore concentrate on the polysaccharide components. These can be separated into the three groups by their different solubility properties (see Figure 6.29), and the isolated components can then be studied further.

Comparatively mild treatments, such as extraction with hot water, remove the pectin. This is a gelling polysaccharide made up of linear polymers with numerous side branches. It provides flexible support and traps large amounts of water within a net-like structure.

The second component, which dissolves in alkaline solution but not in water, is hemicellulose; there are several types of hemicellulose present and they provide a means of binding microfibrils of cellulose together into stacks and may also provide additional binding between individual cellulose molecules within each microfibril.

The residue that is left is the fibrous polysaccharide, cellulose. This provides the main structural component of the cell wall.

◇ What features of the chemical structure of cellulose make it ideal as a structural support molecule?

◆ (i) individual polymers of cellulose are composed of chains of glucose molecules linked by β-1,4 glycosidic bonds providing a strong linear molecule and (ii) microfibrils of cellulose are formed by parallel stacking of large numbers of individual polymer molecules, held by inter-chain hydrogen bonds*. (Chapter 1, Section 1.3.1).

Models of cell wall structure can be proposed using the results of these analyses. One rather speculative model is shown in Figure 6.30, in which each shape represents a different polysaccharide.

One type of hemicellulose is shown as a comb-like structure lying alongside each cellulose microfibril, and another type is represented by a wavy line between cellulose polymers within each microfibril. The hemicelluloses and the pectin act as a matrix binding the stacks of cellulose microfibrils together. This interaction is stabilized by hydrogen bonds or covalent bonds.

This general structure of strong microfibrils embedded in a more flexible matrix is typical of most cell walls. The very different stretchiness of walls prepared from actively growing tissues and from mature tissues results in part from different proportions of the three basic components present in the cell walls. The walls of expanding cells contain relatively large amounts of pectin and little hemicellulose; the hemicellulose molecules appear to form the links between the cellulose microfibrils, and these links have to be broken if the microfibrils are to move relative to each other. There are relatively few of these links in immature walls, and so the force needed to cause expansion is quite small.

Material incorporated into the wall after expansion is complete is much richer in hemicellulose and contains less pectin. The increased cross-linking leads to a much more rigid structure. The wall becomes thicker and in some cells the rigidity is further increased by filling the spaces between the molecules with hard substances such as lignin rather than with water molecules trapped in the pectin gel.

* The chemical structure of cellulose is described in Chapter 3 of Norman Cohen (ed.) (1991) *Cell Structure, Function and Metabolism*, Hodder & Stoughton Ltd in association with The Open University (S203 *Biology: Form and Function*, Book 2).

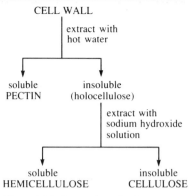

Figure 6.29 Separation of cell wall components according to their solubility.

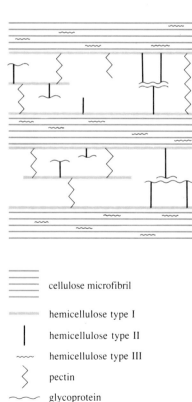

Figure 6.30 Model of maize coleoptile cell wall structure.

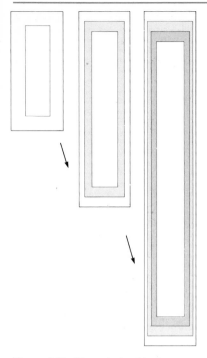

Figure 6.31 The relationship between cell expansion and cell wall synthesis. The wall present before expansion starts is shown in the lighter tone and the newly synthesized wall material is shown in darker tone.

6.4.4 Wall synthesis and cell expansion

So far we have considered how and why existing cell walls can be stretched.

◇ What would happen if this stretching was not accompanied by synthesis of new wall material?

◆ The wall would become thinner and thinner (as shown in Figure 6.31).

The spherical cells of some unicellular algae expand equally in all directions. When electron micrographs of their cell walls are examined, cellulose microfibrils can be seen running in all directions. Similar pictures are obtained from tissues such as apple fruit, which also expand in all directions. In this case, each cell is constrained by contacts with other cells around it and in cross-section is therefore a polygon rather than a circle.

In elongating tissues, the microfibrils tend to run in particular directions, and this regular orientation can be shown in living cells. Figure 6.32 shows the pattern of microfibrils at the inner surface of the cell wall (next to the cell membrane) and on the exterior surface. Those nearest to the cell membrane are the most recently laid down (look back at Figure 6.31 if you are not sure why this is so). The newer microfibrils are laid down at right angles to the direction of growth, but as the cell elongates the microfibrils become re-orientated in line with the direction of elongation (shown by the arrow).

The explanation of how this re-orientation occurs is shown in Figure 6.33. When either turgor pressure or wall extensibility increases, the protoplast will

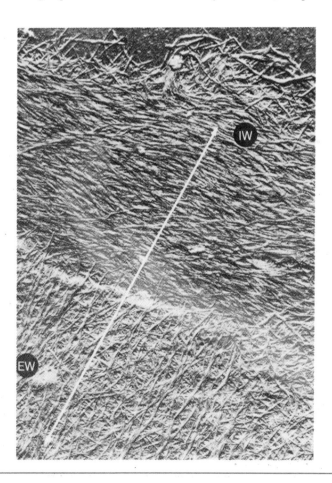

Figure 6.32 Orientation of cellulose microfibrils in the inner (IW) and outer (EW) part of the primary wall.

push against both the side walls and the end walls. The force exerted on the side walls is about three times that on the end walls. The arrangement of microfibrils is rather like a coiled spring so there is considerably more resistance to expansion in width than in length. The net result is that the cell expands more in length, and as it does so the microfibrils become re-aligned in the direction of expansion.

Any account of cellulose synthesis therefore has to explain how the microfibrils are laid down in a particular direction. It also has to explain the fact that electron micrographs show the microfibrils weaving over and under each other (Figure 6.33a is a simplified representation of this interweaving). How are cellulose microfibrils produced *outside* the cell when the cell's synthesizing machinery usually operates *inside* the cell? Other electron micrographs show that the ends of newly laid down cellulose microfibrils are associated with particles embedded in the cell membrane; these particles are cellulose-synthesizing enzymes whose movement may be guided by microtubules (part of the cytoskeleton of the cell) bound immediately beneath the cell membrane. Further evidence of the role of microtubules comes from experiments where cells are treated with colchicine, which disrupts the microtubule structure of the cell. This results in a random orientation of cellulose microfibrils, and so the cells swell much like the spherical cells of unicellular algae. (This is why the colchicine-treated cells in the experiment described in Section 6.2.1 and Figure 6.8 are larger than the untreated cells.) However, little is known about the mechanism of the interaction between cellulose-synthesizing enzymes and microtubules that results in the orientation of the cellulose microfibrils.

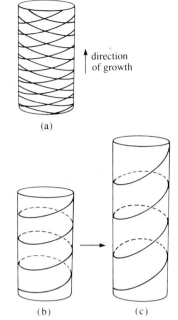

direction
of growth

(a)

(b)　　　　(c)

Figure 6.33 The orientation of cellulose microfibrils in cell walls. (a) Adjacent to the cell membrane. (b) In the walls of an elongating cell.

Summary of Section 6.4

1 The rate of expansion of cells can be described using either the water relations equation $J_v = L_p \times (-\Psi_{int})$ or the growth equation $G = E_x(P - P_{crit})$.

2 The rate of cell expansion can be altered by changing wall extensibility, E_x, turgor pressure above a critical value (by changing either P or P_{crit}), or the hydraulic conductance of the cell membrane, L_p. Current evidence suggests that while changes in P can account for large fluctuations in growth rate, the only factor that is affected by PGRs is wall extensibility E_x.

3 Cell walls consist of cellulose, hemicellulose, pectin and glycoprotein. Cellulose fibres cannot be stretched, but can slip past each other when placed under strain, resulting in wall extension. Elastic expansion of the wall can be explained by the reversible stretching of cross-linking bonds between cellulose and the other components of the wall; plastic expansion requires the breaking of such bonds.

4 Cellulose is laid down on the outside of the cell membrane by cellulose-synthesizing enzymes. For elongating cells, initially cellulose microfibrils lie at right angles to the direction of enlargement, but become re-orientated to lie parallel to the direction of growth as the cell enlarges.

Question 8 (*Objectives 6.1, 6.7 and 6.8*) Which of the following statements are true and which are false?

(a) Isolated cell walls cannot be stretched by applying a force to them.

(b) Reversible cell expansion must involve the breaking of cross-links in the wall.

(c) Chemicals that increase wall extensibility will not stimulate growth when applied to plasmolysed tissues.

(d) In rapidly elongating cells, cellulose microfibrils are initially laid down parallel to the direction of cell expansion.

(e) The synthesis of cellulose microfibrils takes place outside the cell membrane or at its outside surface.

Question 9 (*Objectives 6.7 and 6.8*) Cell walls were isolated from two tissues, L and M. Those from tissue L consisted of 20% cellulose, and each cellulose molecule was on average made up of 2000 glucose residues. The cellulose from tissue M made up 35% of the wall and had an average chain length of 8000 residues. Which tissue is more likely to have been growing rapidly? Give reasons for your answer.

Question 10 (*Objectives 6.2 and 6.7*) The average cell length in a particular region of a pea stem is 94 µm in a dwarf variety and 224 µm in a tall variety. After injection of a solution of a sugar that is absorbed but not metabolized by pea cells into the vascular tissue in a stem of the dwarf variety, the average cell length increased to 183 µm. What can you deduce about the differences in turgor pressure and wall extensibility of the two varieties?

Question 11 (*Objectives 6.2 and 6.7*) The alga *Nitella* consists of chains of giant cells, each up to 2 mm wide and 20 mm long. How would you expect the length of a cell wall isolated from a single *Nitella* cell to change (i) if a weight was suspended from one end of the preparation and (ii) if the weight was subsequently removed? Would this experiment give a more reliable value for E_x than one in which a section of killed pea stem were used?

6.5   STUDYING PGRS AND THEIR ACTION

We indicated earlier that much less is known about the mechanism of PGR action than animal hormone mechanisms. In this section we briefly review some of the proposals and hypotheses that have been put forward to explain the action of one PGR, the auxin IAA. The first case study examines how IAA might cause an increase in cell wall extensibility, and the second discusses experiments that have implicated IAA as the controlling factor in the response of a whole organ — the phototropic response of coleoptiles.

6.5.1 Auxin and cell wall extensibility

In Sections 6.3 and 6.4 we established certain facts about the effect of some PGRs on cells, about the structure of the cell wall and the way cell expansion might be controlled.

◇ Recall two examples where a PGR promotes cell expansion.

◆ (i) Addition of IAA to one side of a growing coleoptile causes curved growth (apparently mimicking a phototropic response); (ii) the flood response shown by the petioles of some amphibious plants is caused by increased levels of ethylene in the presence of auxin. (Sections 6.3.2 and 6.3.3)

Several factors are involved in determining the rate of cell expansion

◇ List three such factors

◆ (a) hydraulic conductance, L_p, of the cell membrane;
(b) turgor pressure above a critical value $(P - P_{crit})$;
(c) cell wall extensibility, E_x (Section 6.4.1).

Of these three, altering the value of E_x provides the most likely means by which the rate of cell expansion could be controlled in growing tissues. Figure 6.34 shows the result of an experiment where the elongation and the plastic extensibility of coleoptile cells were measured under different concentrations of IAA. IAA increases the growth rate of the coleoptile and it also increases the plastic extensibility of the cell walls. This effect of IAA on the extensibility of cells (particularly the outer epidermal cells) has important implications for the growth of the coleoptile, because the epidermal cells have to be passively stretched as the coleoptile grows (Section 6.2.3).

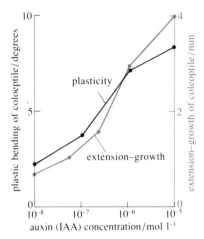

Figure 6.34 The effects of auxin (IAA) on cell wall plasticity and extension growth.

How is cell extensibility altered by the presence of IAA? One early suggestion (first proposed in the 1930s) was that IAA caused the production of new proteins ('loosening enzymes') which acted on the cell wall.

◇ When this idea was first proposed, little was known about the details of cell wall structure. Look at Figure 6.30 again and suggest possible sites of action of such enzymes.

◆ The covalent links (i) between pectin and the cellulose fibres (ii) between hemicelluloses (those within stacks of cellulose and those lining each stack) and (iii) between adjacent glycoproteins are all possible sites.

However, two findings seemed inconsistent with this proposal: (i) IAA acts within 15 minutes of being added to growing tissue and this was considered to be too short a time interval for new proteins to be synthesized, and (ii) adding inhibitors of protein synthesis did not always abolish IAA-stimulated growth.

In the search for a rapid-response mechanism, attention turned to the cell wall and the solution within it (the apoplastic micro-environment). The pectin component of the cell wall traps large quantities of water within the wall (Section 6.4.3) and effectively separates the apoplastic environment from the surroundings. The way the cell controls the conditions of this micro-environment seemed likely to provide the key to understanding the mechanism of controlling wall extensibility.

It had been known since the 1930s that acid solutions have an effect on growth. Figure 6.35 shows the results of an experiment in the late 1960s in which sections of oat coleoptile were transferred from buffer at pH 7 (a) to a buffer solution at pH 7 containing IAA or (b) to a buffer solution at pH 3 with no IAA.

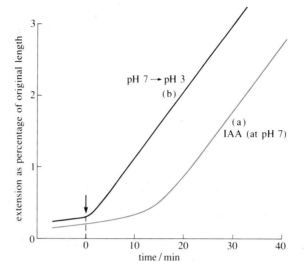

Figure 6.35 The effect of IAA on growth of pieces of oat coleoptile incubated in buffer at pH 7 compared with the effect of transfer from buffer at pH 7 to buffer at pH 3.

The results of this and other experiments suggested that acid conditions mimic the effect of auxin. This led to the proposal that cell wall extensibility was increased by acidification of the wall, and IAA acted by promoting low pH (i.e. acid conditions) in the apoplastic environment. Lowering pH either caused certain bonds in the structure of the wall to become unstable or stimulated the activity of enzymes within the wall (such enzymes would have a low pH optimum); either mechanism would cause an increase in wall extensibility. This suggested mechanism is known as the **acid growth hypothesis** and has been the subject of much debate over the last ten years or so.

The earliest models proposed a membrane-bound receptor for IAA, which, when stimulated, caused the extrusion of H^+ ions from the cell across the plasmalemma into the apoplastic space. A later modified version suggested that an IAA-driven redox chain was located on the plasmalemma and this pumped H^+ across the membrane to the apoplast (using a similar mechanism to that described in Chapter 3, Section 3.4.1).

To test the hypothesis, it would first be necessary to determine whether auxin is capable of stimulating the extrusion of H^+ ions across cell membranes.

◇ Suggest an experiment that would test this. (Hint: in the above experiments, buffers were used to ensure that there would be no changes in pH in the medium surrounding the cells.)

◈ If tissue is incubated in an *unbuffered* (or weakly buffered) medium, the addition of IAA should lead to a lowering of the pH of the medium.

Such experiments can be carried out using a sensitive pH meter provided that there is a very small volume of solution in direct contact with the apoplast. The only way to achieve direct contact is to abrade the tissue to remove cuticle so that extruded protons can get into the solution where measurements are taken. Results of such experiments are similar to those shown in Figure 6.36.

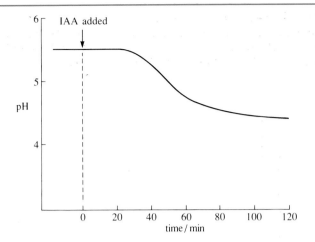

◇ What does this experiment show?

◆ IAA does indeed cause secretion of H^+ ions into the medium, but a change in pH cannot be detected until more than half an hour after application, whereas growth starts within 10–15 minutes.

So a major problem for the acid growth theory is the timing of the events that are supposed to be cause and effect. However, it is extremely difficult to obtain accurate measurements of the pH within the cell wall matrix, and so the delay in the onset of pH changes in such experiments may simply be due to these technical difficulties. A further difficulty for the theory is that wall loosening occurs optimally at pH 3–4, but experiments such as those summarized in Figure 6.36 indicate that the addition of IAA does not produce such low wall pH values.

◇ What effect would you expect if you repeated the experiment in Figure 6.36 but used a neutral buffer solution instead of water?

◆ Any H^+ ions pumped across the plasma membrane to the cell wall matrix would be 'mopped up' by the buffer, so maintaining a steady pH of the cell wall. You could predict that there would be no increase in cell wall extensibility and growth would not be promoted.

Early experiments of this type seemed to show the effect predicted by the hypothesis, and for a while the acid growth hypothesis held sway; however, more recent experiments have cast doubts on this important prediction of the hypothesis. IAA has been shown to cause some cell elongation even when the apoplastic environment is kept at a high pH by the addition of appropriate buffers.

◇ Suggest a further test of the hypothesis.

◆ Do the reverse of the buffering experiment, by introducing an external promoter of proton extrusion.

An agent that promotes proton extrusion in many systems is fusicoccin, a toxin produced by certain fungi. When added to plant cells it increases the rate of proton pumping across the cell membrane to the cell wall matrix. Both IAA and fusicoccin promote elongation of tissues, and when applied at the same concentration (about 1 μmol l^{-1}) they have a similar effect. However, fusicoccin is a much more powerful promoter of H^+ extrusion than IAA, so when adjustments are made to ensure that growth rates can be compared under conditions of identical increase of H^+ extrusion, different rates of growth occur in tissues treated with IAA and fusicoccin. This suggests that something other than H^+ extrusion is causing the difference in growth rate under these two conditions.

So while the acid growth hypothesis seemed to provide a promising framework for explaining the mechanism and control of cell expansion, it does not provide a complete explanation. An interesting feature of the hypothesis was its similarity to some animal hormone systems: it involved receptor molecules which when activated generate a 'second messenger', in this case H^+ ions (in animal systems the second messenger is often the nucleotide cyclic AMP). According to the hypothesis, second messenger was responsible for promoting the observed effect. We take up the issue of the validity of adopting mechanisms derived from animal systems and applying them to plants in Section 5.4.3.

So if predictions that arise from the acid growth hypothesis are not supported by recent (1990) evidence, where do we go from here?

It has been demonstrated recently that a further effect of IAA on cells is to induce the production of messenger RNAs before elongation occurs. So far, this finding has been confirmed only in a few tissues; however it has also been demonstrated that inhibitors of mRNA synthesis abolish IAA-induced cell elongation.

◇ What is the implication of these findings for the earlier hypothesis that IAA promotes the production of wall loosening enzymes?

◆ These findings provide support for the idea that new proteins are produced due to the presence of IAA; the nature of the proteins produced as a result of the production of these mRNA molecules needs to be determined before they can be classed as 'wall loosening enzymes'. A range of new proteins would be expected to be produced as the cell grows.

But what about the major objection to the notion of producing new wall-loosening enzymes when it was first proposed? The facts that IAA acts within a short time period and that inhibitors of protein synthesis often have no measurable effect still have to be countered.

Recent experiments have compared the effect of adding mRNA inhibitors to abraded tissue and to intact tissue; the inhibition is shown only in abraded tissue! Presumably, early reports that inhibitors of protein synthesis did not always suppress IAA-induced growth could be due to such inhibitors not being taken up sufficiently by the cells. So again it seems important to make the cells fully accessible to treatments, for example by removing the cuticular barriers of plant cells under study. In experiments investigating H^+ extrusion this was necessary to ensure that protons could get out, but in this case it ensures that the agents used can get in. Ironically, the techniques developed during the experiments that investigated the acid growth hypothesis have led to a reassessment of the alternative theory.

6.5.2 Phototropism: A case study of the control of cell expansion

Having considered some possible ways by which the rate of growth of an individual cell might be controlled by the extensibility of the cell wall, this section examines a response of a whole organ — the bending of a stem or coleoptile towards a source of light. This response is called **phototropism**. Other bending (or tropic) responses occur in other plant organs — gravitropism (bending towards gravity in the case of roots or away from gravity in the case of shoots) and thigmotropism (bending of a shoot in response to touching an object) are two common examples. The bending response is due to unequal rates of cell expansion on the two opposites sides of the structure, and so the mechanisms that control cell expansion described in Section 6.4 would seem to provide a means of controlling these whole-structure tropic responses.

One of the first documented reports of phototropism was by Charles and Francis Darwin in 1880. They observed that when coleoptiles bent towards the light, the curvature occurred in the growth region below the tip. However, if the tip of the coleoptile were removed the response was abolished (see Figure 6.20). There was also reduced curvature when the tips were covered. Although the Darwins did not propose any specific model to account for their observations, other workers inferred that there must be some 'influence' passing from the tip (the 'site of perception') to the growth region (the 'effector').

◇ What forms could this 'influence' take?

◆ The signal could be transmitted by cell–cell contact (as in animal nerve connections); another possibility would be by a chemical substance diffusing from the site of perception to the effector.

Some 30 years after these initial observations, a series of experiments were carried out by several investigators, working independently, to elucidate the nature of this 'influence'. The results of some of these experiments are summarized in Figure 6.37.

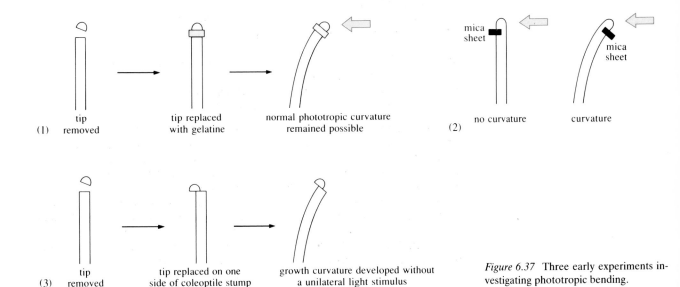

(1) tip removed → tip replaced with gelatine → normal phototropic curvature remained possible

(2) mica sheet — no curvature / mica sheet — curvature

(3) tip removed → tip replaced on one side of coleoptile stump → growth curvature developed without a unilateral light stimulus

Figure 6.37 Three early experiments investigating phototropic bending.

Experiment 1 The tip of the coleoptile was cut off, and a small amount of agar (or gelatine) placed on the cut surface; the tip was then replaced and the coleoptile subjected to sideways (unilateral) light.

Experiment 2 A small sheet of mica (a glass-like material) was inserted part way into the coleoptile between the tip and the growing region. The coleoptiles were then placed in unilateral light, either from the side on which the mica sheet had been inserted or from the opposite side.

Experiment 3 The tip of the coleoptile was cut off and then replaced asymmetrically on the cut surface; the coleoptile was placed in normal (not unilateral) light.

◇ What conclusions can be drawn from the results of experiments 1, 2 and 3?

◆ Experiment 1 shows that the bending response can be re-established by replacing the excised tip on an agar-covered cut surface of the coleoptile; the 'influence' inferred from the Darwins' experiment can clearly diffuse through the agar block, and so must be chemical in nature.

Experiment 2 suggests that this chemical is a growth promoter. Where the chemical can diffuse down the 'shaded' side of the coleoptile, the bending response occurs, apparently due to increased growth rate. Where the chemical is prevented from diffusing down the shaded side, the bending response does not occur.

Experiment 3 also supports the deduction that the chemical involved is a growth promoter. Where it diffuses down on one side only, even in all-round light, that side seems to grow at a faster rate, producing curved growth similar to that observed in phototropism.

The search then began for the nature of the growth promoter, but it was not until 1926 that Fritz Went obtained from cut coleoptile tips a substance which seemed to fit the bill. The substance was collected from cut tips of *Avena* (oat) coleoptiles placed onto agar blocks (Figure 6.38a). When these blocks were placed asymmetrically on decapitated coleoptiles the bending response was similar to that described in experiment 2. Moreover, the angle of curvature was shown to be dependent of the amount of 'hormone' collected in the agar block. Indeed the angle of the curvature induced in the coleoptile became the means of measuring the amount of 'hormone' obtained from other tissues (one of the first examples of biological assay, or **bioassay**, see Figure 6.38b).

Figure 6.38 Experiments which suggested that tropic curvature of oat coleoptiles is due to unequal distribution of IAA. (a) Outline of the experimental procedure. (b) Results obtained using different numbers of coleoptiles on the agar blocks. (c) Results obtained using different concentrations of IAA on the agar blocks.

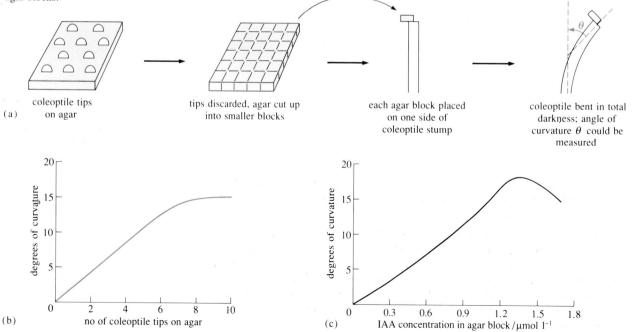

(a) coleoptile tips on agar → tips discarded, agar cut up into smaller blocks → each agar block placed on one side of coleoptile stump → coleoptile bent in total darkness; angle of curvature θ could be measured

(b) degrees of curvature vs. no of coleoptile tips on agar

(c) degrees of curvature vs. IAA concentration in agar block / μmol l^{-1}

The curvature induced by the 'hormone' seemed to match results obtained by adding IAA to agar blocks which were then placed asymmetrically on decapitated coleoptiles; the more IAA added to the block the greater the angle of curvature produced, as shown in Figure 6.38c. It was not until much later (early 1970s) that other techniques showed that IAA was present naturally in coleoptiles.

◇ From the above experiments and information, suggest a mechanism by which IAA could control the phototropic response.

◆ Curvature of the coleoptile or stem could be caused by the asymmetrical distribution of auxin produced in the tip and transported down to the growth zone to a greater extent on the shaded side than on the lit side. Where IAA is present in larger quantities, the cell walls would be more plastic, resulting in these cells expanding more. This would cause uneven growth rates, resulting in curvature towards the light source.

This idea was proposed by Went (working on phototropism) and independently by a Russian, Cholodny (working on gravitropism) and became known as the **Cholodny–Went theory**. A main preoccupation with workers exploring this theory was to establish whether the asymmetrical distribution of the auxin occurred by (A) lateral transport from the lit side to the shaded side or (B) breakdown of the auxin on the 'inside' bend region. (Figure 6.39).

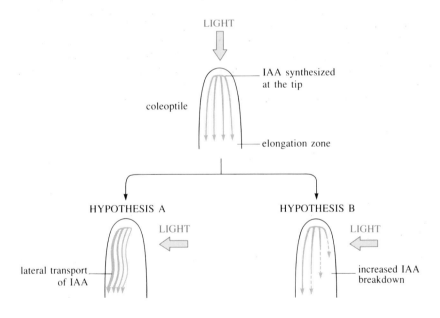

Figure 6.39 Two hypotheses explaining how unequal levels of IAA occur on the light and dark sides of phototropically stimulated maize coleoptile.

◇ Examine Figure 6.40, which summarizes a set of experiments designed to investigate these two hypotheses. Can either hypothesis be *discarded* as a result of these experiments?

◆ Hypothesis B. This predicts the increased breakdown of auxin as it moves down the illuminated side of the coleoptile during phototropic stimulation; but the data in Figure 6.40 show the total amount of IAA collected in the agar blocks (adding together the two values in each of (b), (e) and (f) is almost the same irrespective of whether the tips were in darkness or lit from one side. So IAA is not destroyed by illumination and hypothesis B must be rejected.

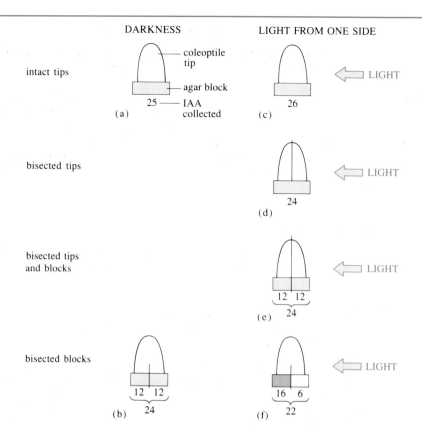

DARKNESS   LIGHT FROM ONE SIDE

Figure 6.40 The results of collecting IAA over a 3 h period from maize coleoptile tips treated in various ways. Each number is an IAA concentration which is the mean of at least 40 measurements.

◇ Do the data support the other hypothesis?

◆ The data provide some support for hypothesis A. (f) shows that auxin is distributed towards the dark side. This redistribution does not occur in darkness (b). In (d) and (e) mica sheets bisect the coleoptile tips and prevent lateral movement to the dark side. So the data are consistent with the hypothesis, but it would be unwise to say that they 'prove it'.

The accumulation of this sort of evidence spanning over 50 years work supported the Cholodny–Went theory that a PGR (here identified as IAA) controlled the tropic responses. Note the similarities with certain animal hormone control systems:

(a) A hormone is produced at a specific site (here, the coleoptile tip).

(b) The hormone passes from a site of production to target cells.

(c) Arrival of increased levels of the hormone elicits a response at the target cells, resulting in the observable change.

Now consider the following observations concerning phototropism.

Figure 6.41 shows the results of measuring growth rate on two opposite sides of a coleoptile both before and after stimulation by unilateral light.

◇ What happens to the growth rate of the shaded side after stimulation has occurred? Is this predicted by the Cholodny–Went model?

◆ The growth rate (shown by the slope of the line) on the shaded side remains virtually identical to the rate before stimulation; the Cholodny–Went theory would predict an increase in growth rate due to the presence of increased levels of IAA.

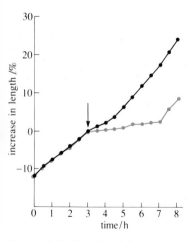

Figure 6.41 Growth of both sides of *Avena* coleoptiles before and after unilateral phototropic stimulation (indicated by the arrow). The illuminated side is shown in green and that on the shaded side in black. (Increase in length is expressed as a percentage of the length measured at the start of unilateral illumination.)

◇ What happens to the growth rate on the illuminated side?

◆ There is an almost immediate 'stop' to growth on the illuminated side, followed by a restoration of original growth rate some time later.

Similar graphs have been obtained from experiments on a wide range of coleoptiles and stems; phototropic and gravitropic responses all have the common feature of a 'stop' to growth on what becomes the 'inside bend'. This may sometimes be accompanied by an increase in growth rate on the 'outside bend' (by up to 10% of pre-stimulated growth rate), but the 'stop' (reduction in growth rate by 90% of the pre-stimulated rate) on the inside bend is the universal feature.

◇ How does the timing of the response shown in Figure 6.41 fit with a hormone-based theory?

◆ The graph suggests that the response to unilateral light is almost immediate; the production and redistribution of IAA with this time span would be impossible.

◇ Look at Figure 6.42 showing the measurement of growth rate along zones of a stem subjected to unilateral light. Do these Figures support a hormonal redistribution theory?

◆ A hormonal redistribution model would predict a 'wave' of differential growth passing down from the top zone to lower zones as the redistribution takes effect; the fact that each zone responds simultaneously to the unilateral light conflicts with this sort of model. Simultaneous reactions of all zones along the length of the coleoptile or stem strongly suggest a direct effect of the stimulus on the growing cells.

The experiments shown in Figure 6.40f have recently been repeated, measuring IAA activity on the illuminated and shaded side (using the bioassay technique described above) but then comparing this with a direct measurement of the amount of IAA present collected on each half of the agar blocks using thin-layer chromatography (a technique unavailable when the experiment was first carried out). The results of the bioassay confirmed the original finding. The activity of IAA on the illuminated side was about 38% of that on the shaded side. However, the actual *amount* of IAA present in both sides was the same.

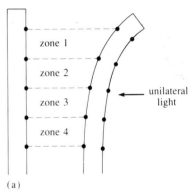

(a)

Figure 6.42 Experiment to measure the curvature of zones of coleoptiles. (a) Plastic beads were placed at equal intervals along two sides of the coleoptile before it was placed in unilateral light. (b) Experimental results.

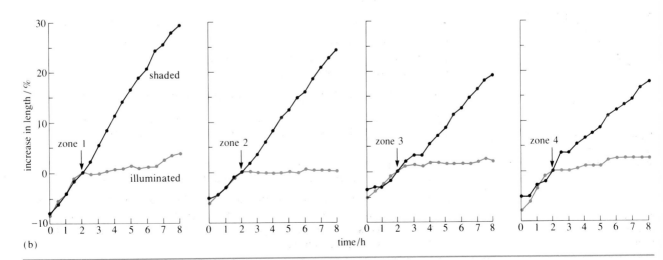

(b)

◇ What does this finding imply for the Cholodny–Went theory?

◆ It causes problems for the original theory! Uneven growth is occurring even though there is no gradient of IAA across the coleoptile.

◇ How could the activity of IAA be different in the illuminated and shaded side even though the actual amount does not change?

◆ (i) Inhibitors that reduce the activity of IAA might be unevenly distributed between the illuminated and shaded sides. There is some evidence that inhibitors are present and unevenly distributed but the chemical nature of these is yet to be elucidated.

(ii) the action of IAA might be different on different cells because some cells might become more (or less) sensitive to an unchanging concentration of IAA.

The debate continues, with proponents of the 'hormonal' theories (divided between those favouring uneven inhibitor distribution and those proposing uneven auxin distribution in different zones of growing tissue) countering the arguments of the 'no hormone needed' school of thought.

Our discussion of this case study illustrates several important points about experiments investigating PGRs in general. First, while it is tempting to explain plant responses in terms of cause and effect mediated through some hormone-like substance, this may not always be the case. In animal systems, responses occurring via changes in hormone levels typically involve an increase in hormone by between 15 and 30 times the base level, whereas this is very unusual in plant systems.

◇ Recall one plant response that is elicited by a sharp rise in a PGR.

◆ The closure of stomata by the guard cells in response to increased levels of abscisic acid.

But the investigations into the role of auxin in tropic curvature illustrate the dangers of assuming this kind of model. Alternative hypotheses, such as direct effects of stimuli on cells which alter their growth rate, also need to be tested if we are to elucidate the mechanisms involved.

6.5.3 Why are PGRs and their action so elusive?

These two studies of investigations into PGR action have illustrated some approaches that have been used to elucidate PGR action in plants. The first approach tried to untangle the biochemical events that lead to increasing cell wall plasticity, and the second approach attempted to correlate a change of PGR concentration with a physiological response. These two examples represent only a tiny fraction of the vast amount of work undertaken in the study of PGRs. However, in both cases the interpretation of the results of the experiments leads more to a debate of how the underlying system might be working rather than providing definitive answers. This is fairly typical of many such investigations!

A similar approach to the study of animal hormones has yielded a detailed knowledge of their mode of action. Animal physiologists and biochemists have been successful in unravelling the complexities of a wide range of hormones and can describe in detail the role of each hormone and its mode of action. Figure 6.43 summarizes the way a typical animal hormone operates.

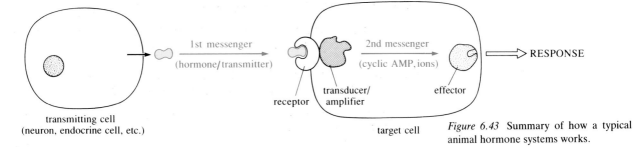

Figure 6.43 Summary of how a typical animal hormone systems works.

One reason why our knowledge of PGR action lags behind that of animal hormone systems is that plant cells present very different technical problems to the investigator than do animal cells. It is much more difficult to get at the system under study without damaging it. For instance, trying to measure the changes in the cell wall micro-environment requires abrasion of the cuticle at the surface, and who knows what effect that has on the system? Animal hormone systems typically have many different hormones each with a specific effect on identifiable target tissue. Plants have relatively few types of PGR, each of which can have many different effects on a wide range of tissues. PGRs operate at very low concentrations (IAA concentration in plant material is about 1 part in 10^9). New techniques in identifying PGRs and determining their concentration in plant material should help to lift this restriction on investigations. An important weapon in the armoury of the animal physiologist is the range of antagonists to determine the biochemical mechanisms and binding properties of hormones. With a few exceptions (such as silver ions as antagonists of ethylene action) such antagonists are virtually unknown for PGRs and this imposes severe limitations on the types of experiments that can be carried out.

But the relative lack of progress in our understanding of PGR systems cannot be blamed entirely on such technical limitations. Note that in both case studies there is an implicit assumption: 'target cells' respond to changing levels of a hormone-like substance (IAA). But are we justified in giving IAA and other PGRs hormone-like qualities?

One definition of an (animal) hormone is a substance produced in minute amounts in one tissue that serves to control the activity of another tissue.

Consider Figure 6.43. Note that a typical animal hormone is synthesized by endocrine cells, which are usually grouped together in glands.

◇ What experiments would you carry out to test whether a physiological response was due to a hormone produced by a particular gland in an animal?

◆ (i) Remove the gland and see if the response is abolished.
(ii) Replace the gland in the body and see if the response is restored.
(iii) Replace the gland by an extract of the gland or an artificial supply of the hormone and see if the response is restored.

The problem for plant physiologists with this sort of approach is that there are no equivalents to endocrine glands in plants, so the source of a PGR production cannot be isolated. While it is likely that some types of cell produce more of a particular PGR than others, most cells have a capacity for PGR production. In most cases it is certainly not possible to identify cells that have a higher capacity for PGR production from their structure. Moreover, the cells that produce PGRs may also respond to the PGRs that they produce, so there is the further problem of identifying target cells. In animals, however, the hormone-producing cells are usually conveniently gathered into an easily recognizable and discrete tissue or organ, and are separated from the target tissues.

These problems mean that plant physiologists are left with a limited version of the experiment sequence described above: only a modified step (iii) is possible — to investigate the effect of an artifical supply of PGR on intact plants or plant tissues. Up to the early 1980s this was the main feature of plant 'hormone' investigations. The general framework for such investigations is:

(a) Substance X occurs in certain plants.

(b) We have a supply of substance Y which closely resembles the structure of substance X.

(c) When Y is added to a plant it causes a specific response (e.g. stem elongation).

(d) It is therefore likely that substance X has a role in controlling stem elongation.

◇ List as many shortcomings of this general approach as you can

◈ Amongst the items in your (possibly very long!) list should be:
(i) Y is in addition to levels of X which may continue to be produced during the course of the experiment.
(ii) Y may be added externally in similar concentrations to the levels of X. But it would be difficult to demonstrate that it reaches any target tissue at the appropriate concentration; methods of improving uptake (such as removing cuticle layers) themselves alter the properties of the tissues being investigated.
(iii) Y may not be transported to sites of action in the same way as X.
(iv) Y may exert an effect on the tissue, but in quite a different way from the mechanism of X.

You probably listed a good many more. The basic point is that this approach to investigating PGR action provides merely circumstantial evidence. To establish whether a PGR is operating as a 'hormone', it must meet a far more stringent set of criteria.

First, in a plant system, it must be possible to demonstrate that the site of *perception* is separate from the site of *response*. This is not easy to establish in many cases, for the same reasons that it is difficult to locate the site of production of a PGR. There is a similar problem in locating the 'target tissue'. Most cells, for instance, respond in some way to increasing IAA levels. In the phototropism case study, the underlying assumption behind the series of investigations was that perception of light (at the tip) was remote from the response zone (below the tip); alternative hypotheses suggest a direct effect of light on the responding cells, and do not need to invoke the concept of a 'hormone' at all. As pointed out by one experimenter, you don't need a postal system to mail yourself a letter!

The *mechanism of response* must be shown to be identical whether the added chemical or the chemical produced by the cells is invoking the response. Adding IAA to one side of a coleoptile will certainly induce curved growth by accelerating growth on the treated side. However, analysis of a tropic curvature demonstrates that the mechanism relies on the cessation or slowing down of growth on one side compared to the other.

The *timing of the response* must be appropriate. If the response is suspected to be caused by an increase in level of a particular chemical, then this increase must be measurable before the effect is observed. This criterion is met in cases such as ethylene production prior to rapid extension of the stem and petioles in amphibious plants, but has been the source of considerable dispute in other cases. In tropic responses, some workers argue that any detectable shift in the amount of IAA across the responding tissue occurs after the response has started (i.e. it could be a *consequence* not a *cause* of the response).

The *magnitude of the response* must correlate with the magnitude of the change in the level of the chemical suspected of controlling the response. This is particularly important where the large changes in concentration of the chemical are required to produce relatively small changes in the responding tissue. Some responses are induced only when artificially high levels of a PGR are added. Few cases seem to meet this criterion: ethylene production can rise dramatically within tissues when submerged, and ABA levels in leaves also rise prior to stomatal closure. However, in other cases the changes in PGR levels that are thought to accompany a physiological response are really quite small (typically 3 times the base level). This is quite different from a typical animal hormone-based response system, where the concentration of the hormone may rise by 15–30 times the base-level.

The fact that PGR levels in general do *not* change dramatically in the way that some animal hormones do has led some workers to suggest that the mechanisms involved with PGR action are fundamentally different from those of animal hormone systems. The tissues may not be responding in the 'animal-like' manner of target tissues to changes in hormone level. Instead, the cells may be changing their *sensitivity* to fairly static PGR levels. The fact that in many experiments artificially high levels of PGR need to be added suggests that the plant system is geared not to respond to the small changes in internal PGR concentration.

The sensitivity of a cell to a PGR could be altered in several ways. First, the amount of PGR transported to the receptor sites on the cell membrane or within the cell could be altered. Secondly, the number of receptor sites could be increased or decreased, so altering the receptivity of the cell to unchanging PGR levels in much the same way as receptor regulation occurs in animal cells. A third possibility is that the events occurring after the PGR has bound to a receptor could be altered. Cells could have a different capacity to respond at different times during their development. These possibilities are summarized in Figure 6.44.

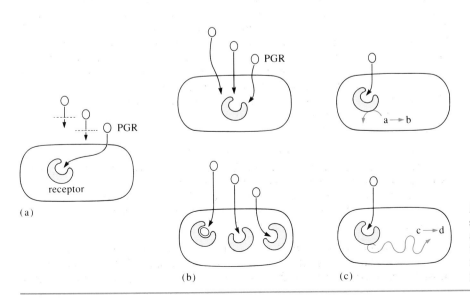

Figure 6.44 Three ways in which the sensitivity of a cell to a PGR could be altered. (a) Different amounts of PGR transported to receptor sites. (b) Different numbers of receptors. (c) Different post-binding effects.

The concept of response capacity could help to explain the wide range of responses shown by different cells to the same PGR. Different cells may have different receptors capable of binding with a particular PGR; in each case the formation of a PGR–receptor complex would stimulate a different sequence of events within the cell, so producing a different response (Figure 6.44c).

So one reason why the results of many investigations of PGR action are difficult to interpret is that the experiments are often designed under the assumption that plant cells respond to changing PGR levels, much like animal cells to hormones. The challenge to the plant physiologist is to design experiments that will distinguish between responses due to changing PGR levels within the plant and responses due to changing sensitivity of cells to the PGR. It could be that most systems lie between these two extremes; add to that the further complexity that cells may respond to PGRs they themselves produce, and we have a fairly complex tangle to unweave.

PGRs have certain features in common with animal hormones at the biochemical level. Both bind to receptors in cells (or on their surface) and as a consequence a sequence of events is generated within the cell that produces the response. However, the physiology of the response system in plants seems very different from that in animals.

Summary of Section 6.5

1 Many theories of how growth events are controlled assume that growing cells respond to changing PGR levels, but it is difficult to establish this hormone-like role of PGRs experimentally. Problems include technical ones and often having to work with artificial PGRs to elucidate mechanisms of action. In practice, few PGRs meet the criteria necessary to be classed as hormones in the animal sense.

2 It has been suggested that cells may be able to change their sensitivity to PGRs (which do not usually change in concentration to any great extent in the plant). This suggests a different control system in plants from typical animal control systems that involve hormones.

3 IAA-induced cell expansion is thought to be controlled either by acidification of the cell wall (the acid growth hypothesis) or as a result of the generation of new gene products, such as wall-loosening enzymes. The latter now seems more likely.

4 Tropic curvature, once thought to be caused by uneven distribution of IAA causing uneven growth, appears to be controlled either by growth inhibitors or simply by the direct effect of the stimulus on the cells that respond.

Question 12 (*Objectives 6.1, 6.2 and 6.9*) Which of the following results are consistent with the acid growth hypothesis for auxin action?

(a) IAA does not stimulate the elongation of oat coleoptiles in the presence of substances that make membranes permeable to H^+ ions.

(b) When IAA is added to oat coleoptiles, the tissue grows at an increased rate for at least 12 hours. After treatment with acid, growth is stimulated, but it declines to the control value after only 2 hours.

(c) In many tissues, inhibitors of protein synthesis reduce the auxin-induced stimulation of growth.

(d) The addition of IAA stimulated elongation and acid secretion in sections of pea stem. When abscisic acid is added at the same time as IAA, both growth and acid secretion are less than in the absence of ABA.

Question 13 (*Objective 6.11*) Which of the following statements help to explain why there are few reports of the successful isolation of PGR receptors in plants?

(a) PGR-producing cells are not grouped together in recognizable areas.

(b) The plasmalemma tends to remain associated with fragments of cell wall during purification.

(c) No tissues are known where very large numbers of PGR receptors might occur.

(d) All PGRs have effects on most plant tissues.

(e) PGRs are lipid-soluble.

Question 14 (*Objectives 6.2 and 6.10*) Genetic variants of *Arabidopsis* plants have their tropic responses abolished. These plants have also been shown to be insensitive to IAA. How would supporters of each of the following theories about phototropism view this information: (i) the Cholodny–Went theory (ii) inhibitor redistribution theory and (iii) a theory that proposes a 'direct effect' of light on cells.

6.6 PGRS AND DIFFERENTIATION

The techniques and experiments described so far have concerned the control of one aspect of growth — growth in length due to individual cell expansion. In this section we review the role of PGRs in the development of plant organs. Information about such roles comes from two main sources; the study of cells developing in tissue cultures and the study of the development of cells derived from cambium in intact plants.

6.6.1 Tissue culture

Tissue culture is a technique that exploits the ability of many plant cells to revert to a meristematic state. The healing of damaged plant tissues often involves the proliferation of a mass of undifferentiated cells. A similar growth of undifferentiated cells can be artificially induced by placing a small amount of tissue in an appropriate growth medium. Such a medium can maintain cells in the meristematic state and extend their proliferation indefinitely. The same term, **callus**, is used to describe both the natural and artificial growths.

Tissue culture cells have a relatively unspecialized structure (see Figure 6.45a), and the development of such cells can be easily followed.

◇ What limitations are there to this approach to studying how cells develop?

◆ The organization of cells in an artificial callus is quite different from that in intact plants, so the interactions between the cells may also be different.

Almost all tissues will form a callus when placed on a medium containing a suitable mixture of salts, sugars, vitamins and PGRs. However, you should remember from Section 6.2 that most tissues already contain unspecialized but inactive cells, and these, rather than more differentiated cells, may be the source of the callus cells. Sometimes a callus, and eventually a whole plant,

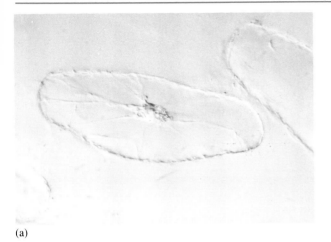

(a)

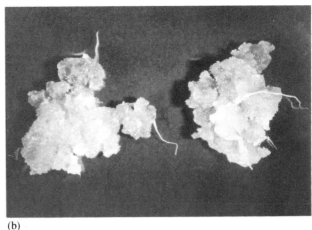

(b)

Figure 6.45 (a) A typical tissue culture cell seen in a light microscope using phase contrast optics. (b) The differentiation of a culture to form roots.

can be formed from a single cell in culture, which demonstrates that such cells are **totipotent** (i.e. that each cell contains a complete set of genes to generate a complete individual). For some reason certain plants, notably carrot and tobacco, are particularly amenable to this sort of manipulation, and whole plants have been regenerated from cultured cells derived from a variety of tissues. With other cultures, it has only been possible to induce the formation of particular organs (usually roots as shown in Figure 6.45b) or certain cell types (usually xylem and phloem). It may simply be that the chemical requirements for the differentiation of particular cells have not yet been satisfied, but at least we can use the cultures to investigate the factors involved in the various component parts of plant development.

Early work on callus generation showed that it was quite easy to sustain a culture of cells in a medium containing auxin, but not so easy to ensure that the cells divided. This prompted the search for a 'division factor' and lead to the discovery of cytokinins. Most cultures need quite a high concentration of added auxin and cytokinin in order to grow. It is interesting to note that in crown gall disease (an example of a naturally occurring callus growth), invading bacteria induce the infected cells to make their own auxin and cytokinin. The result of this PGR excess is a tumour-like callus growth. Other organisms, including the nematode worms that cause 'root knots' to form and the symbiotic bacteria in nodules on the roots of peas and beans, also stimulate growth, probably by secreting PGR-like chemicals.

Tissue culture techniques were first developed in the 1950s, and the early experiments were concerned with developing the right mixture of components in the culture medium. In one classic experiment the experimenters found that by adjusting the amounts of various PGRs they could promote either the formation of roots or of shoots. Some of their results are shown in diagrammatic form in Figure 6.46.

Some treatments lead to the formation of roots, others to shoot initiation, while in the remainder the tissues continue to grow in a disorganized manner. The figure shows the effects of different concentrations of the artificial cytokinin, kinetin, at only one auxin concentration, but a similar range of results is obtained at other concentrations. From the complete set of results, it is apparent that the type of differentiation observed depends not on the absolute concentration of either of the PGRs but on the ratio of the two.

◇ Work out from Figure 6.46 auxin:kinetin ratios that favour root formation, shoot formation and disorganized growth.

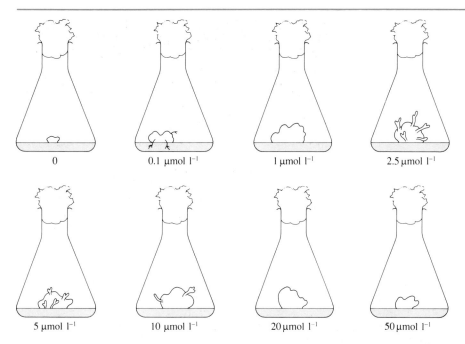

0

0.1 μmol l⁻¹

1 μmol l⁻¹

2.5 μmol l⁻¹

5 μmol l⁻¹

10 μmol l⁻¹

20 μmol l⁻¹

50 μmol l⁻¹

Figure 6.46 The effect of various concentrations of the artificial cytokinin, kinetin, on the growth of tobacco callus in the presence of $10\ \mu mol\ l^{-1}$ IAA.

◆ Roots are formed when there is a lot more auxin than cytokinin (100:1 is shown), shoots when there is only slightly more (4:1, 2:1 and 1:1 are shown). Disorganized growth predominates at intermediate values (10:1 is shown) or when there is more cytokinin than auxin. With no cytokinin there is no growth.

◇ What would you expect to happen if callus was grown on medium containing $0.1\ \mu mol\ l^{-1}$ kinetin and $0.25\ \mu mol\ l^{-1}$ IAA?

◆ The medium described has an auxin to cytokinin ratio of 2.5 to 1, and shoot initiation is therefore the most likely result.

Comparatively few generalizations can be derived from tissue culture studies, but auxin:cytokinin ratios do often seem to be important in differentiation of callus to form either tissues or specialized cells such as xylem. Many studies consider only the effects of these two PGRs. Other factors are undoubtedly involved, and the work with tobacco callus was extended to show that the ratio of gibberellin to cytokinin controlled both the number of shoots formed and their shape (in the presence of GA_3 there are fewer shoots and they have elongated leaves).

Nor are PGRs the only regulators involved: in lilac and bean callus, both xylem and phloem elements are formed at appropriate ratios of auxin to cytokinin. The relative amounts of the two types of vascular tissue depend not on PGR concentration but on the concentration of sucrose in the medium. It is tempting to suggest that a high sugar level stimulates the development of the cell type (phloem) that is responsible for its transport.

Fragments of tissue do not necessarily need to be persuaded to form callus before they can form other organs. Layers of tissue from two to four cells thick can be removed from many stems or leaves and kept alive in culture. In tobacco (again!), roots, shoots and flowers can be formed as a result of PGR manipulation. This has the advantage that the origin of the new tissues and their relationships to structures such as vascular tissue can be easily followed under the microscope.

These techniques now form the basis of a rapidly growing micropropagation industry. A wide range of ornamental and horticultural plants (from roses to oil palms) are propagated in this way. A single short section of stem of one plant containing the meristematic zone can be cut into about fifty segments and each segment can then be cultured in an appropriate medium to generate complete stems; changing the medium of the culture stimulates the stems to produce roots. Tobacco plants are produced when shoots initiated from callus are transferred to a root-inducing medium and then to soil. Other cultures produce whole plants directly from small groups of cells (embryoids) that develop in a remarkably similar way to the embryo in a developing seed. Although whole plants have been developed from single cells, the first step is usually the formation of a small aggregate of cells, and once triggered, the process then continues without further modification of the medium. The potential for such propagation techniques is enormous. For instance, where a complete plant can be generated from a single cell, one callus culture is a source for over 50 000 plants!

These techniques have important applications in several areas. First, they provide the possibility of isolating agriculturally useful mutants rapidly. For instance, it is much easier to select for salt-tolerance from a million cells than from a million plants. Secondly, micropropagation techniques produce virus-free plants (see Section 6.2), and this has had an enormous impact on the production of many ornamental plants such as orchids as well as crops. Third, cultured cells can multiply extremely rapidly; grown in bulk, some can also be induced to synthesize important drugs such as morphine or digitalin. It is easier to obtain drugs in this way than to extract the substances (found perhaps in a few tissues only), from large quantities of plant material gathered in the field.

A further application is in genetic manipulation. When immature pollen grains are cultured, they may give rise to callus and thence to plants whose cells are haploid (they have only half the normal number of chromosomes). Totally homozygous diploid plants which will breed true for every character, can be produced from doubled-up haploid cells formed as a result of treatment with colchicine (see Section 6.2.1). There is also a possibility of mixing genetic material other than by normal cross-breeding, by fusing haploid cells after digesting away their protective walls. Hybrid tobacco plants have been produced in this way, and there has been some success in fusing unrelated cells, although not with their subsequent regeneration. Another possible carrier of new genetic information is the crown gall bacterium we mentioned before (Section 6.3.3). The bacterium contains a small piece of DNA (called a plasmid) in addition to its main chromosome, and this becomes inserted into the host cell DNA after infection. The tumour persists even after the infecting bacterium has been killed by heat treatment. This suggests that the factor or factors which promote tumorous growth must be encoded in the plasmid. One effect of these factors is to promote PGR synthesis by the host cell.

Modern biotechnology provides a means by which, theoretically at least, we can replace the tumour-inducing portion of the plasmid DNA with more useful genes, such as those coding for particular enzymes and then infect plants with these 'beneficial' bacteria. The possibility of developing a wheat plant that does not need fertilizer because it has enzymes that fix atmospheric nitrogen or the hybrid tomato-potato that bears crops above and below ground are intriguing. Current research is already developing tomatoes that are insensitive to ethylene (see Chapter 7). This is done by inactivating the gene that codes for one of the enzymes (polygalacturonase) that softens the fruit, so these tomatoes have a much longer shelf-life. The possibilities are endless. But the feat of biotechnology in inserting or deleting genes has limited value unless such plants can be propagated. So micropropagation techniques will play an increasingly important role in the generation of genetically engineered plants.

6.6.2 PGRs and cambial activity

Is the development of vascular elements in callus comparable with the process in an intact plant? It is encouraging (at least to those who work with tissue cultures) that there are similarities.

Figure 6.47 shows the effects of auxin on the growth of cucumber roots. One of the effects was to stimulate cambial activity, thus producing more vascular tissue and making the root thicker. Roots removed from pea seedlings and transferred to a nutrient solution respond in the same way to added $5\ \mu mol\ l^{-1}$ auxin. If the tip is removed and transferred to fresh medium, it will continue to grow, and the tip transfer can be repeated several times over. However, when $5\ \mu mol\ l^{-1}$ auxin is added to the transferred roots, there is no stimulation of vascular thickening.

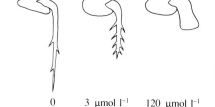

$0 \qquad 3\ \mu mol\ l^{-1} \qquad 120\ \mu mol\ l^{-1}$

Figure 6.47 Germination of cucumber seeds in various concentrations of the synthetic auxin NAA and the effect on subsequent root growth.

◇ What does this experiment suggest about the normal source of growth factors for the root?

◆ It appears that some factor needed for vascular thickening is normally supplied to the root by the shoot tissue. There is sufficient of this factor present for excised roots to respond to added auxin, but it is destroyed or diluted out in cultured roots.

A simple but elegant way of simulating the normal supply routes is shown in Figure 6.48. The medium in the vial (representing the shoot) and the medium on the plate (to simulate absorption through the root tip) were supplemented with various factors.

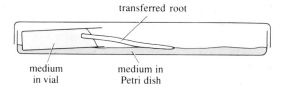

transferred root

medium
in vial

medium in
Petri dish

Figure 6.48 A technique for supplying excised roots with different media via the cut end and the root tip.

Maximum stimulation of cambial activity was observed when the vial contained mineral nutrients, $10\ \mu mol\ l^{-1}$ IAA and 8% sucrose, and the plate contained mineral nutrients and 4% sucrose. This is rather surprising because roots surrounded by medium containing the same mineral nutrients, $10\ \mu mol\ l^{-1}$ IAA and 8% sucrose, do not show vascular thickening. The important missing factor is not the presence or absence of a particular PGR but how it is supplied.

Similar experiments with radish roots showed that in this species, vascular thickening required an additional PGR, cytokinin. Thus, the same three factors — auxin, cytokinin and sucrose — are involved in vascular development in both callus and root tissue, although different tissues may have different requirements.

The importance of the direction from which the stimulus comes can also be seen in the control of cambial activity in the stems of trees such as poplar (*Populus* sp.) and sycamore (*Acer pseudoplatanus*). Growth occurs most rapidly in the spring, and cambial activity is initiated first in the region just below the expanding buds. It gradually spreads down the twigs and branches into the trunk and eventually to the roots. One simple explanation would be that PGRs produced by the expanding buds are transported down the stem.

◇ How would you test whether the buds were the source of PGR?

◆ By observing the activity of the cambium in disbudded twigs.

Disbudded twigs show very little growth, and the next step is to identify the active ingredient or ingredients supplied by the buds. Since auxin is required for cambial activity in both callus and roots, it should not surprise you that the next experiment was to apply IAA to disbudded twigs. This is done by applying a dab of lanolin (a greasy substance like petroleum jelly) containing IAA to the points where buds have been removed.

◇ What control experiment should be done?

◆ A disbudded twig should be treated with plain lanolin to make sure that this substance has no effects.

It was found that IAA-treated stems did become thicker, but not as thick as intact stems. Four questions therefore need to be answered:

(i)   How is IAA transported to its site of action?

(ii) What effect does IAA have on cambial cells?

(iii) What factors other than IAA are involved?

(iv) Is IAA involved in the intact plant?

Consider first the transport of IAA:

◇ What can you deduce about the route by which IAA is transported from the buds? (Consider the direction of flow in the vascular tissues.)

◆ The transpiration stream in the xylem is upward from the roots, and the organic nutrients needed for growth will be supplied to the bud in the phloem. Neither flow is in the required direction, and some other route must be used.

It is thought that the IAA transport depends largely on transfer from one cell to another (although it does also move in the phloem). Does it move down a concentration gradient, or is the direction of movement controlled in some other way? Suppose that you have been given a number of equal lengths of stem tissue, some blocks of agar jelly containing radioactively labelled IAA and some pure agar jelly. You are asked to test whether IAA can be transported in either direction along the stem, down a concentration gradient.

◇ How would you use these materials to measure the amount of IAA transported along the stem?

◆ The basic technique is to supply labelled IAA from a donor agar block at one end of the tissue and place a receiver block at the other end. After a period of time, the blocks are removed and the amount of labelled IAA present in each is measured (see Figure 6.49).

◇ How would you estimate the amount of IAA present in the blocks?

◆ Because labelled IAA is used, a quick check on the radioactivity in each block would tell you if anything has been moved. If there were some

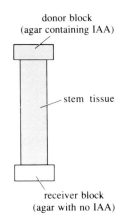

donor block
(agar containing IAA)

stem tissue

receiver block
(agar with no IAA)

Figure 6.49 A technique for measuring the rate of auxin movement through stem tissue.

radioactivity in the receiver block, you would then need to show that it was due to IAA (rather than some substance formed by metabolism of IAA) by chemical assay or bioassay.

◇ How would you investigate the direction in which IAA can move?

◆ Apply IAA to the apical end of a stem section, and measure the amount transported to an agar block at the basal end. The experiment should be repeated with a similar piece of tissue applying IAA at the basal end with a receiver block at the apical end.

Figure 6.50 shows the results of this experiment. The orientation of the stem is important, because the differences between the two sections on the left could simply mean that IAA cannot be moved against a gravitational force. This interpretation can be eliminated by placing the stem horizontally, or by comparison with stems turned the wrong way up. As you can see, this makes no difference to the direction of transport, which is always from apex to base. This **polar transport of auxin** only from the apex to the base (described as basipetal) is a general property of plant stems.

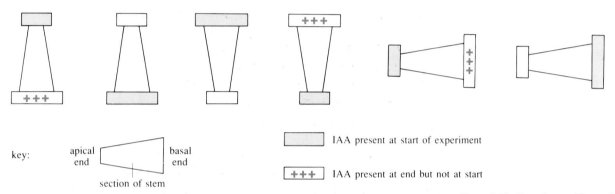

key:

apical end | basal end

section of stem

▭ IAA present at start of experiment

⊞ +++ IAA present at end but not at start

Figure 6.50 Experiments illustrating that the transport of auxin is always from the apex to the base of stem tissue.

◇ Can you suggest a further refinement to test whether a concentration gradient is needed to drive basipetal transport?

◆ Apply the same concentration of IAA at both ends and see if the concentration in the block at the basal end increases. (It does.)

Stems therefore have a powerful mechanism for transporting auxin towards the roots, and roots transport auxin towards the tip (acropetally), but what effect does auxin have on the cambium? To answer this, the experiments on poplar twigs were continued in the laboratory by taking lengths of disbudded twigs, placing the basal end in water and sealing the upper end with lanolin. Figure 6.51 shows sections of twigs treated with pure lanolin and with IAA in lanolin, together with twigs treated with gibberellin alone and in combination with IAA.

It is clear that both auxin and gibberellin have an effect: detailed examination shows that auxin causes the development of some cells into xylem elements, while gibberellin causes division of the cambial cells without subsequent differentiation. The action of gibberellin and auxin together is most definitely more than the combination of the actions of the two PGRs alone — there is virtually normal development of a complete ring of new xylem. This synergistic action of two PGRs is extremely common. In this case, we can suggest that gibberellin primarily acts to promote cambial division and auxin to induce differentiation, but in most cases there is no such obvious separation.

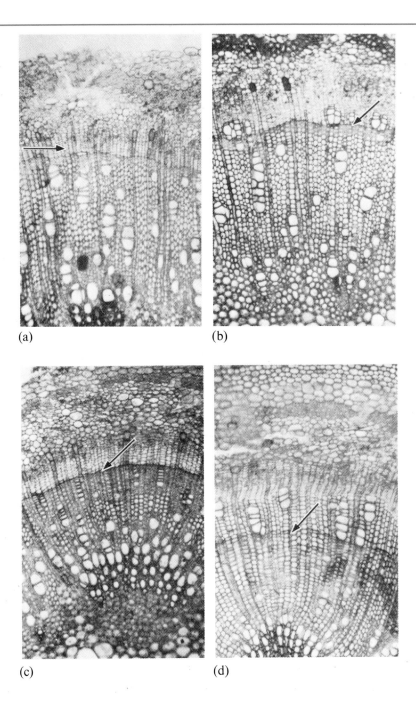

(a)

(b)

(c)

(d)

Figure 6.51 Sections through twigs treated with PGRs applied in lanolin from the apical end. (a) Lanolin only. (b) Lanolin containing IAA. (c) Lanolin containing GA$_3$. (d) Lanolin containing both IAA and GA$_3$. The arrows indicate the start of new growth.

It was also found that new phloem was formed mainly in response to gibberellin. However, do not be misled into believing that this is the whole story. It is fine as far as it goes, but the 'phloem' formed is almost entirely phloem parenchyma rather than the sieve tubes formed in whole twigs.

Do experiments such as these help to explain normal development in plants? It has been shown that expanding buds contain high concentrations of auxin and gibberellin, but it is difficult to measure the PGR concentration in the region of the cambium. In the absence of contrary evidence, the idea that these two PGRs have an important role in normal development is the best explanation available.

6.6.3 Other interactions

You should appreciate from the examples we have chosen that many developmental steps are controlled by a combination of two or more factors. We have considered PGR ratios and synergistic interactions between PGRs. In other processes too, the effect of two PGRs together may simply be the sum of the effects of each PGR alone (an additive effect). In yet other examples, the interaction may be antagonistic, and one PGR will counteract the effect of the other. PGR interactions are involved at various stages in the plant life cycle. Many of them are studied by observing comparatively major changes in the growth of the plant rather than at the cellular level.

Summary of Section 6.6

1 Tissue cultures can be used to study the stimuli needed for cell differentiation in isolation from the whole plant.

2 The effect of two PGRs applied at the same time is often not the same as the sum of the effects of the two alone. Differentiation can be controlled by the relative amounts of each PGR involved rather than just by absolute concentrations.

3 Cambial activity is controlled by PGRs in both roots and shoots, although different conditions are needed for the two organs.

4 IAA is transported in a polar way, basipetally from cell to cell towards the base of the stem, and acropetally along roots.

5 Micropropagation techniques, developed from tissue culture expertise, have important commercial applications, particularly in the production of large numbers of virus-free and genetically-engineered plants.

Question 15 (*Objectives 6.2, 6.12 and 6.13*) The growth of a callus requires oxygen and a supply of auxin. If one end of a segment of tobacco stem were inserted into nutrient medium containing 5 μmol l^{-1} 2,4-D (a synthetic auxin) would you expect to get a better growth of callus if the apical end or basal end were uppermost?

Question 16 (*Objectives 6.2 and 6.13*) In an experiment designed to measure the effects of a growth inhibitor on the gibberellin-induced growth of a plant, the increases in length of treated plants were:

no gibberellin and no inhibitor	50%
gibberellin alone	100%
gibberellin plus inhibitor	50%.

What can you conclude about the effect of the inhibitor on the response of the tissue to gibberellin?

6.7 CONCLUSIONS

At the start of this chapter, we pointed out that there were a number of external and internal factors that affect plant growth and development. This chapter has concentrated on the internal factors, particularly the role of PGRs on the growth and differentiation of individual cells. Chapter 7 will examine growth and development in a wider context — responses of whole plants or plant organs to external stimuli, and how these responses appear to be mediated by PGRs.

The current state of knowledge about PGRs presents an incomplete and somewhat unbalanced overall picture. On the one hand, applications of PGRs are widespread and are of increasing importance in agriculture and horticulture, particularly with regard to micropropagation techniques which depend on the manipulation of appropriate PGRs in the growth media. These applications represent an enormous potential for controlling plant growth and generating large numbers of plants with beneficial characteristics. Yet our understanding of the fundamental processes underlying such control is frustratingly limited.

New techniques are now emerging that may break through some of these limitations. For instance, many early experiments on PGRs had to rely on the application of artificial analogues of natural PGRs; the discovery that plant cells can be induced to produce more of their own PGR as a result of infection by certain agrobacteria enables a whole range of new experiments to be devised. Another limitation has been the difficulty in measuring the extremely low PGR levels in plant tissue; new analytical techniques now enable researchers to identify the presence of minute traces of any PGR reliably and routinely. So at last a chemical assay method can be compared with the traditional bioassays in an attempt to distinguish between presence of a PGR and its activity. Information about both is necessary in order to distinguish between responses due to changing PGR levels and those due to changing sensitivity of cells. Most responses probably lie between these two extremes.

It is generally assumed that PGRs operate by binding to receptors within the target cells. The difficulty has been to determine whether binding leads to other events that result in the physiological response. Emphasis is now being placed on investigating the formation of new gene products when a cell is stimulated. Techniques borrowed from genetic engineering are beginning to make an impact here. Genes and gene switches can be inserted in and deleted from the genetic make-up of particular cells; such experiments will enable us to establish whether new gene products are required to produce a response to a stimulus. Advances such as these should soon produce information that will fill the gaps in our knowledge of how PGRs operate at the biochemical level.

But however good the techniques become, advances in our understanding of the role of PGRs in whole plants will require experiments that test *appropriate* hypotheses about the way the activity of cells is co-ordinated. The assumption that PGRs work in a similar way to animal hormones and devising experiments under this assumption has not produced the advances expected in our understanding of how the co-ordination system as a whole works in plants. One of the recurring themes in this chapter is the danger of assuming that PGRs operate in a similar way to animal hormones; there are examples of PGR action that do have such parallels, but the ability of cells to change their sensitivity to background levels of PGRs provides an additional dimension to the way in which growth and developmental processes could be controlled. It is also becoming clear that there is no single mechanism that satisfactorily explains all the different actions of PGRs on cells. It is this multiplicity of action that distinguishes PGRs from animal hormones.

As one (somewhat frustrated!) worker commented; 'Multicellular plants are not green animals; the sessile and autotrophic lifestyle may reflect a distinctively evolved approach to communication between cells and from the environment'.

OBJECTIVES FOR CHAPTER 6

Now that you have completed your study of this Chapter, you should be able to:

6.1 Define and use or recognize definitions and applications of all the terms set in **bold** type. (*Questions 7, 8 and 12*)

6.2 Design, interpret and evaluate experiments that relate to the processes of cell enlargement and cell differentiation and the roles of PGRs in these processes. (*Questions 4, 5, 6, 10, 11, 12, 14, 15 and 16*)

6.3 Describe the locations of cell division, cell expansion and cell differentiation in a growing plant. (*Questions 1, 2, 3 and 4*)

6.4 Describe and compare the way in which growth occurs in stems, roots, leaves and coleoptiles. (*Questions 2 and 3*)

6.5 List the five main classes of PGR and name one representative of each and briefly describe one developmental process where it is a controlling factor. (*Question 7*)

6.6 Give examples of some commercial applications of PGRs or their antagonists. (*Question 7*)

6.7 List three factors that determine the rate of cell expansion and predict how altering the value of each would alter the rate of expansion. (*Questions 8, 9, 10 and 11*)

6.8 List three components found in most plant cell walls and briefly describe the role of each. (*Questions 8 and 9*)

6.9 Explain how the auxin IAA is thought to increase cell wall plasticity (i) according to the acid growth hypothesis and (ii) by inducing the production of new protein. Assess whether data from experiments support or conflict with each hypothesis. (*Question 12*)

6.10 Assess whether data from experiments on tropic curvature support or conflict with hypotheses that involve (i) uneven distribution of PGRs or inhibitors (ii) differential sensitivity of cells (iii) direct effects of stimuli on responding cells (*Question 14*)

6.11 Explain why it is more difficult to study PGR responses in plants than hormonal responses in animals. (*Question 13*)

6.12 Give a simple account of the use of tissue cultures and give examples of the commercial exploitation of this technique. (*Question 15*)

6.13 Evaluate experiments concerned with (i) the interaction of PGRs and (ii) the role of IAA in controlling cambial activity in the plant. (*Questions 15 and 16*)

◆ CHAPTER 7 ◆ THE REGULATION OF PLANT GROWTH

7.1 INTRODUCTION

The division, expansion and differentiation of plant cells (in other words plant development) are subject to strict control. Whilst some of the processes involved appear to occur whatever the growth conditions, others are initiated or modified by changes in the environment.

The seeds of some annual plants, when grown under constant laboratory conditions, will complete their normal life cycle of germination, vegetative growth, flowering and senescence without prompts from external signals. However, even in these plants, the onset and duration of each developmental phase can be subtly influenced by stimuli such as light, temperature, nutrients and water availability. Environmental influences such as these are said to affect the developmental flexibility or **phenotypic plasticity** of plants. It is this flexibility which ensures that plants can adapt to the conditions in which they are growing. In many plants, changes in their external environment have a profound effect on their subsequent development, and we will see in this chapter that some phases in the life cycle of plants are under the control of specific environmental signals.

It has been proposed that the signals which plants perceive from the environment are transduced by the family of molecules termed plant growth regulators (PGRs — see Chapter 6). Evidence to support this proposal is difficult to obtain because, as you will recall from the previous chapter, the response to a PGR can be influenced by either the concentration of the molecule or the sensitivity of the plant to it. You should not, therefore, be surprised that the evidence for some of the control mechanisms described in this chapter is sparse or equivocal. However, with the advent of new techniques in molecular biology and the discovery of mutants which develop in abnormal ways, plant physiologists are moving towards a position of greater confidence with regard to the role of PGRs in certain developmental events. A number of stages in the life cycle of flowering plants have therefore been chosen to illustrate different aspects of the control of plant development.

Both seed germination (Section 7.2) and the development of buds in spring (Section 7.3) require the resumption of growth in dormant apical meristems. Other processes involve changes in the pattern of growth at active meristems leading to their vegetative (Section 7.4) or reproductive (Sections 7.5 and 7.6) growth. The final events we consider deal with the terminal phases of plant development such as senescence, fruit ripening and abscission (Section 7.7). Although these processes ultimately lead to the death of an organ or whole plant, we will see that they do not occur in a chaotic fashion but are regulated in a highly co-ordinated way — in other words, death is an active process in plants.

7.2 SEED GERMINATION

The life-cycle of a flowering plant starts when haploid male and female gametes fuse to form the single-celled diploid zygote that eventually gives rise to the mature plant. Fertilization takes place within the ovule, and the embryo continues to develop in this protected environment. By the time the mature seed is shed from the parent plant, the embryo is already well developed and embryonic root (or **radicle**) and shoot (or **plumule**) portions can be recognized. The structure of the seed of a typical dicot plant, the French bean (*Phaseolus vulgaris*) is shown in Figure 7.1. As you can see, the shoot- and root-forming portions take up a relatively small proportion of the seed, which mainly consists of food reserves — in this case proteins and polysaccharides stored within the seed leaves or **cotyledons**. In mature seeds of monocots, such as barley (*Hordeum vulgare*) or wheat (*Triticum aestivum*), the food reserves are stored in a tissue which surrounds the embryo called the **endosperm** (Figure 7.2). One of the earliest events in germination is the mobilization of reserves providing the energy and raw materials necessary for the establishment of the seedling.

During the development of the seed, the embryo usually enters a period of inactivity or **dormancy** during which time the average water content of the seed falls to below 15%. This phase is important because it reduces the likelihood of germination taking place on the mother plant. The duration of the dormant phase can vary from species to species. In cereals, it may be so brief that farmers become concerned about rain around harvest time because of the danger that wheat and barley grains will germinate within the ear, making them useless for either flour milling or brewing. In other species, the period of dormancy may last many months or years. For instance, seeds from the pernicious weed wild oat (*Avena fatua*) have been known to lie dormant in the soil for up to seven years from the time of shedding.

Four major factors determine whether, and when, a seed will germinate. These are: water availability, light, temperature and the age of the seed.

Water availability

Seeds need a supply of water because of the very low moisture content they have when they are shed. However, few seeds will germinate if totally submerged in water, because the growth of new tissues requires metabolic energy and this is usually obtained by aerobic (i.e. oxygen-requiring) respiration. The seeds of desert plants must respond rapidly to rainfall if they are to exploit the short-lived conditions favourable for growth, and heavy rain has been shown to wash out *inhibitors of germination* which are present in these seeds. Chipping away a portion of the seed coat or **testa**, is often recommended to promote germination in sweet peas (*Lathyrus odoratus*), and this strategy can be effective in other species as well. A common explanation for the success of this treatment is that the testa acts as an impermeable barrier and restricts uptake of water by the embryo.

◇ In what other ways might the testa inhibit germination?

◆ It might limit gaseous exchange between the embryo and its environment; it might produce an inhibitor, or prevent the leaching of an inhibitor as described previously; or it might act as a physical barrier to radicle emergence — the first external sign of germination.

Light

One reason why light may act as a signal for germination is that, under natural conditions, its perception would indicate that the seed is at or near the surface

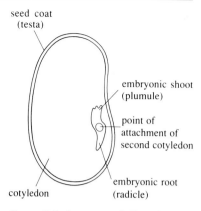

Figure 7.1 Structure of French bean seed, observed by removing the upper cotyledon.

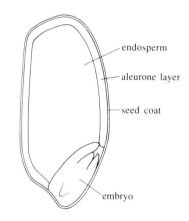

Figure 7.2 Structure of a grain of barley, observed by cutting it in half.

of the soil. This information would be of particular importance for species having small seeds with few stored reserves. For instance, many weed seeds germinate as soon as the soil in which they may have lain for many years, is turned over. On the other hand, germination is inhibited by light in a few species that grow in arid regions. Presumably, in these species it would be advantageous if seed germination, and hence root establishment, took place some way below the soil surface thus reducing the risk of desiccation.

Temperature

In temperate regions, many seeds are shed towards the end of the growing season. The requirement of a minimum period of chilling before germination becomes possible (known as *after-ripening*) helps to ensure that germination does not take place in the autumn. Conversely, seeds need warmth to germinate, and the requirement for a certain minimum temperature prevents ripened seeds from germinating before the arrival of spring.

Age

The fourth factor, age of the seed, is of critical significance for germination. Seeds stored under conditions of low relative humidity may remain viable for many years. Indeed viable *Canna compacta* seeds found in an excavated tomb in Argentina have been estimated by carbon dating to be approximately 600 years old. In contrast, some hydrated seeds can lose their viability in just a few months. Seeds of cultivated plants often have a particularly short lifespan if kept under conditions where their moisture content is in excess of 15%.

7.2.1 The germination of barley grains

Although barley does not require chilling before it will germinate, germination is more rapid after a short period of dry storage than it is immediately after harvest. This partial inhibition of germination disappears within a month of storage and the only factors that seems to influence germination thereafter are the availability of water and adequate temperature. Because barley grains require such straightforward stimuli to promote their germination they are good subjects for the characterization of some of the processes associated with this developmental event.

As stated previously, food reserves in the mature barley grain (Figure 7.2) are stored in the endosperm tissue (rather than in the cotyledons as in French bean), and are mainly in the form of starch together with some protein. A layer of metabolically active cells, the **aleurone layer**, is located between the endosperm and the seed coat.

Water uptake occurs because the seed has a lower water potential (Ψ) than its surroundings. Dry seeds contain a lot of colloids (in this case, the starch grains), and these contribute to m, the matric pressure, and therefore lower the water potential (Chapter 5, Section 5.2).

◇ Would you expect seeds in which the embryo is dead to take up water?

◆ Yes, because if water uptake is due to matric pressure, it will depend on the presence of colloids, rather than on the presence of living cells. Of course, the passive uptake of water in these non-viable seeds is not followed by the normal processes associated with germination.

In many varieties of barley the first visible sign of germination is the emergence of the radicle from the seed. This event, however, is preceded by a series of biochemical reactions including the breakdown of the stored starch to yield the glucose necessary for use both as a respiratory substrate and as a source of the carbon skeletons of the molecules needed for growth. The enzyme responsible for the mobilization of the food reserves, **α-amylase**, was first extracted from germinating barley grains in 1833 — although the site of its production was not finally pin-pointed until the 1960s. An early test for the production of α-amylase was to cut grains in half transversely and place them, cut surface down, on a thin sheet of agar (a jelly-like material) containing a small amount of starch. After an appropriate time, the half-grains with and without an embryo were removed and iodine solution (which stains starch blue) was poured onto the agar. Where α-amylase had diffused from the tissues into the agar and broken the starch down, the agar failed to turn blue, resulting in the appearance of clear spots against a blue background (Figure 7.3).

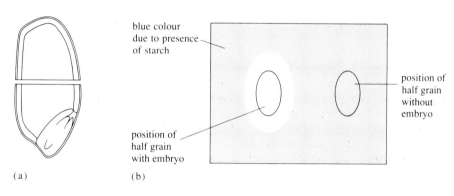

Figure 7.3 (a) position of half grain with embryo, blue colour due to presence of starch, position of half grain without embryo (b)

Figure 7.3 The effect of the presence of the embryo on the production of α-amylase by barley grains. (a) After transverse halving, the grains were placed cut surface down on a sheet of agar containing a small amount of starch. (b) 48 hours later the grains were removed and the agar stained with iodine solution.

◇ Which of the following explanations are consistent with the results shown in Figure 7.3?

(i) Absorption of water activates α-amylase, which is present in endosperm cells.

(ii) Absorption of water causes the embryo to produce α-amylase.

(iii) Absorption of water causes the embryo to secrete a substance that induces α-amylase production in other parts of the seed.

◆ Only explanation (i) can be discounted (because half-grains without embryos fail to produce α-amylase). Although (iii) is more complex than (ii), both explanations are equally plausible.

◇ Suggest an experiment to distinguish between explanations (ii) and (iii) by the starch gel technique.

◆ Embryos could be dissected from endosperm tissue and placed on the starch gel to see whether they could produce α-amylase.

It is not easy to remove all traces of endosperm from isolated embryos; but when the experiment was carried out in the 1890s, a clear area was found round the shoot end of the embryo. Explanation (ii), that the embryo secretes an enzyme that releases substrates for its own growth, seemed to be valid. So the story remained for another fifty years despite the observation by Haberlandt that, in germinating seeds, the starch grains disappeared just as rapidly from cells just under the surface of the testa as from those adjacent to the embryo.

The next development came when Japanese researchers incubated barley grains without embryos in a culture medium in which isolated barley embryos had previously been cultured. By this time, it was possible to measure the activity of α-amylase accurately and, although traces of the enzyme were found to be present in the medium after incubation with isolated embryos, α-amylase activity increased dramatically when the embryo-less portions of the barley grains were added. Earlier workers had not realized that isolated embryos did not produce enough α-amylase to account for what happened in the whole grain. The Japanese researchers subsequently identified the ingredient from the embryos that induced α-amylase synthesis as the gibberellin, GA_3 (Chapter 6). At the same time, it was shown in Australia that GA_3 (which had by then been purified from other sources) induced α-amylase production when it was added to embryo-less grains.

Haberlandt's microscopical observations had suggested that cells near the surface of the grain might be important. This is the position of the aleurone layer (Figure 7.2), the only metabolically active tissue in the seed apart from the embryo, and it was suggested that these cells were induced by gibberellin to synthesize α-amylase.

If the seed coat is removed and the surface of the seed scraped very carefully, the aleurone cells can be separated from the endosperm.

◇ What experiments could be performed to test whether gibberellin acts on the aleurone cells in the way described?

◆ Incubate the different parts of the seed (embryo, endosperm and aleurone) in media with and without added gibberellin. If the description is correct, only the medium containing gibberellin and aleurone cells will have high α-amylase activity.

Gibberellins, particularly GA_1 and GA_3, do indeed stimulate aleurone cells to secrete α-amylase into the surrounding medium.

The levels of gibberellins in whole seeds do increase at the start of germination, immediately before an increase in α-amylase can be detected, and gibberellin does induce α-amylase production by isolated aleurone cells when applied at physiological concentrations. The sequence of events summarized in Figure 7.4 is therefore well supported by experimental evidence (even if conclusive proof is missing). Uptake of water causes the embryo to secrete gibberellin. This PGR diffuses to the aleurone cells which are then induced to secrete α-amylase. This hydrolyses the starch grains in the underlying endosperm and the glucose produced is presumed to diffuse back to the embryo where it is used to fuel growth.

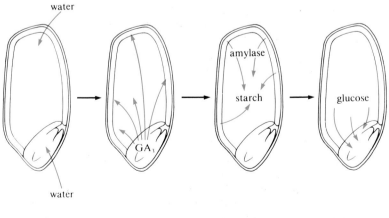

Figure 7.4 Mobilization of stored reserves in a barley grain.

The aleurone layer can be removed from barley grains which have previously been imbibed in water. It can then be treated with cell wall-degrading enzymes to release aleurone protoplasts (i.e. wall-less cells). These protoplasts respond to gibberellin just like intact tissue and secrete α-amylase (see Figure 7.5).

◇ Explain the observation that aleurone protoplasts secrete α-amylase in response to lower concentrations of GA_3 than intact seeds.

◈ Much of the GA_3 applied to intact seeds fails to get to the aleurone layer and therefore a higher concentration of the PGR must be applied. This is a general criticism of experiments where PGRs are applied to whole plants but may act on only a few cells.

However, many questions remain unanswered. For instance, does water uptake induce the synthesis of gibberellin or release it from some previously inactive form? Does gibberellin stimulate RNA synthesis, protein synthesis, protein export or some combination of the three?

◇ What can be deduced about the involvement of RNA synthesis from the results presented in Table 7.1?

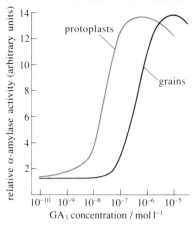

Figure 7.5 Dose–response curve of barley grains and aleurone protoplasts to GA_3.

Table 7.1 The effect of inhibitors on α-amylase production by barley aleurone layers. Isolated aleurone layers were incubated in 10^{-3} mol l^{-1} GA_3 for 48 hours. α-amylase activity in the medium was then scored as present $(+)$ or absent $(-)$. Actinomycin D is an inhibitor of RNA synthesis; cycloheximide inhibits protein synthesis.

Treatment	Inhibitor added	Time inhibitor present	α-amylase production
1	none	—	+
2	actinomycin D	0–48 h	−
3	actinomycin D	10–48 h	+
4	cycloheximide	0–48 h	−
5	cycloheximide	10–48 h	−

◈ Gibberellin stimulates RNA synthesis (treatment 2 causes inhibition), but the RNA is stable once formed (treatment 3 does not cause inhibition).

Actinomycin D inhibits the synthesis of all types of RNA — messenger (mRNA), ribosomal (rRNA) and transfer (tRNA). So GA_3 could stimulate (i) RNA synthesis generally, (ii) the synthesis of α-amylase mRNA specifically or (iii) the synthesis of either rRNA or tRNA in a system in which the levels of these RNAs limit the rate of protein synthesis. Experiments in which mRNA extracted from barley grains is translated *in vitro* suggest that gibberellin specifically affects the synthesis of α-amylase mRNA (i.e. transcription of the α-amylase gene).

Barley germination and gene expression

Over the last few years it has become clear that the coordination of developmental events relies heavily on precisely regulated changes in the expression of specific genes. We have seen in the preceding section that α-amylase plays a crucial role in the germination of barley grains, and we will discover later in this chapter that other enzymes play critical roles in regulating developmental events such as ripening and abscission. With the advent of the revolutionary techniques of molecular biology it is now possible to find out not only *when* a particular gene is expressed but also *what* regulates its expression. In the near future it is hoped to be able to explain *why* a gene exhibits tissue-specific expression. That is, in the case of barley grains, why gibberellin dramatically stimulates α-amylase synthesis in the aleurone cells but only slightly in cells of the embryo. This information not only has scientific value but also has significance in any goal to manipulate crop growth and development by genetic engineering procedures.

Like many recent scientific developments, molecular biology is cloaked in its own jargon and there is not space in this book to go into the details of experimental procedures. Nevertheless it is important to be acquainted with some of the approaches used in molecular biology to study gene expression and these are outlined in Box 7.1.

Box 7.1 Some genetic engineering techniques

Cloning

mRNAs from a tissue are extracted and purified. Using an enzyme called *reverse transcriptase*, DNA copies of the mRNAs are synthesized, and then inserted into the bacterium *Escherichia coli*. Each bacterium takes up a different DNA copy (or cDNA) of the mRNA molecules present. The cDNA is replicated by the bacterium each time it divides, and in this way a 'library' of cDNAs coding for different mRNAs extracted from the tissue is constructed. cDNAs of mRNAs which are specifically associated with a developmental event can subsequently be identified.

Sequencing

The nucleotide sequences of the cDNAs are determined. Since each triplet of bases codes for a particular amino acid, once the amino acid sequence of a protein is known the cDNA which codes for it can be identified.

One of the main aims of these techniques is to sequence the gene which codes for a particular protein. Since the cDNAs are direct copies of

mRNAs, these *do not* represent the sequences of the genes themselves since the mRNAs are processed (see Figure 7.6). Many genes are split by *introns* and these are lost during post-transcriptional modifications. Moreover, sequences outside or *upstream* from the mRNA coding regions themselves are now known to play important regulatory roles and these are not included on the mRNAs. These sequences are known as *promoter* regions. Since the cDNAs will have a similar structure to the genes from which the RNAs originated, it is possible to use those cDNAs that have been identified as coding for particular proteins as aids to isolate the genes themselves. The genes can then be sequenced.

Once the gene coding for a particular protein has been identified, its expression can be monitored during a developmental event. It is then possible to study its promoter regions for clues as to how its expression is regulated, and its structure can be altered to examine the impact that this has on the developmental process.

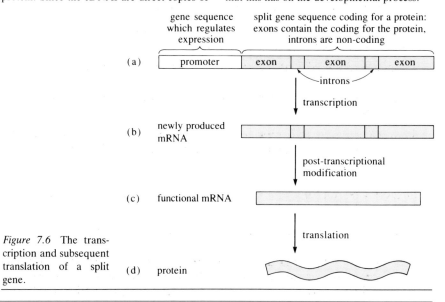

Figure 7.6 The transcription and subsequent translation of a split gene.

The techniques of molecular biology have enabled the biochemical events which are loosely termed germination to be unravelled. There is now good reason to believe that, in barley, the PGR gibberellin has a crucial role to play. It acts both to initiate and coordinate the mobilization of the stored food reserves which are central to seedling growth and development. But what induces the synthesis and release of the gibberellin from the embryo in the first place? Although there is not yet an answer to this question, other germination systems are also being studied in detail, and in some of these, answers may prove to be more forthcoming.

7.2.2 The germination of lettuce seeds

Lettuce (*Lactuca sativa*) 'seeds' are actually dry fruits in which the true seed is surrounded by a layer equivalent to the shell of a nut. We shall refer to this extra layer *plus* the very thin seed covering as the *coat*. Unlike barley grains, seeds of the variety of lettuce known as 'Grand Rapids' exhibit dormancy. The duration of this condition is strongly influenced by environmental conditions (Table 7.2).

Table 7.2 The effect of experimental conditions on the germination of 'Grand Rapids' lettuce seeds. For each treatment, approximately 50 seeds were placed on moist filter paper in a Petri dish. Dishes for dark treatment were immediately wrapped in silver foil. Where applicable, the seed coat was damaged by gently abrading the seeds before sowing. The numbers of germinated and ungerminated seeds were counted after 30 hours.

Treatment	Illumination	Temperature	Seed coat	Number of seeds germinated	ungerminated
1	dark	25 °C	intact	10	38
2	light	25 °C	intact	47	3
3	dark	2 °C for 4 h, then 25 °C	intact	40	9
4	dark	25 °C	damaged	46	5

◇ What are the three experimental treatments that can be used to stimulate the germination of 'Grand Rapids' lettuce seeds?

◈ 1 Place the seeds in the light rather than in complete darkness.
   2 Chill the seeds for a few hours initially.
   3 Damage the seed coats before the start of the experiment.

◇ What is the percentage germination in the control treatment and in each of the others?

◈ The percentage germination in the control experiment is 21%; when the seeds are placed in light it is 94%, when they are chilled it is 82% and when the seed coat is damaged it is 90%.

◇ Suggest a reason for the lower percentage germination of the cold-treated seeds.

◈ Metabolism and growth will be slower during the cold treatment.

This is confirmed by using different combinations of treatments. Cold plus light and cold plus coat damage do not stimulate germination any more than does cold alone. It therefore looks as if control of germination in 'Grand

Rapids' might be like a single on/off switch that can be turned to the 'on' position by light, by chilling or by coat damage. Most of the work on 'Grand Rapids' has been concerned with how this switch might work and in particular how it is triggered by light.

The effect of light

The results shown in Table 7.2 are from an experiment which was set up in a well-lit laboratory. When the dishes were wrapped in aluminium foil, unchilled seeds with an intact coat did not germinate readily. However, if the foil cover was taken off for as little as one minute two hours after the seed had been imbibed with water, most of the seeds germinated.

◇ Are dry seeds sensitive to light?

◈ Apparently not, since no special precautions were needed in setting up the dark control in Table 7.2.

Sensitivity to light seems to develop as the seeds begins to take up water. Once sensitivity has developed, a brief exposure to light stimulates germination, and this stimulatory effect of light is not reversed by returning the seeds to darkness.

In order to exert an effect on plant development, light must first be absorbed by some *pigment* within the plant. One way to ascertain the identity of the pigment responsible is to determine the wavelengths of light which are most effective at stimulating germination (Figure 7.7).

◇ Which wavelengths are most effective?

◈ Light in the 600–700 nm wavelength region induces maximal germination. The receptor is therefore likely to be a pigment that absorbs light strongly in this region of the spectrum, i.e. red light.

The proportion of seeds that germinate in complete darkness varies from batch to batch. It was about 21% in the results presented in Table 7.2, but about 50% in the results shown in Figure 7.7. In addition to the stimulation by red light, Figure 7.7 shows that there is a marked inhibition of germination by light of wavelengths around 720–760 nm. This region is right at the limit of the visible spectrum and is called '*far-red*' light. Although the stimulatory effect of red or white light on germination is not reversed by darkness, it can be abolished if the light treatment is immediately followed by a brief exposure to far-red. Table 7.3 shows the effect on germination of alternating 5 minute periods of red and far-red light.

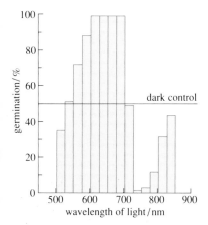

Figure 7.7 The effect of different wavelengths of light on the germination of 'Grand Rapids' lettuce seeds.

Table 7.3 Control of germination of 'Grand Rapids' lettuce seeds by red and far-red light. After being allowed to absorb water in the dark, seeds were given consecutive 5 minutes of illumination with red and far-red light and the percentage germination determined.

Treatment	% germination
red	70
red/far-red	6
red/far-red/red	74
red/far-red/red/far-red	6
red/far-red/red/far-red/red	76
red/far-red/red/far-red/red/far-red	7
red/far-red/red/far-red/red/far-red/red	81
red/far-red/red/far-red/red/far-red/red/far-red	7

◇What factor determines whether a high proportion of the seeds germinate?

◆The wavelength or *quality* of the light to which they were last exposed. A high percentage of the seeds germinate if this was red, but a much lower percentage if this was far-red.

The original explanation of the results shown in Figure 7.7 was that there were two light receptors: one that stimulated germination as a result of the absorption of red light and another that inhibited germination after the absorption of far-red light. However, the fact that the effect of red light could be completely and repeatedly reversed by far-red, suggested to the two American scientists, Borthwick and Hendricks, who characterized the response in the 1950s, that there were two forms of the same pigment, which could easily be interconverted. The existence of such a pigment (which became known as **phytochrome**) remained unproven for some time.

◇In view of the wavelengths of light phytochrome absorbs, why should it be difficult to detect small amounts of the pigment in tissues such as leaves?

◆The presence of phytochrome will be masked by the large amounts of chlorophyll, which also absorbs red light (Chapter 2).

The presence of a pigment that absorbs light of the wavelengths expected for phytochrome was therefore first detected in the coleoptiles of dark-grown oat (*Avena sativa*) seedlings, which contain no chlorophyll. Figure 7.8 shows the absorption spectrum of purified oat phytochrome.

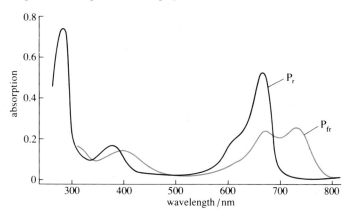

Figure 7.8 The absorption spectrum of solutions of oat phytochrome that had previously been illuminated with far-red light (green line) and with red light (black line). The light intensity used to obtain the absorption spectrum is too low to effect the conversion.

The two interconvertible forms of phytochrome, P_r and P_{fr}, have different absorption spectra. P_r predominantly absorbs red light and P_{fr} predominantly absorbs far-red light. As predicted, absorption of red light converts P_r to P_{fr}, and absorption of far-red light does the reverse. This is usually written as an equilibrium:

$$P_r \underset{\text{far-red light}}{\overset{\text{red light}}{\rightleftarrows}} P_{fr}$$

In other words, red light converts phytochrome P_r (the red-absorbing form) to P_{fr} (the far-red absorbing form). Far-red reverses the effect of red light because far-red converts phytochrome P_{fr} to P_r.

◇Which form of phytochrome predominates after illumination with red light?

◆P_{fr} (the far-red absorbing form).

Phytochrome cannot be detected in dry lettuce seeds, although it appears rapidly once water uptake has begun. This is further evidence in favour of the involvement of phytochrome in light-stimulated germination because the sensitivity to light develops at about the same time that phytochrome becomes detectable. However, because of the difficulty of detecting it, a demonstration that a response shows red/far-red reversibility is often the strongest evidence available for the involvement of phytochrome.

An observation that must be explained, however, is that white light has a similar effect on germination as red light. Figure 7.9 shows the relative amounts of different wavelengths of light in natural daylight. The effect of white light will depend on which form of phytochrome it will cause to predominate.

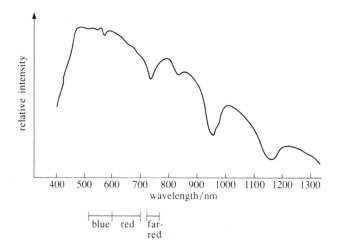

Figure 7.9 The distribution of wavelengths in natural daylight measured on a sunny day.

◇ Which form of phytochrome will predominate in plants exposed to natural light?

◈ Both forms will be present, but there will be more P_{fr} than P_r on a sunny day because daylight then contains more red light (600–700 nm) than far-red (720–760 nm) and P_{fr} is formed (from P_r) by absorption of red light.

It is evident from the preceding discussion that both the amount of phytochrome and the form it is in can play an important role in regulating developmental events such as germination. Furthermore, these features can be influenced by both the quality and the quantity of light. For instance, it is now known that phytochrome is synthesized in the form of P_r and that this process is itself inhibited by high levels of P_{fr} generated by red light. In addition, P_r is the stable form of the pigment while P_{fr} can be destroyed in the presence of light or revert back to P_r in the dark. These reactions therefore contribute to the complexities which surround phytochrome-mediated developmental events:

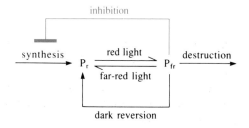

The light-induced reactions are very rapid compared with synthesis, destruction and reversion, which explains why short periods of illumination are extremely effective.

◇ From the description so far, can you decide whether germination is stimulated by the presence of P_{fr} or by the absence of P_r?

◆ No. The results described so far are equally consistent with either alternative.

How then might it be possible to distinguish between these alternatives? Different light sources produce irradiation containing different proportions of red and far-red wavelengths. Consider two light sources A and B, which are expected to convert 5% and 1% respectively of the phytochrome in plant tissues to the P_{fr} form. A sample of 'Grand Rapids' seeds was kept under each of the two sources.

◇ If germination is controlled by the amount of P_{fr} present, would you expect the percentage germination to be very different in the two samples?

◆ Since there will be five times as much P_{fr} present under A as under B, a large difference in the percentage germination might be expected.

The difference in amount of P_r (95% compared to 99%) is proportionately much smaller. In fact, many plant tissues do show a much greater response to light sources like A than to sources like B. It is difficult to see why such a proportionately small change in P_r should produce a large response. So, in the absence of contradictory evidence, it is assumed that effects triggered by phytochrome depend on the amount of P_{fr} present. In lettuce seeds, germination is stimulated by light and therefore by high P_{fr}. Other developmental processes which are also known to be mediated via phytochrome, such as increased stem length in dark-grown seedlings (*etiolation*), are inhibited by light and occur only when P_{fr} levels are low.

The precise amount of P_{fr} needed to stimulate germination in 'Grand Rapids' is not known. Under continuous illumination most of the phytochrome will be in the P_{fr} form throughout the course of the experiment, but what about seeds given only a short light treatment and then returned to the dark?

◇ What will happen to P_{fr} levels in these seeds.

◆ P_{fr} levels will be high immediately after illumination, but will decline as P_{fr} reverts to P_r or is destroyed.

When seeds are exposed to a continuous sequence of red followed by far-red light treatments, the percentage germination is determined by the quality of the light to which they were last exposed (Table 7.3). What happens to the percentage germination as the interval between the red and the far-red light treatments is increased? Table 7.4 shows the effect on germination of giving 5 minutes of far-red illumination various times after 5 minutes of red light.

◇ What can be deduced from these results?

◆ Far-red light, given up to an hour after red light, completely reverses the latter's effect, suggesting that high P_{fr} levels must be maintained throughout this period to stimulate germination. After 8 hours, far-red has no inhibitory effect, suggesting that the processes which lead to germination have been irreversibly triggered. Between 1 and 8 hours, the percentage of seeds germinating increases despite far-red treatment.

Table 7.4 The effect on the germination of 'Grand Rapids' lettuce seeds of 5 minutes of red illumination followed by 5 minutes of far-red illumination separated by dark intervals of various lengths.

Treatment	% germination
red only	100
red/no dark interval/far-red	15
red/1 h dark interval/far-red	15
red/2 h dark interval/far-red	20
red/4 h dark interval/far-red	40
red/6 h dark interval/far-red	60
red/8 h dark interval/far-red	100

The dormancy-breaking mechanism in lettuce

Light is not the only stimulus that will stimulate germination of 'Grand Rapids' lettuce seed. In addition to chilling and abrasion of the seed coat (Table 7.2), germination can also be promoted by gibberellin.

The question that therefore needs to be answered, is: do all these stimuli promote lettuce seed germination by the same mechanism?

Earlier in this chapter the properties of the testa were examined, and it was concluded that it might affect germination by restricting the leakage of an inhibitor out of the seed. One way in which we could investigate this hypothesis is to carry out the series of experiments outlined in Figure 7.10. This figure shows the impact on germination of making particular incisions on seeds of 'Grand Rapids' lettuce.

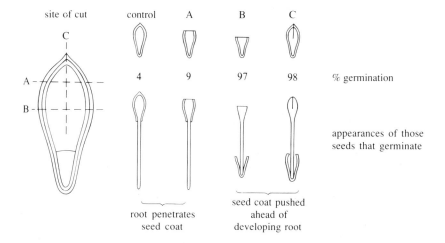

Figure 7.10 The effect on germination of cutting particular regions of 'Grand Rapids' lettuce seeds in the dark.

◇ If dormancy is due to the presence of a water-soluble inhibitor, would you expect seeds with the top quarter removed (treatment A, Figure 7.10) to germinate in the dark?

Yes, because the inhibitor should have diffused from the cut surface. In fact only 9% germinated, which implies that dormancy of 'Grand Rapids' seeds is not due to the presence of an inhibitor.

Seeds with the top half removed (treatment B) and those split vertically (treatment C) do germinate readily in the dark, although the growing root pushes the seed coat ahead of it rather than penetrating it. This suggests that

'Grand Rapids' seeds may exhibit dormancy because they are unable to overcome the physical restraint imposed by the seed coat, even after they have absorbed water.

The ability to overcome the restriction of the seed coat may be part of the mechanism responsible for light-stimulated germination in 'Grand Rapids'. Although water is absorbed in the dark as a result of the matric pressure of the endosperm, it is insufficient to break the seed coat. However, the 'pushing force' of the embryo appears to increase after illumination. Embryos from seeds that have not been exposed to light do not grow in mannitol solutions more concentrated than $0.4 \, \text{mol} \, l^{-1}$ (because the water potential of the cells is less negative than that of the mannitol solution and so water flows out of the cells). After 5 minutes illumination with red light, growth is possible in mannitol solutions of up to $0.46 \, \text{mol} \, l^{-1}$, so the water potential of the cells must have decreased — becoming *more* negative. This can only have happend because the cells developed a higher osmotic pressure (Π). This means that they can develop a higher turgor pressure (P) when fully turgid (since $\Psi = P - \Pi$; Chapter 5, Section 5.2).

It seems most likely that illumination leads to an increase in turgor pressure within the cells of the embryo, possibly as a result of rapid changes in membrane permeability to ions such as K^+ and H^+. Illumination is known to alter the electrical resistance of artificial membranes consisting of purified lipids and phytochrome. It is therefore tempting to suggest that the light stimulation of germination is mediated through a direct effect of membrane-bound phytochrome on ionic permeability. The stimulation of germination by chilling could operate in a similar way, with membrane properties being affected either directly or indirectly through changes in levels of PGRs such as gibberellins.

Although reasonably coherent stories about the control of germination have emerged from studies on barley and on 'Grand Rapids' lettuce, the same systems may not be operative in other species.

7.3 DORMANCY IN PERENNIAL PLANTS

Just as seeds can enter a period of dormancy, so too can vegetative organs of some herbaceous plants. It is therefore pertinent to pose the question whether the resumption of growth from the overwintering structures formed by perennial plants (e.g. bulbs, corms, tubers, winter buds) can be triggered in the same way as seed germination? Removing the outer covering layers quite often stimulates seed germination, but rarely affects the growth of the dormant buds of trees. However, there are some similarities, in that light can stimulate bud break in certain trees, and in many species growth of buds can be stimulated after a period of chilling or by the application of gibberellin. In fact, at least in temperate climates, the major factor involved in triggering development of both the dormant buds of trees and other dormant organs is cold. For instance, neither young sycamore (*Acer pseudoplatanus*) trees nor *Gladiolus* corms will resume growth if the dormant plants are kept in a warm greenhouse.

◇ How would you explain the observation that bud break commonly takes place in March or April rather than in January, although the buds should already have been well chilled by then?

◈ The outside temperature is too cold for growth of buds until the spring. Twigs brought indoors from January onwards will, however, develop quite rapidly.

Lettuce seeds germinate following just 4 hours at 2 °C (Section 7.2.2), although seeds of other species require more prolonged cold treatment. However, chilling for at least a week is often required to break bud dormancy. The response to light is also rather different. Seed germination may be triggered by a single short exposure to light, but buds of trees such as beech (*Fagus sylvatica*) (which will be illuminated during the day throughout the winter months) break only in response to the *lengthening* of days in the spring. This latter difference may be more apparent than real, since phytochrome has the property of being able to act as a clock as well as a simple switching mechanism (see Section 7.6).

The subsequent internal processes that result in bud growth are no better understood than those leading to germination. The limited amount of evidence from the application of PGRs and the measurement of their levels is consistent with the control of bud dormancy by a balance between growth promotion by gibberellin and inhibition by abscisic acid (see Chapter 6, Section 6.3).

◇ How would you expect the levels of these two PGRs to change in sprouting potato tubers?

◈ Gibberellin would increase or abscisic acid would fall (or both). Alternatively, the sensitivity of the tubers to these PGRs might change (see Chapter 6). These changes might be restricted to the tissues comprising the 'eyes' of the tubers since these are the dormant buds of the potato.

In fact, the concentrations of both PGRs in the tuber tissues of potato vary in the predicted direction towards the end of the dormant period. However, not all treatments that stimulate sprouting are accompanied by an increase in gibberellin concentration, while others do not seem to lead to a fall in the level of abscisic acid. It is therefore impossible to identify one major internal factor involved in the breaking of dormancy. It is also difficult to separate the initiation of bud development from the subsequent growth of the bud. Finally, although all the buds on the plant may have been exposed to the same dormancy-breaking treatment, growth of the apical bud inhibits development of buds lower down the stem in many species — a phenomenon known as **apical dominance** (Chapter 6).

In general then, the principles controlling dormancy appear to be similar in all organs, although the balance between internal factors (such as PGRs) and the environment may be different in different species and different tissues.

Summary of Sections 7.2 and 7.3

1 Plants form a number of types of dormant structures, which can survive adverse conditions.

2 Breaking of dormancy in seeds may be affected by one or more of the following factors: water availability, integrity of the seed coat, light and chilling (followed by higher temperatures). Growth of other dormant structures is usually in response to a chilling period, although increasing day length may also act as a trigger.

3 Application of gibberellin often stimulates the initiation of growth of dormant structures. Evidence for the importance of changes in level of this PGR *in vivo* is limited to a few species.

4 Mobilization of starch stored in barley endosperm to provide the glucose needed for growth of the embryo is brought about by the enzyme α-amylase. Synthesis of this enzyme by cells in the aleurone layer is triggered by gibberellin secreted by the embryo in response to the uptake of water. Gibberellin stimulates transcription of the α-amylase gene.

5 In 'Grand Rapids' lettuce seeds, germination can be triggered by light, chilling, or rupture of the seed coat. Restrictions by the testa may be a major factor in imposing dormancy of the seed. Changes in ion transport as a response to external stimuli could increase the internal (turgor) pressure of the embryo.

6 The effect of light on seed germination is mediated by the pigment phytochrome. Phytochrome can exist in two forms: the red absorbing form, P_r, and the far-red absorbing form, P_{fr}. These are interconverted by light. P_{fr} appears to be the active form, and since this can be converted back to the inactive form P_r by far-red light, this explains why phytochrome responses may be reversible.

7 Buds break in response to lengthening days in the spring. The control of bud dormancy may be a result of a balance between growth promotion by gibberellin and inhibition by abscisic acid.

Question 1 (*Objectives 7.1, 7.2, 7.3 and 7.4*) Which of the following statements are true, and which are false? Give reasons for your answers.

(a) Moistened seeds kept in a dark refrigerator will always germinate more readily than moistened seeds of the same species kept in a warm and dark airing cupboard.

(b) The germination of barley grains involves the action of PGRs on metabolically inactive cells.

(c) Light-stimulated germination is triggered by the absorption of light by the P_r form of phytochrome.

(d) The shape of the action spectrum for a phytochrome-triggered response is similar to the shape of the absorption spectrum of P_{fr}.

Question 2 (*Objectives 7.1, 7.2 and 7.3*) Figure 7.11 shows the effect of light on the germination of seeds of *Phacelia tanacetifolia*. One dish was kept in the light, one was kept in the dark except for 3 minutes red light 6 hours after the start of the experiment, and one was kept in the dark except for 3 minutes red light followed by 3 minutes far-red 6 hours after the start of the experiment. Can you deduce which dish received which treatment? What is the effect of a high level of P_{fr} on the seeds?

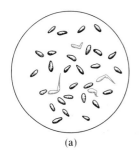
(a)

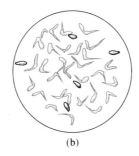
(b)

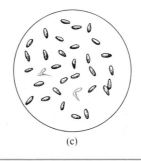
(c)

Figure 7.11 The effect of different light treatments on germination in *Phacelia* (see text of Question 2).

Question 3 (*Objective 7.4*) Potatoes are usually stored under cover in a frost-free place, because potatoes that have been frosted taste unpleasantly sweet. Suggest an explanation for this response.

7.4 VEGETATIVE GROWTH

This section describes how growth is controlled in plants that have emerged from a period of dormancy, but have not yet started to form flower buds or other structures associated with reproduction.

The precise form that vegetative growth takes depends on the balance between a number of internal and environmental factors. If we compare a sunflower (*Helianthus annuus*) plant, which normally produces a single unbranched stem, with a tomato (*Lycopersicon esculentum*) plant, in which the developing side shoots have to be 'pinched out' at intervals of about two weeks in order to concentrate growth into the main stem, it is apparent that the degree of apical dominance varies from species to species. Within a single species, variability is generally less marked, but it can exist and is most readily influenced by both the age of the plant and the conditions under which it is growing. Most plants become more bushy as they get older, but the growth of side shoots in young plants of *Coleus* (or flame nettle, a common pot plant) can be stimulated by increasing the level of illumination or by the application of nitrogen-containing fertilizers. Thus, nutrition can affect the form of growth as well as the growth rate. The most effective means of stimulating outgrowth of lateral buds is by removing the apex. The inhibitory effect of the apex can be mimicked by the application of auxin in lanolin paste to the apical stump (Chapter 6, Section 6.5). Although the precise mechanism by which auxin inhibits lateral growth is not clear, there is good reason to believe that it is responsible for maintaining apical dominance *in vivo*. The capacity of a plant to sustain a supply of replacement apices is of major ecological advantage since it allows plants to recover rapidly when the growing tip is damaged or destroyed by the actions of late frosts or grazing herbivores, through the development of previously repressed buds.

Another example of the effect of interaction between internal and environmental factors on growth is illustrated in Figure 7.12. In this diagram we see what happens when pieces of root of dandelion (*Taraxacum officinale*) are oriented in different ways and then allowed to grow. It is clear that a fixed internal axis must exist which dictates that shoots develop from what was originally the apical end and roots from the basal end, regardless of orientation.

This phenomenon of axis fixation or **polarity** is not yet understood, but it is known to be established at the first division of the zygote and subsequently passed from cell to cell. Adjustment of the positions of plant parts in response to environmental stimuli involves the differential growth of particular cells. These responses are termed *tropisms* and may be induced by such stimuli as light, gravity and mechanical stimulation. It has been proposed that tropic bending occurs in reaction to an asymmetrical distribution of PGRs such as auxins within the responding organs. Whilst there are some observations that support this assertion, convincing evidence is lacking (see Chapter 6, Section 6.5).

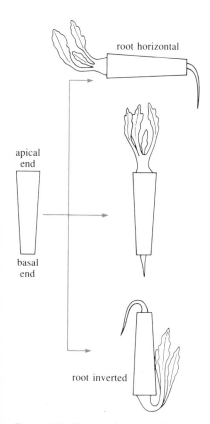

root horizontal

apical
end

basal
end

root inverted

Figure 7.12 The development of shoots and roots from pieces of dandelion root.

7.4.1 Light and vegetative growth

There are at least three ways in which light may vary, each of which can have important effects on plant growth. These are: quantity (i.e. intensity or duration), quality (i.e. wavelength) and direction.

Although phototropism is primarily a response to light coming from a particular direction (Chapter 6, Section 6.5.2), the degree of phototropic bending also depends on the quality of light supplied (Figure 7.13).

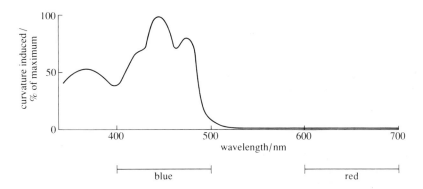

Figure 7.13 The relative efficiency of different wavelengths of light in stimulating phototropic curvature in French beans.

◇ How will French bean seedlings respond to unilateral (one-sided) illumination with (a) red light and (b) blue light?

◆ They will continue to grow vertically in red light, but will bend towards the blue light source.

◇ How would you explain the observation that seedlings bend towards the light in natural environments?

◆ Because daylight contains some blue light, which stimulates the phototropic response. The other wavelengths have no effect on stem orientation.

Although there is much debate about the identity of the receptor pigment involved in phototropism, it is clear from the phototropic action spectrum (Figure 7.13) that this light-induced response is *not* mediated by the pigment phytochrome (compare Figure 7.13 with the absorption spectrum of phytochrome — Figure 7.8).

There is evidence, however, that phytochrome is involved in the unfolding of the hook in the seedling stem and in the first stages of leaf expansion in French beans (Figure 7.14). As in the germination of lettuce seeds, these processes appear to be triggered when the level of P_{fr} rises above a certain level and then continue without the need for further illumination.

Interestingly, whilst a high level of P_{fr} appears to be necessary to stimulate expansion of leaf cells, this form of phytochrome inhibits elongation of cells in the stem. As a consequence of this, plants grow taller in the dark than the light. In other words, different plant cells have the ability to interpret the same environmental signal in different ways. This is a fascinating aspect of plant development and one to which we shall return later.

Much longer periods of illumination (i.e. hours) are required for the expansion of leaves to full size and for the synthesis of chlorophyll than for the induction of lettuce seed germination. The relative effectiveness of different wavelengths of light in inducing these responses to long-term exposure is shown by the green line in Figure 7.15. For comparison, Figure

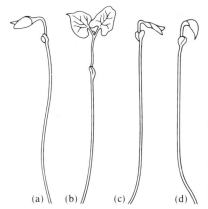

Figure 7.14 The appearance of French bean seedlings (a) grown in continuous darkness, (b) exposed to red light for 2 minutes, (c) given 2 minutes of far-red light and (d) given 2 minutes of red light immediately followed by 2 minutes of far-red.

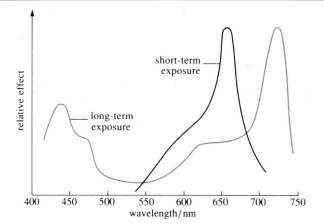

Figure 7.15 The relative effectiveness of light of different wavelengths in stimulating plant development for processes where long-term exposure is required (green line) and where short exposures are sufficient (black line).

7.15 also shows the light wavelengths most effective at inducing rapid responses (black line). Although the wavelengths responsible for both types of event appear very different, we now know that both long-term and rapid responses to light are mediated by phytochrome, but in different ways.

The wavelengths that are most effective in stimulating long-term responses are in the far-red part of the spectrum (Figure 7.15). This observation is paradoxical since these are just the ones that usually inhibit the rapid phytochrome responses. However, since the action spectrum for long-term responses resembles the absorption spectrum for P_{fr} (Figure 7.8), the explanation may be that these responses require either continuously high levels of P_r or low levels of P_{fr}.

In fact, results from experiments with sources that emit different proportions of red and far-red light suggest that it is not the actual amounts of P_r or P_{fr} that are important in the long-term responses, but their relative proportions, i.e. the ratio of $P_r:P_{fr}$. Long-term responses are thought to occur for as long as the proportion of phytochrome in the P_{fr} form in a tissue (i.e. $P_{fr}/(P_r + P_{fr})$) lies between certain limits.

Under most light sources, both reactions $P_r \rightarrow P_{fr}$ and $P_{fr} \rightarrow P_r$ occur, although there is a point (known as the **photostationary state**) at which their rates are equal. At this point the proportion of phytochrome in the P_{fr} form will remain constant. Two different light sources, each producing a photostationary state of 5% P_{fr}, might be expected to induce the same pattern of development in plants grown under them. In fact, increasing the intensity of illumination (i.e. irradiance) tends to induce a more marked response. This observation suggests that plant tissues must be able to monitor the rate of interconversion between P_r and P_{fr} since this will be dependent on the intensity of light whilst the relative proportions of phytochrome in the form of P_r and P_{fr} will not. It has therefore been proposed that plant tissues must contain some mechanism for monitoring the rate of interconversion of P_r to P_{fr} (or vice versa) as well as the proportion of P_{fr}.

Plants grown entirely in the dark elongate faster than normal, their leaves do not expand and their stems are often hooked over at the top. Such plants are described as **etiolated**. They are ideally suited to pushing up rapidly through the soil whilst offering minimal frictional resistance. If leaf expansion was fully triggered by a short exposure to light, the plant would no longer be able to grow rapidly if it was covered over again. Long-term responses therefore appear to enable plants to respond flexibly to changes in the environment in such ways that their patterns of development optimize their growth.

◇ What will happen if the growing tip of a plant is covered over with soil?

◆ The shoot will grow in the manner typical of dark-grown tissue until it once again breaks through the soil surface.

Although the responses of plants to light seem complicated, the stimulus is very important for their autotrophic existence, and plants have developed the capacity to respond to the subtlest of changes in illumination. One example of this is the response of plants to shade. We have already considered how plants react to shading from one side by growing towards the source of illumination. However, how would they respond to shading from other plants? Plants growing under a leaf canopy will not only receive *less* light than those not overshadowed by other plants, but they will also be subject to *different wavelengths* of light. This is because chlorophyll acts as a very efficient absorber of red light (Figure 7.16).

◇ Will the proportion of P_{fr} be higher in plants growing in open air or in shaded habitats?

◆ It will be higher in open habitats because in shade there will be proportionately more far-red light and therefore more P_r.

Stems of the plant fat hen (*Chenopodium album*), a common weed in arable crops, are more elongated when grown mixed with faster growing species that shade them than when mixed with slower growing species. Fat hen plants are also more elongated when grown under lights emitting a high proportion of far-red wavelengths than when grown under an equal intensity light containing little far-red.

◇ Do the results obtained with artificial lights suggest that the response to shading is triggered by the quantity or by the quality of the light?

◆ By quality: the results show that at least part of the stimulation of growth by shading by other plants must be due to the increased proportion of far-red light.

Furthermore, by using a wide range of different light sources, it can be shown that the rate of stem growth in this species increases as the predicted proportion of P_{fr} in the photostationary state decreases. Not all plants exhibit this response to shade. Plants fall into approximately three groups. These are shade avoiders (such as fat hen), shade tolerators such as dog's mercury (*Mercurialis perennis*) and bluebell (*Hyacinthoides non-scriptus*) and shade neutral species such as nettle (*Urtica dioica*). The growth response of representatives of these three groups to elevated far-red light can be seen in Figure 7.17.

As you might predict, shade tolerators do not exhibit a growth stimulation in response to far-red light since they would not be able to pierce the canopy of plants such as trees under which they were growing!

Summary of Section 7.4

1 The form that a mature plant takes is dictated by the degree of dominance that the apex imposes on growth of the lateral buds associated with the main stem of the plant.

2 The inhibitory influence which originates from the apex is thought to be the PGR auxin.

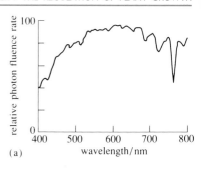

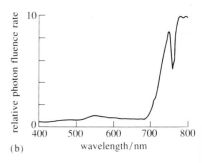

Figure 7.16 The spectral distribution of (a) unfiltered sunlight and (b) sunlight after filtering through leaves within a crop canopy.

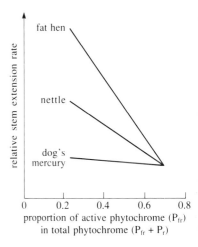

Figure 7.17 The relationship between stem extension rate and the proportion of phytochrome in the P_{fr} form for three species of herbaceous plant.

3 The growth of a plant can be influenced by the quality and quantity of light, and the direction from which it comes. De-etiolation of dark-grown seedlings requires long periods of illumination. These long-term light responses are mediated by phytochrome but are different from most of the rapid responses induced by the same pigment since they are stimulated more by far-red than by red light. Although rapid responses are largely dependent on the *proportion* of phytochrome in the form of P_{fr}, long-term responses are apparently also influenced by the *rate of formation* of P_{fr} from P_r.

4 Phytochrome may play a critical role in mediating the response of plants to shading. Plants growing under a canopy will be subjected to elevated far-red light since the red wavelengths in the natural spectrum will be absorbed by the pigment chlorophyll. The elongation growth of shade avoiders is stimulated by low levels of P_{fr}. Such plants will therefore continue to elongate until they grow through the leaf canopy under which they are shaded.

Question 4 (*Objectives 7.1, 7.3 and 7.5*) Consider the following statements about the effect of light on plant development and decide whether each is true for (i) phototropism, (ii) long-term phytochrome responses, and (iii) responses triggered by phytochrome.

(a) The altered pattern of development persists only for as long as an inducing stimulus is applied.

(b) No response is observed if the plant tissue is lit equally from all sides.

(c) The response can usually be evoked by red light alone.

Question 5 (*Objectives 7.1, 7.2, 7.3 and 7.8*) Seeds of a dock (*Rumex* sp.) were germinated and the seedlings grown for 35 days under bright lights, which (it was predicted) would convert 70% of the phytochrome present to the P_{fr} form (treatment A). Some seedlings were then transferred to two qualities of dim lighting (the intensity of each being 40% of the original), one of which was predicted to give 40% P_{fr} (treatment B) and the other 70% (treatment C). Leaf area and leaf weight were estimated after six days. The results are shown in Table 7.5.

Table 7.5 The effect of light quantity and quality on leaf growth of dock seedlings.

Treatment	Light quantity as percentage of original	Predicted percentage P_{fr}	Total leaf area/cm^2	Leaf dry weight/mg	
				leaf blade	petiole
A	100	70	158	442	135
B	40	40	208	446	194
C	40	70	209	448	141

What are the effects of light intensity and light quality on the growth of (a) leaf blades and (b) petioles (i.e. leaf stalks)? Suggest why these effects may have adaptive value for plants growing in natural environments.

7.5 VEGETATIVE REPRODUCTION

In contrast to vertebrate animals, reproduction in plants can take place without the requirements for the union of male and female gametes. This process of asexual or vegetative reproduction is possible because many plant cells display totipotency (Chapter 6, Section 6.6). Unlike sexual reproduction which results in genetic recombination as a result of the processes of reassortment and crossing-over during meiotic cell division, vegetative reproduction results in the production of new plants that are genetically identical to their parent.

Some of the most pernicious garden weeds such as couch grass or twitch (*Elymus repens*) can spread extremely rapidly by producing long-lived underground stems or rhizomes which periodically produce new shoots which appear above ground. These shoots do not necessarily become separate plants, although they are quite capable of an independent existence if the connecting stems are severed. In the strawberry (*Fragaria* sp.), new plants are formed at the end of **runners**, elongated stems which grow horizontally rather than vertically upwards. The tips of the runners develop leaves and they will also root provided they are in contact with a suitable medium (such as soil).

A potato (*Solanum tuberosum*) tuber planted in the spring may yield many times its weight of potatoes by late summer. As well as being a means of vegetative reproduction, tubers are also dormant structures by which the potato plant can withstand winter conditions. They are formed by swelling of the tip of a short-lived underground stem or **stolon** (Figure 7.18). The central parenchyma tissue expands as the storage polysaccharide starch is accumulated, and each of the immature lateral shoot apices becomes an 'eye', capable of developing into an aerial shoot in the next growing season.

◇ Why do portions of seed potatoes, each with only one eye, grow when planted?

◉ Each contains a shoot apex capable of forming a whole plant.

Potato plants can even be grown from peelings, although the buds will have a slow start since they are deprived of food.

In *Solanum andigena*, the wild ancestor of the cultivated potato, buds on the portion of the stem below ground begin to develop into stolons during the long days of summer. However, the stolon tips do not begin to swell into

Figure 7.18 The formation of stolons and tubers by potato plants.

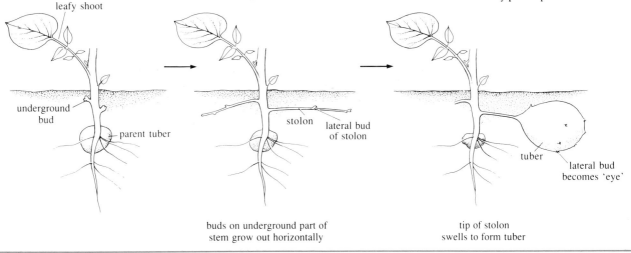

leafy shoot

underground bud

parent tuber

stolon   lateral bud of stolon

buds on underground part of stem grow out horizontally

tuber

lateral bud becomes 'eye'

tip of stolon swells to form tuber

tubers until day length begins to shorten. The presence or absence of stolons appears to be determined primarily by the age of the shoot, whereas **tuberization** is controlled by an environmental factor, day length.

Why do lateral buds develop as stolons, and why is this development restricted to the lower region of the stem? Consider three possible explanations for the localization of stolon development, all of which assume that stolon development is triggered by the action of some factor (such as a PGR) on lateral buds.

1 Stimulatory concentrations of the factor are present throughout the plant. Stolon development is dependent on stem age, and only the older, lower buds have developed sensitivity to the factor.

2 Stimulatory concentrations of the factor are present throughout the plant. Stolon development is dependent on the environment of the bud. Only buds whose surroundings are dark or moist (i.e. underground) will respond to the factor.

3 The concentrations of the factor vary in different parts of the plant. Only the lower buds are exposed to a sufficient concentration of the factor to stimulate stolon growth.

Various experiments designed to distinguish between these three possibilities are described below.

◇ What can be deduced from the fact that when the overground portion of a potato plant is cut off, the tips of the developing stolons turn upwards and develop into new aerial shoots?

◆ This shows that the ability of lateral buds to form leafy shoots is not irreversibly lost as the stem becomes mature. It is not directly relevant to any of the explanations 1–3 (note that 1 states that older buds acquire the ability to form stolons, not that they lose their ability to form leafy shoots).

◇ What can be deduced from the fact that when potato stems are cut below the fourth leaf from the apex and the base of the cutting inserted into potting compost, stolons develop from the lowest lateral buds within a short time?

◆ Since comparatively young buds can develop into stolons, explanation 1 can be discounted.

◇ What can be deduced from the fact that when the cuttings described in the previous experiment are inserted into flasks containing nutrient solution which are kept in either the light or the dark, results similar to those shown in Table 7.6 are obtained?

Table 7.6 The effect of light and dark on stolon development in potato. Cuttings were placed with the lowermost bud below the surface of the nutrient solution and the second lowest bud just above its surface (10 cuttings were used for each treatment).

Treatment	Stolons from lowest bud		Stolons from second lowest bud
	number formed	mean length/cm	number formed
light	10	5.9	1
dark	10	8.5	8

◆ Darkness is needed in order to trigger stolon development in the second from bottom bud; this supports explanation 2. However, light seems to have no effect on stolon initiation from the lowermost bud (although it does inhibit stolon growth); this supports explanation 3.

As you will see from experiments described below, it is comparatively easy to induce stolon development in the light by the application of certain PGRs. However, the practice of 'earthing up' potatoes has a marked effect on the yield obtained, so the effect of the environment cannot be ignored. The 'correct' explanation probably lies somewhere between 2 and 3 — in other words, gradients of some stolon-forming factor are present in the plant, but the responses of the tissues (or the gradients themselves) are affected by the environment. What is known about the nature of the factor (or factors) concerned?

If you look at a potato plant you will see that, although each plant has several stems, each arises from an eye on the tuber and relatively few stems have side branches. In other words, there is strong apical dominance in the potato.

◇ What would you expect to happen if the growing point of a potato stem is pinched out?

◆ Lateral shoots would develop. Figure 7.19 shows that this response is observed.

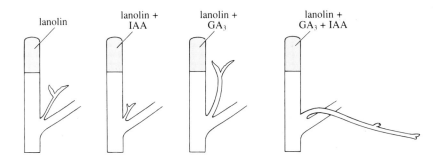

Figure 7.19 The effect on the development of lower buds of potato plants of applying IAA and/or GA$_3$ in lanolin to the cut stump.

As discussed in Section 7.4, auxin produced by the shoot tip is generally considered to be the PGR primarily responsible for apical dominance. In accord with this hypothesis, Figure 7.19 shows that application of the auxin IAA inhibits the growth of lateral shoots in cuttings. In contrast, GA$_3$ stimulates the growth of leafy lateral shoots, but if the gibberellin is applied in combination with IAA, the lateral bud immediately below the point of application is induced to develop as a stolon (even though it is above ground level). Stolon development in the whole plant might therefore be triggered by an increase in the concentration of either auxin, or gibberellin, or both.

One way in which this process could be studied further would be to apply PGRs to growing potato plants and analyse the effect that this has on plant development. Smearing lanolin round a portion of the stem of an intact potato plant does not affect its subsequent growth. However, stolon development is often stimulated if the lanolin contains GA$_3$ — although the stolons are shorter than normal. It is difficult to assess how much gibberellin is absorbed in such treatment, but this result does suggest that stolon development from buds above ground might normally be prevented by gibberellin levels being too low.

◇ What evidence would corroborate this suggestion?

◆ The concentration of gibberellin should be higher in the region of buds that develop into stolons. Unfortunately, there is no evidence that this is so. This does not *necessarily* rule out a role for gibberellins since it might have an impact on stolon development via changes in cell sensitivity.

Tuber formation is stimulated in both growing plants and cuttings when the leaves are exposed to day lengths below a critical value (typically about 12 hours in *S. andigena*).

What about the effects of PGRs on tuber formation? Application of gibberellin to growing plants tends to inhibit tuber formation.

◇ If gibberellin is involved in the normal control mechanism, how would you expect its level to change?

◆ The high level needed to stimulate stolon development should fall before tuber formation.

However, even if gibberellin is needed to trigger stolon development, an increase in its level could be restricted to the stimulated lateral bud itself. Little gibberellin is normally found in tubers, although (you may recall from Section 7.3) their gibberellin content rises and abscisic acid content falls as they begin to sprout. Abscisic acid has been reported as promoting tuber formation in several species, including some varieties of potato. Figure 7.20 shows one experiment that suggests that potato leaves may respond to short days by altering either the production or transport of abscisic acid, which in turn either inhibits stolon growth or promotes tuberization.

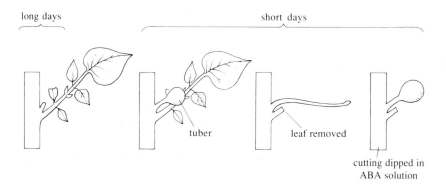

Figure 7.20 The effect of day length and abscisic acid on tuber development from potato cuttings grown in culture.

On the evidence presently available, it appears that the control of vegetative reproduction in potato occurs at two separate points: stolon development and tuber formation. Circumstantial evidence suggests that PGRs could be involved in some way, with gibberellin having opposite effects at the two control points. It is possible that PGRs might act directly to alter the course of development or indirectly by modifying the distribution of other compounds which play a critical regulatory role.

Summary of Section 7.5

1 Plants can propagate themselves by means of vegetative reproduction. This is possible because the majority of plant cells are totipotent.

2 Vegetative reproduction occurs through, for example, the production of rhizomes, runners or, as in the case of potato, stolons. It is the tips of the stolons which swell and develop into tubers.

3 Stolon development is stimulated by gradients of some unidentified factor, although the induction of stolons is also markedly influenced by the environment. Treatment of lateral buds or decapitated shoots with gibberellins can induce stolon development if auxin is applied at the same time.

4 Tuber formation is stimulated in whole potato plants and cuttings by exposing the leaves to day lengths of less than a critical value (typically 12 hours). Tuber formation can be inhibited by application of gibberellin and stimulated by treatment with abscisic acid.

Question 6 (*Objectives 7.1, 7.2 and 7.3*) Four experiments (A–D) that formed part of an investigation into factors affecting the development of strawberries are described below. The strawberry is a 'false' fruit, the fleshy part being actually a swollen modified stem, which carries the individual true fruits or 'achenes'. However, for simplicity, we refer to the fleshy part as the berry and to the achenes as the seeds, although neither term is strictly correct.

Experiment A When all the seeds were removed from strawberries on days 4, 7, 12, 19 or 21 after pollination, growth of the berry stopped completely, although in the last case the berry did turn red at the usual time (28 days after pollination).

Experiment B No growth of the berry was observed in unpollinated flowers. Fertilization of the ovule contained in one seed led to some growth of the berry in the area surrounding that seed; fertilization of three ovules led to three small areas of growth.

Experiment C Seeds were removed from a strawberry four days after pollination so as to leave only three rows of seeds. A long flat strawberry developed if the rows were vertical, whereas a short fat strawberry resulted if the rows were horizontal.

Experiment D Seeds were removed from a strawberry nine days after pollination. The berry continued to develop if it was covered with a lanolin paste containing auxin, but not if it was covered with lanolin paste alone.

Which of the following statements are supported by these experiments?

(a) Developing seeds release growth substances necessary for the growth of the strawberry.

(b) Growth substances released by the seeds trigger the development of the berry, but their presence is not essential at all stages of growth.

(c) Auxin is the only PGR involved in the development of the strawberry.

(d) The effect of growth substances released by the seeds is restricted to the area adjacent to the seed.

(e) Auxin is supplied from seeds to the berry.

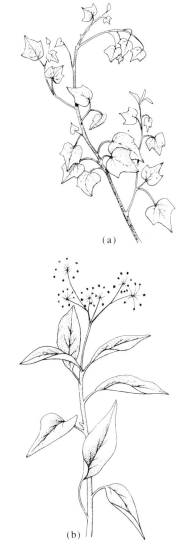

(a)

(b)

Figure 7.21 Ivy. (a) Juvenile form. (b) Adult form.

7.6 FLOWERING

As a rule, plants do not flower until they have achieved a certain size or age. For instance, oak (*Quercus* sp.) trees remain in a juvenile (non-reproductive) phase for up to 30 years, whereas the juvenile period in herbaceous plants commonly lasts only a few weeks — until the seedling has developed half a dozen or more leaves. The transition between juvenile and adult plant is the result of an internal control mechanism, which is unaffected by the environment. Once established, the transition is fairly stable. Thus cuttings from branches of ivy (*Hedera helix*) with the characteristic adult form (Figure 7.21b) flower soon after they become established. However, there are few processes in plant development that are completely irreversible, and the juvenile–adult transition in ivy is no exception. The cuttings gradually revert to forming shoots typical of the juvenile phase (Figure 7.21a), and do not flower again until the plant has attained the larger size normally associated with the production of adult shoots. Reversion must also be induced by events associated with meiosis, since adult plants ultimately produce seeds which germinate into plants with the characteristic juvenile form.

It is also possible to induce some fruit trees to become reproductive while still comparatively young. This is achieved by grafting buds of the required variety at a particular age onto a special ('dwarfing') root stock that favours reproductive rather than vigorous vegetative growth.

A number of complex developmental events have to occur before the opening of a fully formed flower takes place. The earliest of these changes can be seen at the shoot apex. When the tip of a non-flowering shoot of cockleburr (*Xanthium strumarium*) is teased apart, a structure similar to that shown in Figure 7.22a is seen.

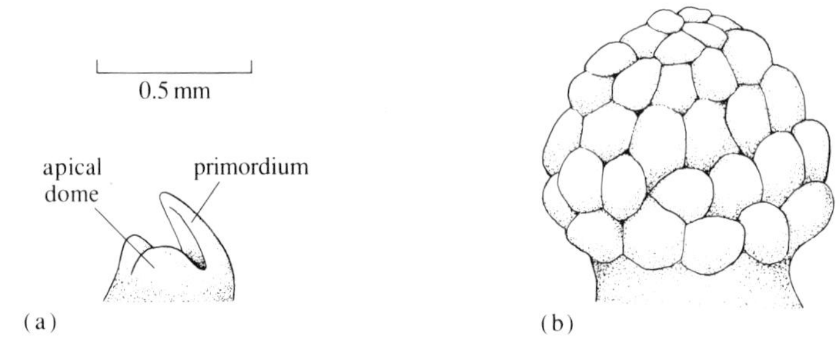

0.5 mm

apical dome   primordium

(a)   (b)

Figure 7.22 The structure of the shoot apices of cockleburr. (a) Before a flower-inducing treatment. (b) After.

◇ What are the structures arising from the sides of the apex?

◆ The tissues or **primordia** that will develop into leaves.

Within a few days of flower induction, the lateral structures formed have a rather different appearance and are destined to develop into flowers rather than leaves. As more flower primordia are formed, the structure of the apex becomes like that shown in Figure 7.22b. Even at this stage, the floral apex is still surrounded by developing leaves formed before the inducing treatment. To the naked eye, induced plants at this early stage look no different from non-induced controls.

In some species, the further development and expansion of each primordium to form a flower follows an inductive treatment without any further stimulus; but, in other species, there is a period of inactivity until flower development and expansion is triggered by a set of conditions possibly very different from those that trigger flower initiation. It is therefore necessary to be very careful when interpreting observations such as 'species X flowers only after treatment Y', in case treatment Y simply serves to trigger the development of preformed floral apices. In this Section, we are concerned only with the control of the *induction* of flower formation.

Assuming that a plant is old enough to flower, what determines when it actually does so? A second generalization about flowering is that many plant species flower at particular times of year. Is this due to an internal annual rhythm in the plant or a requirement for particular environmental conditions? If it were the latter, one should be able to induce premature flowering by adjusting the environmental conditions. Chrysanthemums (*Chrysanthemum morifolium*), which flower naturally in late autumn, are now available in shops throughout the year because it is possible to induce flowering by altering the environment so as to mimic seasonal changes. However, although this demonstrates that the environment can play a key role in dictating the timing of flowering, some plants maintained under conditions favourable to flowering may well continue to flower only about once a year because of the persistence of internal rhythms. Nevertheless, most of what is known about flowering is related to the effects of the environment, and in particular to the effects of varying day length or temperature.

On the basis of how flower formation can be stimulated by altering the growth conditions, plants can be classified as 'long-day' plants, 'short-day' plants, 'chill-requiring' plants (some of which are also 'long-day' or 'short-day' plants) and plants which are environmentally 'neutral' (Table 7.7).

Table 7.7 Flowering behaviour in plants.

Type	Behaviour	Examples
short-day	flower only in short days	soya bean (*Glycine max*) morning glory (*Ipomoea* sp.)
	flower sooner in short days	rice (*Oryza sativa*)
long-day	flower only in long days flower sooner in long days	clover (*Trifolium* sp.) pea (*Pisum sativum*) petunia (*Petunia hybrida*)
chill-requiring	flower when temperature rises after chilling	primrose (*Primula vulgaris*)
chill-requiring long-day	flower in long days after chilling	henbane (*Hyoscyamus niger*)
chill-requiring short-day	flower in short days after chilling	*Chrysanthemum* hybrids
neutral	flowering little affected by the environment	tomato (*Lycopersicon esculentum*) French bean (*Phaseolus vulgaris*)

Long-day plants flower when the day length (number of hours of light) exceeds a certain critical value. Many of the known examples are native to temperate regions where they flower during the summer. The class can be divided into those that will not flower at all in short days, and those that will eventually flower, although much later than they would have done given long days.

Short-day plants flower (or flower most readily) when day length is less than a critical value. They are found both in tropical regions, where day length varies comparatively little during the year, and in temperate regions. The critical value for day length depends on the plant being considered.

◇ When would you expect short-day plants to flower in Britain?

◆ In spring or in autumn. Most British short-day plants are autumn flowering.

Spring-flowering plants are usually found to show a marked response to chilling rather than to short day length. However, when considering chill-requiring plants, it is particularly important to distinguish between the control of flower induction and flower development.

◇ What other developmental processes are triggered by chilling in plants of the temperate zone?

◈ Seed germination (Section 7.2) and the release of winter buds from dormancy (Section 7.3).

Some species do flower solely in response to chilling (e.g. primrose (*Primula vulgaris*)), but others are in fact long-day plants in which floral apices formed during one summer remain dormant until the following spring (e.g. henbane (*Hyoscyamus niger*), daffodil (*Narcissus* sp.)).

Some plants fall into more than one of the four categories. A few examples in which flower induction can be stimulated by particular day lengths only if the plant has been previously chilled are given in Table 7.7. Plants that do not flower until the second season of growth may be showing this type of response.

◇ What is an alternative explanation for this behaviour?

◈ The juvenile (non-reproductive) phase might persist until the second season.

We will now consider some of the simpler mechanisms whereby flower induction may be controlled by temperature (Section 7.6.1) or day length (Section 7.6.2). Those plants with requirements for particular combinations of long and short days will be ignored, along with neutral plants in which flower induction does not seem to be affected by either day length or temperature. It is assumed that internal factors (such as age) are the major controlling influences in the case of day- and temperature-neutral plants.

7.6.1 Chilling and flower induction

A characteristic of plants that form floral apices only after chilling is that they do not normally flower in the same season as the seed was sown. In order to understand what happens during the intervening winter, it is necessary to find out exactly how much chilling is needed to induce flowering, as well as the location of the temperature-sensitive cells and the nature of the changes that occur after chilling.

It is possible to define the necessary stimulus quite precisely. Most cold-sensitive species will not flower unless they have been exposed to temperatures in the range 0–2 °C for at least a week, with longer exposures necessary for maximum effect. One continuous exposure can be replaced by the cumulative effect of several shorter treatments. These requirements are very similar to those needed for overcoming dormancy in winter buds.

Winter wheat is an example of commercial importance; it is sown in the autumn for harvest the following summer.

◇ Why may the crop fail if the winter is either exceptionally mild or exceptionally severe?

◈ In a mild winter, the plants will not be chilled for long enough to induce flower formation. If the winter is severe, the plants may be damaged by the very low temperatures (see Chapter 5, Section 5.7.3).

In temperate climates, the risk of losing the crop is outweighed by the prospect of higher yields than can be obtained from spring-sown varieties (which do not require chilling and therefore flower later in the same season). However, in the Mediterranean region and in Russia, winter wheat crops would be unlikely to succeed without special cultivation.

◇ Suggest how damage by severe winters might be avoided.

◆ The seedlings could be grown under cover with the temperature maintained just above freezing. They need not be planted out until the spring, when the coldest conditions have receded.

The technique, developed in Russia, of soaking seeds for just long enough to initiate germination and then burying them in the snow was called **vernalization** (i.e. 'made spring-like'). This term is now applied to any cold-treatment that triggers floral development.

The next question is: which parts of the plant need to be vernalized in order to induce flowering?

◇ How would you try to find out which part of the plant was sensitive to chilling?

◆ Cool particular areas of the plant and see which treatment induces flowering.

Experiments in which different parts of the plant have been surrounded by cooling coils or by vacuum flasks full of ice have shown that with very few exceptions, *chilling of the shoot apex* is the only treatment which induces flowering. The shoot apex is the part that gives rise to flower buds, so it could be that chilling acts directly on certain cells, causing them to form flower tissue. However, there is evidence from both winter rye (*Secale cereale*) and from henbane (*Hyoscyamus niger*) that some sort of flowering stimulus can be transmitted from the chilled apex to other tissues.

If the tip of a vernalized rye plant is pinched out, apical dominance is relieved and so side shoots develop. Although the apices of these side shoots were not active at the time of chilling, they flower without further cold stimuli. It is difficult to be certain that the first buds that develop were not already formed when chilling took place; but when the pinching out is repeated twice more, side shoots that develop from side shoots from the original side shoots also form flowers. Although these experiments suggest the existence of a flower-forming stimulus, the results are equally consistent with the induction of some change in all chilled cells that is passed from cell to cell at mitosis and causes the cells in active shoot apices to develop flower primordia.

◇ When an extract of vernalized rye plants was applied to a long-day *Rudbeckia* plant growing under short days, the recipient plant flowered. What can be deduced about the flower-forming stimulus?

◆ Only that certain chemical substances in the extract are capable of stimulating flower formation.

In order to make any further deductions, it would be necessary to know the effect of extracts from control (i.e. unvernalized) plants. The experiment certainly does not achieve its original aim of demonstrating the existence of a flower-forming substance, and furthermore, that the same chemical is involved in temperature-sensitive and in light-sensitive plants.

When one shoot of a chill-requiring variety of henbane is chilled, flowers form at the tips of all the shoots, provided that the plant is kept under long days.

◇ Can either of the explanations offered in connection with flower formation in the side shoots of rye be discounted?

◆ It is difficult to see how an effect passed on at cell division can explain these results.

It seems likely that the flower-forming stimulus produced during chilling (which has been given the name **vernalin**) is a chemical that can be transported between different parts of the plant. But what is the nature of this chemical? If a shoot is taken from a vernalized henbane plant and grafted onto an unvernalized plant, the latter flowers without being chilled (Figure 7.23). However, if for some reason the graft does not 'take', there is no transfer of the flowering stimulus to the unvernalized plant. Responses that are thought to be mediated by one of the common types of PGR can usually be mimicked by application of the substance in lanolin or agar, with no need for tissue continuity to be established. There does therefore appear to be something rather different about the flower-forming stimulus, although there is no firm evidence as yet about its nature.

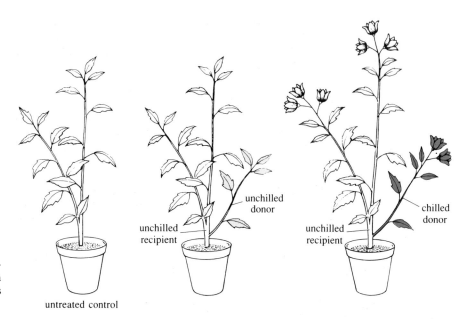

Figure 7.23 The transmission of a cold-stimulated flower-inducing factor across a graft union in henbane (donor tissue is shown in green).

The stimulus does not even seem to have exactly the same properties in all chill-requiring plants. In chrysanthemum, for example, the stimulus has not been shown to be transferred from a chilled shoot either to other shoots of the same (intact) plant, or across a graft union. This might indicate that a transport system for the stimulus is absent in chrysanthemum. On the other hand, it may be too simplistic to consider only the effects of a flower-forming stimulus — a substance which inhibits flowering may also be involved. If so, the process of flowering would be dependent on the balance between the flower-forming and the flower-inhibiting substances.

◇ If this sort of balance was involved in the control of flowering, how would the levels of growth regulators be altered after flower-forming treatments?

◆ Either the level of a flower-forming substance would go up, or the level of a flower-inhibiting substance would fall (or both).

As yet there is no information as to the identity of either the putative flower-forming or flower-inhibiting substances.

7.6.2 Day length and the control of flowering

The induction of flowering by changes in day length (often called **photo-periodism**) involves complex interactions between internal control systems and external stimuli which can only be outlined here.

What stimulus is needed to induce flowering?

Figure 7.24 shows the percentage of plants flowering after the transfer of non-flowering chrysanthemum and spinach (*Spinacia oleracea*) plants to particular 24 hour cycles of light and dark.

◇ Which of the two examples is the long-day plant?

◈ Spinach. It flowers only when day length exceeds 14 hours.

In fact, the transition between flower-inducing and non-inducing treatments occurs at a **critical day length** of 14 hours for both spinach and chrysanthemum. However, because chrysanthemum is a short-day plant, it flowers only when day length is less than this value.

The two examples shown in Figure 7.24 have the same critical day length. However, unlike the response to chilling, where most species have similar requirements, the day length requirements vary widely from species to species. Even within a single species, all of whose members are either long-day or short-day plants, different varieties may show a range of critical day lengths.

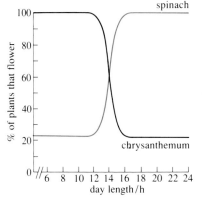

Figure 7.24 The relationship between day length and flowering in chrysanthemum and in spinach.

◇ Two varieties of chrysanthemum, 'Snow' and 'White Wonder', have critical day lengths of 11 and 16 hours respectively. If plants were established in early summer in Britain, which variety would you expect to flower first?

◈ 'White Wonder'. Chrysanthemum is a short-day plant, and day length falls below 16 hours (in July) before it falls below 11 hours (in October).

In tropical regions, the changes in day length through the year are much smaller. Yet tropical species can show photoperiodic control of flowering. Rice (*Oryza sativa*) plants seem to be able to detect changes in day length of only 15 minutes, and similar sensitivity is shown by other species from both tropical and temperate regions. There must therefore be an extremely accurate time-measuring mechanism within the tissues of these plants.

How is day length measured?

Soya bean (*Glycine max*) is usually described as a short-day plant with a critical day length of 14 hours.

◇ Would you expect soya bean to flower given a daily light exposure of (a) 16 hours or (b) 4 hours?

◈ (a) No, because day length is above the critical value of 14 hours.
(b) Yes, because day length is well below the critical value. However, flowering may be markedly reduced with such short periods of illumination.

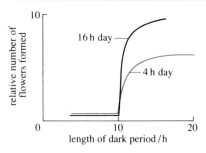

Figure 7.25 The formation of flowers by soya bean plants given a constant 16 hour or 4 hour period of light in association with different lengths of dark period.

Figure 7.25 shows what happens when these day lengths are combined with varying periods in the dark. When 'normal' 24 hour cycles are used (i.e. 16 h light plus 8 h dark or 4 h light plus 20 h dark) the results are as predicted. But Figure 7.25 shows that day length is not the critical factor.

◇ What conditions are needed to induce flowering in soya beans?

◆ The dark period must be at least 10 hours long.

In long-day plants also, it is found that night length is more important than day length. Although no one talks about long-night and short-night plants (rather than short-day and long-day plants), it might be more accurate to do so!

The importance of the dark period is also shown by the effect of a few minutes illumination (a 'night break') in the middle of an inducing dark period. The effects of varying day and night length (with and without a night break) on typical long- and short-day plants are summarized in Figure 7.26. Note particularly that interrupting the day with a brief dark period has no impact on the flowering of either short- or long-day plants (compare (c) with (a)). However, a brief night break in the middle of a long dark period inhibits flowering in short-day plants and stimulates flowering in long-day plants (compare (d) with (b)). The transition to flowering is thus controlled primarily by the length of the uninterrupted dark period.

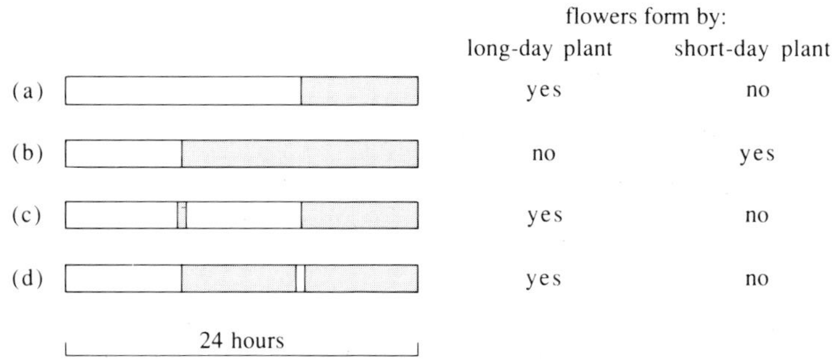

		flowers form by:	
		long-day plant	short-day plant
(a)		yes	no
(b)		no	yes
(c)		yes	no
(d)		yes	no

24 hours

Figure 7.26 The effect of different cycles of light and dark on flowering in long-day and short-day plants.

The results of further investigations of the effects on soya bean of different wavelengths of light in the middle of a long dark period are shown in Table 7.8.

Table 7.8 The effect of a night break with red and far-red light on flowering in soya bean. Plants were grown in a light/dark cycle that would normally induce flower formation, with short consecutive treatments with red and far-red light being given in the middle of each dark period.

Type of night break	Relative number of flowers formed
no night break	8
red	0
red/far-red	3
red/far-red/red	0
red/far-red/red/far-red	2
red/far-red/red/far-red/red	0
red/far-red/red/far-red/red/far-red	1

◇ What is the effect on flowering in soya bean of red light in the middle of a long dark period?

▸ Red light (and therefore, presumably, high levels of P_{fr} — see Section 7.2.2) inhibits flowering in these short-day plants.

Although Table 7.8 shows that the effects of red light can be partially reversed by far-red illumination, the reversibility that is normally characteristic of responses mediated by phytochrome is gradually attenuated. This is because the effect of red light becomes irreversible much more rapidly than it does for lettuce seed germination (see Table 7.3).

Cockleburr is a short-day plant with a critical day length of $15\frac{3}{4}$ hours. It will flower within two weeks of a single cycle of 15 hours light plus 9 hours dark inserted into a 16 hours light plus 8 hours dark regime. When experiments similar to those described for soya bean in Table 7.8 are carried out with cockleburr, it is found that 10 minutes of red light given 4 hours after the start of the single long night completely abolishes the inductive effect.

Recall the relationships between the two forms of phytochrome, P_r and P_{fr}, from Section 7.2.2:

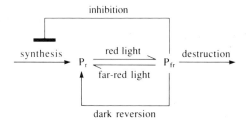

◇ What happens to P_{fr} levels in the dark?

▸ They will decrease. P_{fr} may be destroyed or it may revert to P_r. Phytochrome is synthesized in the form of P_r, and the only reaction that forms P_{fr} from P_r requires red light.

The simplest explanation to account for the results presented so far is that the amount of P_{fr} present gradually falls during the dark period (rather like the level of sand in an old-fashioned egg-timer) and that when it falls below a critical level (the sand runs out), a particular response is triggered. In short-day plants, this response would be flower induction; in long-day plants, it would be an inhibition of flower formation.

◇ Why does a night break with red light abolish the effect of a long night?

▸ Because the conversion of P_r to P_{fr} induced by red light is very rapid compared with the decline in the dark. The P_{fr} level is therefore returned to a high level (the egg-timer is reset).

Figure 7.27 shows the changes in the levels of P_{fr} predicted by this model in (a) short days, (b) long days and (c) short days with a night break. Short-day plants flower only if P_{fr} falls below the critical level (Figure 7.27a) and long-day plants flower only if it remains above the critical level, either because the nights are short enough (Figure 7.27b) or there is a night break (Figure 7.27c).

Although Figure 7.27 illustrates only how the model would account for the response of plants with a critical day length of 12 hours, it is equally applicable to other values.

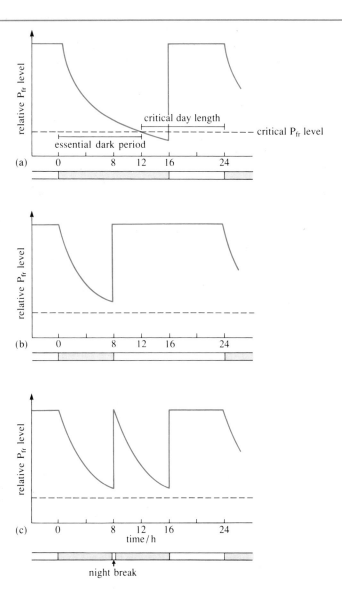

Figure 7.27 Changes in P_{fr} level predicted for a plant with a critical day length of 12 h (i.e. 12 h dark). (a) Given 16 h dark and 8 h light. (b) Given 8 h dark and 16 h light. (c) As (a) but with 5 min light in the middle of the dark period.

◇ Suggest two factors that could determine the length of the essential dark period needed for flowering in short-day plants.

◆ The rate at which P_{fr} levels fall in the dark and the critical level of P_{fr}. A change in either of these factors would result in a different value for the essential dark period and therefore for the critical day length.

Does the model fit all the experimental facts?

One of the purposes of devising models to explain biological processes is that experiments can then be devised to test the assumptions on which they are based and the predictions which follow from them. In other words, models provide a framework within which the question 'Is this how it actually happens?' can be asked.

Do P_{fr} levels actually decrease at a rate consistent with the model? Data from experiments on isolated phytochrome or phytochrome in etiolated seedlings suggest that P_{fr} levels fall during the first 5 hours, but thereafter remain constant. If these results are applicable to flower induction, then the simple model described cannot explain the response of plants which have critical day lengths of less than 19 hours (the majority of day-length sensitive plants, of course).

Many of the experiments designed to test the model involve rather bizarre light regimes to which a plant growing in the wild would not be exposed. These include cycles in which the repeat time is longer or shorter than the normal 24 hour day, and elaborate series of treatments with particular wavelengths of light. Here we consider the results of an experiment with cockleburr that suggest that the model should be modified. As mentioned above, cockleburr is a short-day plant with a critical day length of $15\frac{3}{4}$ hours; it requires only a single night longer than $8\frac{1}{4}$ hours for induction of flowering.

Consider plants treated as shown in Figure 7.28, in which one cycle in a non-inductive regime of 16 hours light plus 8 hours dark is replaced by an inductive one of 12 hours light plus 12 hours dark. The plants flower in this control treatment with 12 hours uninterrupted dark (upper line). The experimental treatment is to give 5 minutes of red light $5\frac{1}{2}$ hours after the beginning of the 12 hour inductive dark period, followed an hour later by 5 minutes of far-red (lower line).

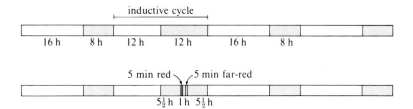

Figure 7.28 Light treatments given to cockleburr plants. The upper line is a control which included one inductive cycle. See text for discussion of the experimental treatment (lower line).

If the model shown in Figure 7.27 is correct, the P_{fr} level would fall during the first $5\frac{1}{2}$ hours of the interrupted dark period. However, it would not reach the critical level before the night break because this takes $8\frac{1}{4}$ hours. The effect of red light is to convert most of the P_{fr} present back to P_{fr}, raising the P_{fr} level. During the next (dark) hour, the level should begin to fall again.

◇ What would be the effect of the far-red light?

◆ Most of the remaining P_{fr} would rapidly be converted to P_r.

◇ Would you expect the plants to flower?

◆ Yes, because the far-red light should decrease the P_{fr} to near the critical level (if not below it).

Even if the P_{fr} level is still above the critical value after the far-red treatment, the remaining $5\frac{1}{2}$ dark hours should be adequate for it to fall to below this level. It is therefore somewhat surprising that plants given this treatment are *not* induced to flower — although, as we have seen in soya beans (Table 7.8), far-red can only partly counteract the night break effect even if it is given immediately after red light.

In order to explain these results and the observations on the rate of decline in P_{fr} levels, a modification of the model has been suggested (Figure 7.29). Although still acting as a 'clock', phytochrome is assumed to operate only during the first few hours of the dark period. As soon as P_{fr} reaches its critical level, a second dark process is initiated. Most night breaks are given at a time when this second process is occurring and are therefore assumed to inhibit it

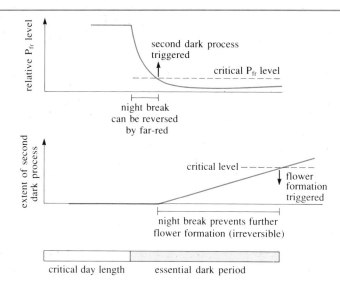

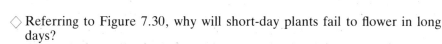

Figure 7.29 A revised model for the control of flowering by two separate processes in the dark.

irreversibly (unless far-red light is given almost immediately, i.e. before inhibition can take effect). The responses that lead to flower formation in short-day plants (or to inhibition in long-day plants) are not triggered until the second dark process is complete.

If this second process involved enzyme-mediated reactions, it is quite possible that it would take several hours to reactivate the enzymes concerned (for instance, by the synthesis of new protein molecules). It is therefore much easier to explain the failure to flower when a normally inductive period follows a night break using the revised model.

Are there rhythms in the sensitivity of plants to light?

Experiments which show that red and far-red light given at various stages of the light period modify the effect of an inductive dark period suggest that plant tissues may show a daily rhythm in their response to P_{fr}.

Rhythmic variations have been observed in a number of processes in plants. One of the best known *circadian* ('approximately one day') rhythms is the movement of the first leaves of runner bean (*Phaseolus coccineus*) seedlings: the primary leaves rise in the morning and fall again towards evening. These movements continue for several days with a cycle of approximately 24 hours even after the plants are transferred to complete darkness. Such observations have led to the development of 'rhythm' hypotheses for the photoperiodic control of flowering. Figure 7.30 shows a possible **endogenous rhythm** for a short-day plant with a critical day length of 12 hours. Each 24 hour cycle is divided into a light-tolerant phase, and a light-sensitive phase during which flower induction is inhibited by high levels of P_{fr}. The length of the light-inhibited phase determines the night length needed for flower induction. Any treatment that results in high P_{fr} levels during the light-inhibited phase will inhibit flowering.

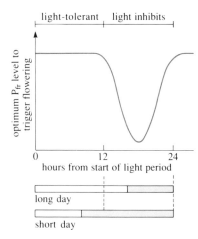

Figure 7.30 A possible 24 h cycle in sensitivity to P_{fr} for a short-day plant.

◇ Referring to Figure 7.30, why will short-day plants fail to flower in long days?

Because part of the light period coincides with the light-inhibited phase.

The requirement for particular levels of P_{fr} is less stringent in the light-tolerant phase, although flower induction may be inhibited unless P_{fr} levels are high at certain points in this phase.

There is considerable disagreement about the importance of endogenous rhythms in flowering. Such rhythms can be observed in both short-day and long-day plants, although they are more important in some species than in others. Even in plants where endogenous rhythms seem to be a major factor in the control of flower induction, phytochrome is involved in the measurement of day length. Whatever the precise mechanism involved, flowering in short-day plants occurs only if P_{fr} falls to a low level during a particular period. Conversely, flowering in long-day plants requires high P_{fr} levels during the sensitive period.

Where is the flowering stimulus received?

If changes in P_{fr} levels do control the production of some flower-inducing substance, which tissues contain the machinery for converting the external stimulus of day length into an internal signal?

Consider the experiment shown in Figure 7.31. The leaves are removed from the upper third of a single-stemmed chrysanthemum plant. By enclosing the remaining leaves inside a light-proof barrier, it is possible to give the upper and lower portions of the plants different day lengths.

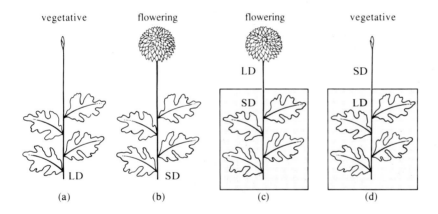

Figure 7.31 The effect of giving a flower-inducing treatment (short days) to the leaves and to the shoot apices of chrysanthemum.

◇ What happens when the whole plant is given either long days or short days?

◆ As expected for a short-day plant, flowers are formed under short days, but not under long days.

◇ Is flowering controlled by the day length given to the shoot apex or to the leaves?

◆ Flowering is induced in plants in which the leaves have been given short days (Figures 7.31b and 7.31c).

The plant in Figure 7.31c flowers even though the apex that develops into the flower has never been exposed to inductive short days. On the other hand, the apex of the plant in Figure 7.31d remains vegetative even though it has been exposed to short days because no mature leaves have been given an inductive treatment. This is quite unlike the situation in cold-stimulated flowering, in which the shoot apex requires the vernalization treatment (Section 7.6.1).

◇ Would you expect completely defoliated plants to flower under favourable day lengths?

◆ No, because there is no light-sensitive tissue to trigger the flower-forming response.

Only small amounts of leaf tissues are needed to induce flower formation. Rye will flower under long days if just 5% of one leaf is left on each plant, and $2 \, cm^2$ of mature cockleburr leaf kept under short days is sufficient to permit flowering of other defoliated branches of the same plant. In another short-day plant, *Perilla*, leaves can be removed from plants grown under long days and then exposed to short days. A single 'induced' leaf grafted back onto a defoliated plant that has never been exposed to short days is sufficient to induce flowering. Indeed, the 'induced' leaf can be removed and regrafted onto a second or even a third plant kept under non-inductive conditions and these will also flower. It therefore seems that the leaves are responsible for the production of some flower-inducing substance, given the name **florigen**, that can convert a vegetative apex into one that develops flowers.

The initiation of flowering can therefore be divided into two distinct processes: first changes in the level of P_{fr} somehow lead to increased levels of florigen within the leaves, and second this florigen acts on shoot apices to convert them from the vegetative to the floral state.

What are the characteristics of florigen?

◇ What can be deduced from experiments such as that shown in Figure 7:31 about the factor that controls flower initiation?

◆ It must be transported from the leaves to the shoot apex in order to exert its effect.

Many experiments have been performed in which tissues from plants given a flower-inducing treatment have caused flowering when grafted onto plants maintained under non-inductive conditions. These range from the induction of flowering in cockleburr plants kept under long days after grafting on a shoot from a cockleburr plant that had been kept under short days, to the bizarre plant shown in Figure 7.32. In this experiment, a long-day *Sedum spectabile* shoot has been grafted onto a short-day *Kalanchoë blossfeldiana* stem and the composite plant kept under short-day conditions.

◇ Explain how it is possible for both species to flower.

◆ *Kalanchoë* normally flowers under short days. It therefore produces florigen, which induces flowering in both *Kalanchoë* and *Sedum* tissues.

◇ Which tissues would flower if the composite plant was kept under long days?

◆ Again, both would flower. Florigen produced by the long-day plant, *Sedum*, would affect both tissues, assuming florigen can be transported downwards from the *Sedum* leaves.

The most important conclusion from these sorts of experiments is that the flower-forming factor appears to be the same in both long-day and short-day plants. Leaves from some light-sensitive plants given an inductive treatment will even promote flowering in the unvernalized tissue of some cold-requiring plants. Although this circumstantially implicates florigen in the induction of

SHORT DAYS

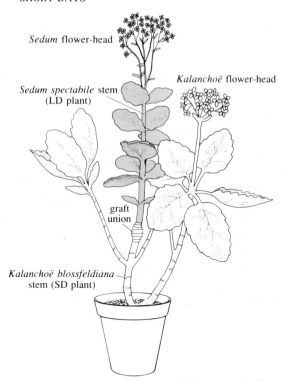

Sedum flower-head

Kalanchoë flower-head

Sedum spectabile stem
(LD plant)

graft
union

Kalanchoë blossfeldiana
stem (SD plant)

Figure 7.32 The result of grafting together portions of long-day and short-day plants (*Sedum* and *Kalanchoë* respectively) and keeping the composite plant under short-day conditions.

flowering by chilling, an alternative explanation is that florigen is able to induce the same sequence of events in a chill-requiring plant as the vernalization treatment itself.

Although florigen can be shown to move readily between tissues (even of different species), the stimulus cannot pass between two tissues separated by a strip of agar.

◇ Does this observation suggest that florigen is one of the common PGRs?

◈ No. Most PGR-mediated processes can be mimicked by application of the PGR in agar or in lanolin.

If contact between an induced donor leaf and a recipient plant is broken at various times after grafting, the recipient flowers only if vascular connections between the two tissues had been established before separation. Similarly, if most of the leaves are removed from a plant and a section of the stem between the remaining leaves and the apex is steamed or girdled, the plant does not flower even if the leaves are given an inductive day length.

◇ What does this suggest about the path of florigen transport?

◈ It suggests that florigen moves in the phloem (Chapter 4) and further support is provided by the rate at which the stimulus is known to travel.

In theory, it should be possible to identify florigen by analysing and comparing the contents of phloem sap in plants before and after flower induction. However, no one has yet managed to extract, let alone purify, a compound with the properties expected of florigen.

Are inhibitors of flower induction involved?

It is generally assumed that florigen stimulates flower formation, and certainly the evidence from short-day plants does suggest that the synthesis of some flower-forming factor is triggered when P_{fr} falls below a critical level. However, when considering long-day plants it is easier to understand how low levels of P_{fr} could trigger the synthesis of an inhibitor than to suggest a mechanism whereby florigen is only made if P_{fr} always remains above the critical level.

◇ Suggest another mechanism that would alter flower-forming ability without requiring changes in the synthesis of either florigen or an inhibitor.

◈ Low levels of P_{fr} might stimulate the destruction of florigen.

We therefore need to consider the possibility of a balance between synthesis and destruction of florigen, as well as a balance between a promoter and an inhibitor. Since no one knows what florigen is, it is hardly surprising that there is little information about the relative rates of its synthesis and degradation. There is, however, quite a lot of indirect evidence that an inhibitor may be involved in the control of flowering.

For example, grafted leaves from a plant given an inductive treatment induce flowering more readily if the uninduced recipient plant has most of its leaves removed before the graft is made. Even in those species in which defoliation is unnecessary, the transferred leaf usually has to be grafted so that it is closer to the apex than are the recipient plant's own leaves. One interpretation of these results is that an inhibitor produced by the recipient's leaves has a tendency to overwhelm the stimulatory effect of the single inhibitor-free leaf. However, another explanation is that the recipient's leaves (particularly those between the graft and the apex) may act as alternative sinks for translocated materials, thus preventing sufficient florigen from reaching the apex. It is also difficult to explain results such as that shown in Figure 7.32 if the level of an inhibitor is the major controlling factor of flower formation. The relative importance of promoters and inhibitors on the initation of floral primordia therefore remains uncertain.

Are any of the known PGRs involved in flowering?

Although quite a number of both long-day and temperature-sensitive plants can be induced to flower by spraying them with a solution of gibberellin, and although gibberellin levels do increase before flowering in these plants, the gibberellins extracted from their tissues do not always induce flowering. Further evidence against this PGR being the universal florigen comes from the failure of short-day plants (as well as some long-day plants) to respond to added gibberellin. An alternative explanation relates to the growth patterns of plants susceptible to gibberellin-induced flowering. They mostly have a rosette of leaves at ground level with the flowers carried on a long stalk. A common effect of gibberellin is to induce stem elongation (Chapter 6, Section 6.3). Thus, while gibberellin is clearly involved in flowering in a limited number of species, there is no evidence that gibberellin is florigen itself, although it may increase as one of the consequences of florigen production.

It would be surprising if other common PGRs were not also involved. One of the first observable events at the apex is a change in the rate of cell division, a process often controlled by cytokinin. The application of cytokinin induces flowering in *Perilla* sp., but this effect is not widespread among plant species.

There are also rare examples of flower induction as a result of treatment with abscisic acid (strawberry) or ethylene (pineapple, *Ananas comosus*).

The nature of florigen therefore remains a mystery. Although it may be a single undiscovered PGR, it is also possible that flowering is controlled by a complex interaction between several substances that no one has yet managed to mimic.

Finally, we should remind ourselves that, although the arrival of a floral stimulus at the apex triggers a series of events that eventually lead to the formation of flowers, these developmental processes may require a particular combination of internal and external conditions quite different from those that induced flowering in the first place.

Summary of Section 7.6

1 Flower induction involves changes in the development of shoot apices and may be triggered by chilling (vernalization) or by changes in day length. Opposite responses to increasing day length are seen in long-day and short-day plants. A non-flowering (juvenile) phase is found in many species.

2 In species that require vernalization before flowering, a week at 0–2 °C is usually sufficient for flower induction. The cold stimulus acts directly on the shoot apex tissue; however, in some species, stimulated tissues appear to form a flower-inducing stimulus that is transported to other apical tissues probably in the phloem.

3 Photoperiodic control of flowering depends on the length of the uninterrupted dark period given. Long-day plants do not flower if the dark period exceeds a particular value, whereas short-day plants do not flower unless the length of the dark period exceeds a value characteristic for the species or variety.

4 The length of the essential dark period is related to the time taken for the concentration of the P_{fr} form of phytochrome to fall to a critical level. This decay may not be the only factor needed to trigger flower induction; there may be interactions with other dark-requiring processes or with endogenous rhythms of sensitivity to light.

5 Changes in day (night) length are detected by mature leaf tissue. After an inductive treatment, some factor or factors must be exported from the leaves to the shoot apices. The chemical nature of this factor, termed florigen, is not known, but it appears to be transported in the phloem and cannot pass between two tissues until a successful graft has been formed. Both promoters and inhibitors of flower induction may be involved.

6 Application of the common PGRs may affect flower induction, but there is little evidence for their involvement *in vivo*.

Question 7 (*Objectives 7.1, 7.3, 7.6 and 7.7*) Which of the following statements are true, and which are false? Give reasons for your answers.

(a) The response to vernalization is mediated by phytochrome.

(b) Day neutral plants kept in a constant laboratory environment are likely to flower at about the same time as plants growing in a natural environment.

(c) Shoots that develop during one summer from chrysanthemum plants (see Table 7.7) will not normally flower until the autumn of the following season.

(d) Flowering in short-day plants is stimulated by high levels of the P_r form of phytochrome.

(e) A brief night-break may trigger flowering in long-day plants.

(f) Flower-inducing treatments may trigger changes in the levels of the common PGRs.

Question 8 (*Objectives 7.1, 7.2, 7.6 and 7.8*) Figure 7.33 shows the results of experiments with two-shoot plants of cockleburr (a short-day species). Why do plants (a) and (b) flower, whereas plant (c) does not? Is flower induction more likely to be limited by the level of a stimulating or an inhibiting factor in this species?

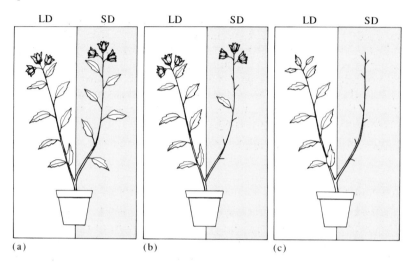

Figure 7.33 The effect of day length on flowering in cockleburr plants with two shoots.

Question 9 (*Objectives 7.1, 7.2 and 7.8*) *Perilla* plants do not flower until they have been exposed to four inductive short days. Flowering occurs at the same time after the fourth day whether two pairs of short days are separated by a week of long days or four consecutive short days are given.

(a) Are these plants likely to be at a disadvantage compared with those that require only a single inductive cycle when growing in natural environments?

(b) What experiments would you carry out to determine whether (i) each short day causes a small amount of florigen to be exported from the leaves but four days are needed before the concentration at the apex is high enough to stimulate flower induction, or (ii) no florigen is exported from leaves until they have received four short days?

Question 10 (*Objectives 7.1, 7.2 and 7.7*) Two species of plants, M and N, both flower if exposed to a cycle of 12 hours light and 12 hours dark. When exposed to 14 hours light and 10 hours dark, M flowers but N does not. What can be deduced about the control of flowering by day length in each of these species? What further experiments would you carry out?

Question 11 (*Objectives 7.1, 7.2, 7.3 and 7.7*) Assuming that the simple model of timing by phytochrome applies, what would you expect to happen if a long-day plant with a critical day length of 11 hours was exposed to cycles of (a) 6 hours daylight, 6 hours dark, 6 hours red light, and 6 hours dark, or (b) 6 hours light, 6 hours far-red, 6 hours red light, and 6 hours dark?

7.7   SENESCENCE, FRUIT RIPENING AND ABSCISSION

In this final section we will consider the factors that lead to the **senescence** (ageing) and death of plant cells. We will also discuss the process of **fruit ripening** which can be considered to be a special type of cell senescence.

7.7.1   Senescence

In the French bean and many other species, flowering is followed by senescence of the whole plant, so that survival of the species depends on the germination and subsequent growth of the seeds produced. In other species, senescence is restricted to particular parts of the plant. For example, the aerial portion of the potato plant dies back after tuber formation, deciduous trees lose their leaves at the end of the growing season, most flower parts (excluding the ovules) die after pollination and senescence may be induced in various organs by disease or damage.

Senescence is usually followed by **abscission** or the shedding of leaves, fruits, etc.

◇ What adaptive value may be conferred by abscission?

◆ It allows the removal of dead or diseased parts and the dispersal of fruits. Also, water loss is reduced and the wind resistance of trees is lowered by the shedding of leaves, which would be of limited use for photosynthesis in the winter anyway. It is likely that the severe storm which struck southern England in the autumn of 1987 would have uprooted far fewer trees if it had come a few weeks later, after leaf fall had taken place.

The post-reproductive death of whole plants appears to be genetically programmed and, once flowering has been initiated, cannot be delayed or prevented by modifying the environment. Since removing the developing pods delays senescence in French beans, a nutritional explanation of senescence has been suggested — the argument being that other tissues die because essential nutrients are normally diverted into the seed tissue. However, an explanation involving PGRs is also possible. Although the nature of the stimulus (termed the **senescence factor**) is not known, one of its effects might be to alter the pattern of nutrient transport. Either way, the end result is a breakdown of cell structure and of macromolecular components. General cellular disintegration is also involved in the senescence of individual organs: protein and chlorophyll levels fall rapidly, nucleic acids become hydrolysed, membranes become leaky and organelles gradually break apart. Although at first sight, these processes might seem to be disorganized in nature, it is clear that senescence and ripening are the consequence of a series of highly coordinated biochemical events.

In some species, roots develop from the base of the petiole of artificially detached leaves. These rooted leaves do not grow, but can be maintained in an apparently healthy condition for a very long time, suggesting that the cells do not have a set maximum life span. However, the life span of an attached leaf is usually limited, with its senescence probably being triggered by environmental cues such as daylength or temperature. The search for an internal signal, the senescence factor, has been little more successful than the search for florigen (Section 7.6).

Much of the evidence for the involvement of PGRs in senescence comes from their effects on detached leaf blades or discs.

◇ List two disadvantages which must be considered when extrapolating from results obtained from these systems to intact plants.

◈ The effects are those of experimentally added PGRs on the properties of tissues separated from the plant.

As a rule, cytokinins and gibberellins tend to retard senescence while abscisic acid and ethylene tend to act as accelerators. However, the results are very variable and dependent on the species considered (e.g. auxin retards senescence in some species and accelerates it in others), and no unifying theory can be presented.

7.7.2 Ripening

The early role of the fruit is to protect the developing seeds. However, once the seeds have reached maturity the fruit changes its role, facilitating the dispersal of seeds. The changes that occur during the process of ripening in a fleshy fruit are often dramatic since they render the fruit attractive and palatable to potential consumers (in the natural environment animals, but in the supermarket ourselves!). The end-product is a fruit with a characteristic colour, texture, flavour and aroma.

Fruits can be classified as *climacteric* or *non-climacteric* on the basis of whether ripening is or is not accompanied by a rise in respiration. A further distinguishing feature between these two groups is that ripening of climacteric fruit is associated with an increase in the production of the PGR ethylene. It is now known that ethylene plays an important role in the ripening of climacteric fruit and will accelerate the process when applied to unripe fruit (Chapter 6, Section 6.3.3).

◇ How does a ripe banana (*Musa* sp.) stimulate the ripening of an unripe kiwi fruit (*Actinidia chinensis*) when placed in the same container?

◈ The ripe banana will produce ethylene. This gas will diffuse into the unripe kiwi fruit and cause it to ripen.

Whilst ethylene is thought to be the natural accelerator of ripening, there is good reason to believe that fruits contain inhibitors which ensure that ripening will only occur once the seeds are mature. Fruit will often ripen more rapidly when removed from the plant. An extreme example of this is the avocado (*Persea americana*) whose fruit will ripen *only* once it becomes detached from the parent plant. Under natural conditions, such fruit would be shed prior to ripening by the process of abscission.

Some of the biochemical changes that accompany fruit ripening have been characterized. Much of the research in this area has been carried out on the tomato fruit, since tomatoes can be grown all year round and are amenable to biochemical analyses. As tomato fruit ripen they stop synthesizing chlorophyll and start producing a red pigment, lycopene. In addition, they convert insoluble starch to soluble sugars, produce a characteristic aroma and they soften. The softening process plays a key role in ripening since it 'opens up' the fruit so that the seeds can be dispersed. Softening is mediated by the breakdown of cell walls, which in tomato is believed to be brought about by the pectin-degrading enzyme *polygalacturonase* (Figure 7.34a). Not all fruit soften to the same extent as tomato, and this may indicate that the process can occur in other ways and be brought about by different hydrolytic enzymes.

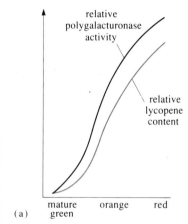

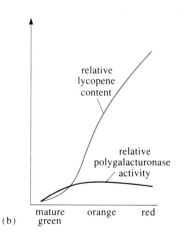

Figure 7.34 Relative polygalacturonase activity (black lines) and relative lycopene content (green lines) during ripening. (a) Normal tomato fruit. (b) Antisense tomato fruit.

The increase in polygalacturonase activity which accompanies tomato ripening is the result of *de novo* protein synthesis brought about by the induction of transcription of the polygalacturonase gene. Using the techniques of molecular biology outlined in Box 7.1 (Section 7.2.1), the gene which codes for polygalacturonase has now been identified and sequenced. Transcription of the gene is specifically promoted by ethylene and inhibitors of ethylene action (such as silver ions) prevent both the appearance of polygalacturonase mRNA and softening.

Another reason why tomatoes are an ideal system on which to study ripening is that the cells of the plant can be induced to accept foreign DNA. This process is known as **transformation** and necessitates the infection of tomato tissue (commonly the stem or cotyledons of a young plant) with a bacterium called *Agrobacterium tumefaciens* (Figure 7.35). This bacterium infects plant tissues and inserts some of its own DNA (called *T-DNA*) into the chromosomes of the host. If the bacterium is *genetically engineered* so that it contains DNA of scientific interest, this can be inserted into the host's chromosomes along with its own. This process is the basis of the major part of the genetic engineering of plants that has been carried out so far.

Unfortunately, many of the major crops of the world (such as the cereals) are monocotyledons, which are not susceptible to infection by *Agrobacterium tumefaciens* and so cannot be genetically engineered by this process.

Tomato plants have now been transformed with a piece of DNA that is the non-coding strand of the gene which codes for polygalacturonase. This piece of DNA is known as an **antisense gene**. Although the precise mechanism is not known, the antisense gene prevents the synthesis of the polygalacturonase enzyme (see Figure 7.34b). One possibility is that the mRNAs produced by

Figure 7.35 Transformation of tomato stem explants with *Agrobacterium tumefaciens* and regeneration and selection of genetically transformed shoots and plants.

1 place *Agrobacterium* containing a foreign gene and an antibiotic resistance gene in the T-DNA onto tomato stem pieces

T-DNA transfer to plant cells
occurs in 2 days

2 place stem pieces onto growth medium containing antibiotic and hormones

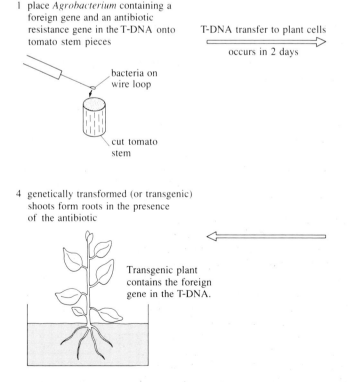

bacteria on wire loop

cut tomato stem

only genetically transformed cells can grow

4 genetically transformed (or transgenic) shoots form roots in the presence of the antibiotic

Transgenic plant contains the foreign gene in the T-DNA.

3 growth on a different medium also containing the antibiotic allows shoots to form

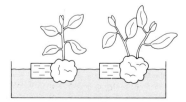

only genetically transformed cells form shoots

the polygalacturonase sense and antisense genes bind together because they have a complementary structure and the duplex that is then formed cannot be translated by the ribosomes.

◇ If polygalacturonase regulates softening of tomato fruit, would you predict that fruit from antisense polygalacturonase plants would be different from normal fruit as they ripen?

◆ Yes. The fruits should change from green to red but they should remain firm.

In fact, fruit from antisense polygalacturonase plants still soften, but appear to do so more slowly than fruit from normal plants. This suggests that other enzymes, in addition to polygalacturonase, may play a role in cell wall breakdown.

◇ What are the commercial implications of regulating the softening process?

◆ Firstly, it will allow growers the opportunity to get more of their produce to market before they spoil. A good example of this problem is mango (*Mangifera indica*) that softens so rapidly that farmers in Africa and South America can only get a small proportion of their produce from the field where they grow to the market place in a saleable state. Secondly, it will increase the 'shelf-life' of fruit in supermarkets.

7.7.3 Abscission

So far in this section, we have outlined the changes that are associated with ripening and senescence. Both these events commonly culminate in the shedding of the fruit, leaf or flower. This process occurs at discrete sites which can be readily predicted. The developmental responses characteristic of abscission are restricted to cells in a small area, known as the **abscission zone**. In many cases, such as at the base of the petiole of the mature leaf of *Coleus* (Figure 7.36), the cells in the abscission zone can be distinguished morphologically prior to the induction of shedding. However, even when structurally

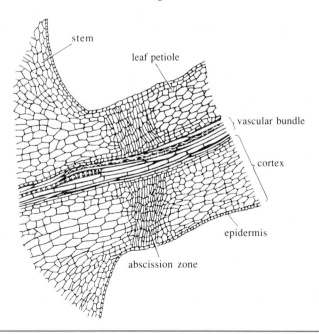

Figure 7.36 The structure of the base of the stalk of a mature *Coleus* leaf.

distinct cells are not evident, abscission always involves the breakdown of cell walls in a localized region by polysaccharide-hydrolysing enzymes. Intercellular contacts in this region are loosened to such an extent that a very weak force is sufficient to separate the organ concerned from the parent plant.

Petioles often detach rapidly after the green leaf blade has been removed, although the abscission-inhibiting effect of the blade can be mimicked by immediate application of lanolin paste containing the auxin IAA. Later auxin application, when the leaf blade is turning yellow, is ineffective, suggesting that IAA may prevent the onset of senescence, but that once triggered, senescence will proceed and abscission will occur regardless of auxin levels. Despite its name, changes in abscisic acid do not correlate well with abscission.

So what PGRs (if any) are involved in abscission? In the days before natural gas, one of the effects of a gas leak was that trees in the immediate vicinity lost their leaves. The constituent of the gas which promoted abscission was subsequently identified as ethylene. The observation that ethylene can promote abscission does not in itself prove that ethylene plays a role in regulating leaf, flower or fruit fall *in vivo*. If this simple hydrocarbon is important, from what tissues does it originate at the time of abscission? The most likely site must be from cells adjacent to the abscission zone since ethylene, being gaseous, will diffuse away rapidly once it has been synthesized. We have already described how many ripening fruits produce ethylene and this could clearly promote their abscission, but is the same true of senescing leaves and flowers? The data in Figure 7.37 show the rate of ethylene production by some senescing leaves and flowers. Both tissues exhibit a marked rise in ethylene production prior to shedding and this provides some further evidence to link the gas with abscission. Further support for this hypothesis has come from the application of inhibitors of ethylene biosynthesis to leaves and flowers, and the use of silver ions that selectively inhibit ethylene action. Both these treatments have been shown to delay senescence of leaves, flowers and fruits, and the abscission of these organs. The use of silver ions to retard senescence of flowers is now routinely used by horticulturalists to enhance the 'vase-life' of flowers such as carnations (*Dianthus* sp.).

Although auxin and ethylene appear to act antagonistically at the onset of senescence, in that auxin can prevent the effect induced by ethylene, auxin applied once senescence has been initiated may actually accelerate the processes leading to abscission (possibly because high concentrations of auxin stimulate ethylene synthesis in many tissues). Cells in the abscission zone are particularly sensitive to ethylene. As little as one part in 10 million (10^7) of air (by volume) will trigger the rapid abscission of mature leaves. Younger leaves take longer to respond, but will eventually fall, even at this minute concentration.

Figure 7.38a shows the time course of abscission in portions of orange (*Citrus sinensis*) consisting of a single leaf attached to a small piece of stem. The

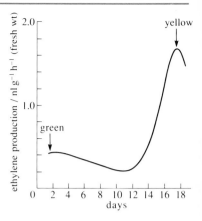

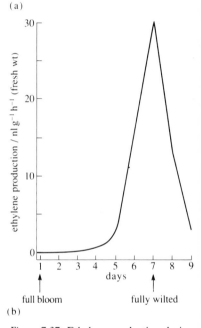

(a)

(b)

Figure 7.37 Ethylene production during the process of senescence. (a) By tobacco leaf tissue. (b) By carnation flowers.

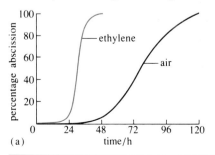

(a)

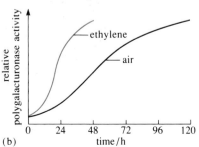

(b)

Figure 7.38 (a) Time course of abscission in detached cuttings of orange in ethylene and in air. (b) Corresponding polygalacturonase activity.

importance of including a control is emphasized by the fact that excision alone is sufficient to trigger abscission. Nevertheless, ethylene further accelerates abscission in this case.

◇ Are the results shown in Figure 7.38b consistent with the suggestion that abscission is due to the weakening of cell walls?

◆ Yes. An increase in activity of the polysaccharide-hydrolysing enzyme polygalacturonase can be detected before abscission in both air and ethylene.

◇ Does abscission have any features in common with the process of ripening?

◆ Yes, both can be promoted by ethylene and involve the breakdown of the cell wall between adjacent cells.

The enzyme whose activity is shown in Figure 7.38b is polygalacturonase, which was also implicated in the softening process which accompanies tomato ripening. Polygalacturonase hydrolyses pectin which is the major component of the middle lamella region of the wall shared by adjacent cells (see Chapter 1). It is this region that shows the first signs of weakening when abscission is studied under the microscope. Another enzyme commonly reported to increase during abscission is *cellulase*, which may hydrolyse the cellulose components of the cell wall. Studies on the molecular biology of abscission have not yet proceeded as far as such studies on the ripening process. However, in the near future it should become clear whether the processes of abscission and ripening which share so many features, have similar regulatory mechanisms. In addition, it may be possible to manipulate the process of abscission using genetic engineering techniques such as the insertion of antisense genes. The artificial regulation of leaf, flower and fruit fall could be of great commercial importance, for instance in enabling much of an oil palm (*Elaeis guineensis*) crop to be harvested at one time, while the prospect of Christmas trees which do not shed their needles in the house is certainly appealing!

We have described in this section the signals which promote abscission and some of the biochemical events which lead to shedding. At the outset, it was stated that these changes were restricted to particular sites. This observation raises the important question of why the changes in enzyme activity induced by ethylene do not occur in cells adjacent to the abscission zone? Although the answer to this question is not yet known, it may turn out that only abscission zone cells contain receptors for ethylene, or have the capacity to bind the gas and transduce the signal into the transcription of the polygalacturonase and cellulase genes. If this is indeed the answer, it will be necessary to find out how these cells differentiate at precise sites so that the abscission events occur only at strategic locations. Knowledge of this aspect of development is at only a rudimentary stage in plants, although it is at a much more advanced stage in animals.

Summary of Section 7.7

1 The final phases that a plant or plant organ undergoes in its life cycle are senescence, ripening and abscission.

2 Senescence is a degenerative process but is the culmination of a strictly coordinated series of biochemical events. It can be affected by all the common PGRs, although the evidence that any one of these is the primary regulator or senescence factor is lacking.

3 Ripening can be considered to be a particular type of senescence restricted to fruits. It results in a series of changes often rendering the fruit palatable to a potential consumer. In general these include changes in colour, texture, aroma and taste. The softening process in tomatoes is mediated by the pectin-degrading enzyme polygalacturonase.

4 Genetic engineering of some plants can be carried out by transformation with the bacterium *Agrobacterium tumefaciens*. Tomato plants have been engineered with a DNA strand which is complementary to the polygalacturonase gene. This antisense polygalacturonase gene inhibits synthesis of polygalacturonase. Antisense fruit change colour but do not soften to the same extent as normal fruit during ripening.

5 Abscission of leaves, flowers and fruits occurs at precise positions. The process can be promoted by ethylene and initially retarded by auxin. The shedding process is mediated by cell wall breakdown and this is restricted to the sites of abscission. Dissolution of the cell wall is precipitated by the action of hydrolytic enzymes including polygalacturonase and cellulase.

Question 12 (*Objectives 7.1 and 7.9*) Which of the following statements are true, which are false? Give reasons for your answers.

(a) Ripening and abscission involve similar anatomical and biochemical changes.

(b) Senescence is the result of a random process of cell autolysis.

(c) The PGR ethylene plays an important role in senescence, ripening and abscission.

(d) Antisense genes can inhibit the synthesis of specific proteins.

Question 13 (*Objectives 7.1 and 7.9*) How do you account for the observation that leaf fall occurs at discrete sites in plants? What mechanisms might be responsible for limiting the abscission response to these sites?

7.8 CONCLUDING REMARKS

Senescence, ripening and abscission are processes which either precede the onset of dormancy, or are associated with the death of the plant. Either way, they can be considered to complete the annual cycle of growth of a typical flowering plant. Having considered the regulation of several stages of this cycle, can any general conclusions be drawn?

There are internal mechanisms that determine when certain processes can occur. Some changes then proceed in the absence of environmental triggers, but most are modified by external stimuli to a greater or lesser extent. Nutrition and water supply have effects on growth and development, particularly where internal controls play a major part. Light and chilling are frequently involved in major changes in development and serve to coordinate growth with seasonal variations. The perception of temperature changes is not well understood, although it may act directly on the cells concerned. Many aspects of the response to altered illumination appear to involve phytochrome in one way or another.

◇ Recall three mechanisms whereby phytochrome can act as a control mechanism in plant development.

◈ High or low levels of P_{fr} may trigger particular developmental processes such as seed germination (Section 7.2). The level of P_{fr} in the photostationary state may affect the form of growth (Section 7.4). Phytochrome may act as a time clock, for example in the photoperiodic control of flowering (Section 7.6).

If the environment has an important influence on both the time scale and duration of developmental processes, it is necessary to consider how these effects may be mediated. The evidence that one or more of the five known classes of PGR (auxins, gibberellins, cytokinins, abscisic acid and ethylene) acts as a transducer of environmental factors remains at best circumstantial. For processes such as vernalization, flowering, and senescence the evidence is so weak that we have invoked the action of hypothetical chemical messengers such as vernalin, florigen or the senescence factor. It remains for us to consider how we are going to identify chemical transducers and probe in more detail how developmental processes are regulated.

As can be seen from the previous sections, the revolutionary techniques of molecular biology are having a dramatic impact on our understanding of some developmental processes. Over the coming years, more information on these will be obtained, and work will begin to unravel some of the mysteries of other phenomena which remain comparatively unexplored at the molecular level. Advances can then be made in the understanding of how the expression of genes playing critical roles in the developmental events are regulated, why they are switched on or off only at certain times in certain tissues, and what factors regulate their expression and how these change. Information such as this will enable plant scientists to work back from the response towards the environmental trigger, and will ultimately expose the identity of the transducer molecules themselves. Whether the transducers turn out to be one or more of the common PGRs, may perhaps remain a mystery until then, but workers have already started along the track to unmask them — it is just a matter of discovering how long the track is!

Another approach which is also proving promising, is the characterization of plant mutants. Such plants have either been generated spontaneously and saved by breeders, or more recently, generated by mutagenic techniques. **Isogenic** or single gene mutants are useful because by studying the phenotype of the mutant plant it is possible to gain an insight into how a developmental process is regulated. Mutants are now available which exhibit abnormal developmental characteristics associated with all of the developmental processes that we have discussed. Moreover, there has been a significant investment of both time and money in recent years into the creation of new mutants in a member of the cabbage family (Cruciferae) called *Arabidopsis thaliana* (thale cress). This plant has been chosen for study because it has a relatively small and simple genome and is thus the subject of considerable research effort. It is hoped that by studying *Arabidopsis* more information will be obtained about how the growth and development of the major crop species are regulated.

Another way in which plant mutants can be induced is to insert a piece of DNA called a **transposon** into their chromosomes by the process of transformation (described in Section 7.7.2). If this DNA interrupts a gene which codes for an important protein such as an enzyme, then a mutant phenotype could be created. By knowing the structure of the inserted DNA, the gene which has been interrupted can be identified, and its sequence determined. This approach may pave the way to isolating genes that play important roles in development and ascertaining how they are regulated.

Plant scientists are on the threshold of making important breakthroughs into the mechanisms of how plant growth and development are regulated. Discoveries in this area will be exciting not only because they will have a dramatic impact on our understanding of plant biology, but also because, armed with this information, it may be possible to manipulate the way that crops grow and interact with their environment.

OBJECTIVES FOR CHAPTER 7

Now that you have completed this chapter, you should be able to:

7.1 Define and use, or recognize definitions and applications of, each of the terms printed in **bold** in the text. (*Questions 1, 2, 4, 5, 6, 7, 8, 9, 10, 11, 12 and 13*)

7.2 Design or interpret experiments relating to topics discussed in the chapter. (*Questions 1, 2, 5, 6, 8, 9, 10 and 11*)

7.3 Write out an equation to show the reactions of phytochrome, and describe the characteristic properties of a phytochrome-mediated response. (*Questions 1, 2, 4, 5, 6, 7 and 11*)

7.4 List the factors that commonly affect the emergence from dormancy, and describe briefly the control of germination in a named seed. (*Questions 1 and 3*)

7.5 Compare and contrast the phototropic response with responses mediated by phytochrome. (*Question 4*)

7.6 List the factors that commonly control flower induction, outline the nature and location of the receptors for external stimuli and describe the characteristics of responses to these stimuli. (*Questions 7 and 8*)

7.7 Describe a simple model that explains how phytochrome could be responsible for the control of flowering by day length, and name one other process with which phytochrome control may interact. (*Questions 7, 10 and 11*)

7.8 Indicate the relative importance of internal and external factors on plant development, and suggest how these factors may be of adaptive value for the particular plants. (*Questions 5, 8 and 9*)

7.9 Describe the main anatomical and biochemical features of senescence, fruit ripening and abscission. (*Questions 12 and 13*)

ANSWERS TO QUESTIONS

CHAPTER I

Question 1 (a) True (Section 1.3.1).

(b) False, because the fibrils are initially orientated perpendicular to the direction of growth, i.e. transversely (Section 1.3.1).

(c) True. The primary wall is cellulose within a matrix of pectins, hemicelluloses and proteins, while the secondary wall is cellulose which is sometimes impregnated with lignin.

(d) False, because, for example, surface cells (e.g. in leaves) with the outer wall thickened with cutin are living.

(e) True (Section 1.3.1).

(f) True. Parenchyma cells are found throughout plants, are thin-walled and pick up stains which attach to the cellulose in their cell walls, but they are not necessarily rounded in shape (Section 1.3.2).

Question 2 A: parenchyma. The cells are thin-walled, rounded and unthickened and stained pale green. These cells function as packing tissue. (Section 1.3.2)

B: collenchyma. The cells are thick-walled but stained green so are not lignified. These cells function as supporting tissue. (Section 1.4.2)

Question 3 See Figure 1.49.

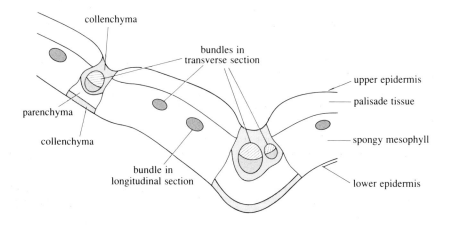

Figure 1.49 Outline diagram of tissues in a section of a Hellebore leaf (answer to Question 3).

Question 4 It is a dicot leaf because two vascular bundles are clearly in transverse section while some smaller ones have been cut longitudinally (Section 1.4.1). In fact the section is a hellebore — you will probably be familiar with at least one species, the Christmas rose (*Helleborus niger*).

Question 5 Plate 5 shows two saxifrages (described by gardeners as mossy (a) and aizoon (b) types). The leaves of mossy saxifrages are usually small and dissected and bright green in colour and the plants branch to form a tight mossy hummock. The aizoon types are often grey coloured and the leaves are in flattened rosettes which lie close to the soil. The grey colour is a result of reflection of light by waxes in the cuticle overlying the epidermis so the plant is not only protected against water loss due to high winds but will also reflect excessive light from the leaves. Both types of saxifrages tend to be found growing in rocky scree or in vertical cracks on rock faces down which water is channelled.

Question 6 (a) Xerophyte leaves may have a thick cuticle, sunken stomata and incurved lower margins. They may also have a low surface area to volume ratio. All these anatomical characteristics serve to limit water loss.

(b) Succulents have highly vacuolated mesophyll tissue for water storage. They also have a thick, waxy cuticle and sunken stomata, which reduces water loss.

(c) Hydrophytes have parenchyma tissue with large air spaces (aerenchyma), which increases buoyancy. They also have a thin cuticle (there is no need to conserve water), and stomata in the upper epidermis only (stomata could not function on the lower, aqueous, surface) and very little lignified tissue (strengthening is not needed because the leaf is supported by water).

Question 7 A is a sclerenchyma fibre. This (incomplete) cell is long and thin and appears thick walled with oblique end walls but without any sign of the special thickening normally found in tracheids and vessels.

B is a vessel element. It is short and squat with a thickened wall and square ends.

C are sclerenchyma fibres in cross section. Note the thick walls (if this was a colour plate they would be stained red) and the angularity of the inside of the walls of the cells.

Question 8 Reticulate thickening (Figure 1.19).

Question 9 (a) A is a xylem vessel. It is thickened with a wide lumen to the cell. It is also located at the centre of the root and is therefore the oldest vessel present. It functions in the transport of water and mineral nutrients from the soil, up the root to the shoot.

B is a cortical parenchyma cell. It is rounded and unthickened and located in the cortex. Its function is packing and possibly food storage.

C is a protoxylem element. It is located at one point of the four-pointed xylem 'star'. It is relatively small in cross-section (compared to A, the xylem vessel). It functions in the transport of water and mineral nutrients.

D is a root hair cell. The cell is within the epidermis with the hair-like extension projecting outward. Its function is absorption of water and mineral nutrients from the soil.

(b) E is epidermis. It is the external layer. Its function is protection of internal tissue and absorption (root hair cells are specialized epidermal cells).

F is cortex. It lies between the epidermis and the stele. Its function is packing and food storage.

G is endodermis. It is the ring of cells surrounding the stele and it is reasonably clear that it is continuous (this is not always the case). It is involved in transport, the waterproof band preventing diffusion of water and ions along the endodermal cell walls between the cortex and the stele. Thus it waterproofs the stele.

H is pericycle. This cell layer lies between the vascular tissue and the endodermis. Its function is partly packing but it is also the site where lateral roots are initiated.

I is phloem. It is an obvious cluster of cells lying between the xylem points. They function in transport of sugars and other organic nutrients from shoot to root.

(c) Four of each.

Question 10 See Figure 1.50.

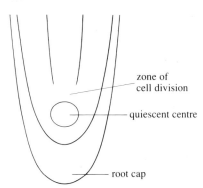

zone of
cell division

quiescent centre

root cap

Figure 1.50 Outline diagram of tissues in the root apex (answer to Question 10).

Question 11 (a) False. The second and third layers may divide at right angles to the surface (like the surface layer), but the central cells only divide *parallel* to the surface, thereby producing growth in length (Figure 1.33).

(b) True. See X and Y, respectively, in Figure 1.31 for examples of the two directions of cell division.

(c) False. The root cap is added to continuously throughout life of the plant.

(d) False. If the cells of the dividing region are damaged the quiescent centre can become meristematic.

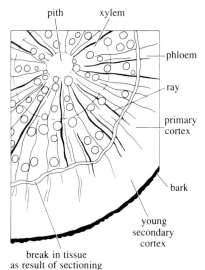

pith   xylem

phloem

ray

primary cortex

bark

young secondary cortex

break in tissue as result of sectioning

Figure 1.51 Outline diagram of a woody stem of apple (*Malus* sp.) (answer to Question 12).

Question 12 (a) See Figure 1.51.

(b) There are two cambial areas. You may not be able to see the cells easily but can infer their presence from the formation of secondary xylem and phloem in the one case and the two layers of cortex and the bark to the outside of the stem.

(c) One year old; there is no evidence of an annual ring.

(d) The reasons for deciding that this is a section of a stem are the small width of the cortex and especially the presence of a pith at the centre of the tissues.

Question 13 (a) True. Monocot roots never develop secondary thickening; instead new roots are formed. Monocot stems may develop additional bundles, but not true secondary thickening, following development of meristematic zones of limited activity.

(b) True. This is one of the ways in which increase in width occurs in monocotyledonous shoots.

(c) False. A Casparian strip occurs in the root endodermal cells of both monocots and dicots. Thickened endodermis is confined to monocots, particularly grasses.

Question 14 The differences between monocot and dicot stems and roots are summarized in the table below.

	Monocots	Dicots
Stems	1 Cortex narrow	1 Cortex wider than in monocots
	2 Vascular bundles randomly distributed or occasionally arranged in one or two rings	2 Primary vascular bundles arranged in a ring
	3 True secondary thickening does not occur (but see answer to Question 13b). Vascular bundles are closed (no cambium) and may be surrounded by fibres	3 Secondary thickening universal. Cambium is first initiated within vascular bundles but soon spreads between them. Secondary vascular tissue is thereby produced resulting in a continuous vascular cylinder or (in trees) a solid stele
	4 Pith usually wide	4 Young pith may be wide but soon reduced or eliminated as secondary xylem fills the centre
Roots	1 Have many vascular bundles, arranged in a ring	1 Commonly, roots have between two and six vascular bundles (i.e. xylem points separated by discrete regions of phloem)
	2 No secondary thickening occurs; instead the root system is strengthened by producing *more* roots	2 Cambium first develops in arcs inside the phloem and outside the xylem. Later it forms a ring round the root and so old roots have a solid stele
	3 Cortex not usually as broad as in dicots	3 Cortex very broad in young dicot roots
	4 Endodermal cells may have U-shaped thickening in older roots	5 Endodermal cells never have U-shaped thickening
	5 Pith wide	5 Pith not usually more than a few cells wide, soon disappearing as the root ages

Question 15 (a) The special characteristics you can see are the clusters of fibres at the corners and also in a broken ring outside the vascular tissue. There is also a very thick cuticle, a very wide cortex and a layer of palisade-like cells just inside the epidermis.

(b) This is a transverse section of the stem of a xerophyte (in fact it is broom; Figure 1.7). Recall that xerophytes have very thick cuticle and often have photosynthetic stems with ridges strengthened by fibres (Section 1.4 and 1.5).

CHAPTER 2

Question 1 (a) and (b) are generally true (See Equations 2.2 and 2.3 and Section 2.1). (c) is false: 'dark' respiration goes on all the time in green tissues but photorespiration occurs only in the light (Section 2.2). (d) is true in some (in fact most) situations but not all. It is not true, for example, when plants are using as their main source of energy food stored in seeds or storage organs.

Question 2 (a): (v); this is the only method listed that is suitable for long-term measurements.

(b): (i) or (iv) (provided that a fairly long incorporation period is given); both of these measure *NP*.

(c): (i) but *not* (iv) in this case because the two tissues being compared are physiologically very different and may store or respire different proportions of the CO_2 fixed.

(d): (i) or (ii); both would measure respiration since there is no photosynthesis in roots.

(e): (iii) or (iv) (provided that a short incorporation period is given); (iii) is necessary for measuring *GP* since respiration (gas exchange measured in the dark) must be added to gas exchange in the light (*NP*) to determine *GP* (Equation 2.1).

Look again at Section 2.2.1 if you had difficulty with any of these questions.

Question 3 The values are all in units of $\mu mol\ CO_2\ g^{-1}$ fresh wt h^{-1}

(a) 75 at 25 °C and 65 at 35 °C (line 2). With a 10 min incubation, *NP* would be measured and these are the 'real' values, when photorespiration is occurring.

(b) 15 at 25 °C and 115 at 35 °C. The amount by which *NP* increases when photorespiration is inhibited (line 3 − line 2) gives the values of photorespiration. Note the large increase with temperature.

(c) 100 at 25 °C and 200 at 35 °C (line 3 + line 1 *or* line 1 + line 2 + (line 3 − line 2)). Here you are applying the equation $GP = NP + R$ where line 1 gives 'dark' respiration and photorespiration is then also added on. If the data in line 3 were not available, i.e. no measurement of photorespiration, then *GP* would have apparent values of 85 and 85 (lines 1 + 2) at 25 °C and 35 °C. It would, therefore, be grossly underestimated at the higher temperature.

Question 4 (a) is false: an increase in a rate-limiting factor *does* cause an increase in the rate of the process, up to saturation (Section 2.3.1).

(b) is false: any plant (whether sun or shade type) will have lower rates of *NP* in shade compared with sun, but shade plants have higher rates than sun plants when *both* are in low irradiances (Section 2.3.2).

(c) is true: the high CO_2 compensation point of C_3 plants is because of photorespiration, which declines as $[O_2]$ decreases. (Conversely, the CO_2 compensation point of C_4 plants, which do not show photorespiration, is insensitive to $[O_2]$.)

(d) is partly false: C_4 species are indeed (with very rare exceptions) sun plants but C_3 species may be sun or shade plants.

(e) is false: some plants show little or no ability to acclimate; for example some always grow in shade or at constant high or low temperatures (see Figure 2.12c and d).

Question 5 (a) The most likely factor is $[CO_2]$ since C_3 plants become light-saturated and limited by CO_2 at quite low light levels.

(b) Probably light; C_4 plants do not become light-saturated until high irradiances, which are higher in the tropics than in the temperate zone.

(c) Initially, $[CO_2]$ is likely to be limiting since all shade plants are C_3 and the situation is as for (a) above. Thereafter (and paradoxically) light is likely to become limiting as photoinhibition, to which shade plants are very sensitive, inhibits gross photosynthesis.

(d) Probably temperature; even for plants that grow at low temperatures, the temperature optimum is usually in the range 15–20 °C.

Question 6 The data show clearly that C_4 grasses and sedges predominate at low altitudes in this tropical region but C_3 types at high altitudes. Since all are in open habitats and receiving high irradiances, the only explanation must be the decrease in temperature with altitude. The C_4 strategy has no advantages when temperatures are low for significant periods. The C_3 plants at high altitudes are probably limited by CO_2 for most of the time (since irradiance will be high), but temperature could be limiting early in the morning before plants warm up.

Question 7 (a) You might expect both species to show a similar adaptation to high temperatures with respect to gross photosynthesis (similar temperature optima and thresholds — see Figure 2.14a and b). However, because of increasing photorespiration, the C_3 species would show much less increase in net photosynthesis with increasing temperature than would the C_4 species.

(b) This difference would disappear if the concentration of either O_2 were decreased or CO_2 were increased, both of which effectively abolish photo-respiration.

Question 8 (a): (vi) and (i); (b): (ii), (iv) and (vii); (c): (i); (d): (iii), (v) and (viii).

If you found this question difficult, look again at Section 2.4.1.

Question 9 (c) applies to (i); (b) and (e) apply to (ii); (a) applies to (iii); (d) applies to (iv) — this is an incorrect statement because it is not electrons but *protons* that flow through coupling factors.

Question 10 (a) and (c) are true. (e) is partly true, the first part of the statement wholly true but not the second part: a high ratio (i.e. high PQ) favours PSI and a low ratio PSII (see Section 2.4.3, Optimizing the use of light, short-term changes).

(b) and (d) are false: only green plants carry out both cyclic and non-cyclic photophosphorylation, photosynthetic bacteria do not carry out the non-cyclic process (see Figure 2.17). Shade plants are more efficient only at low light levels — not at high.

Question 11 (a) Your figure should look something like Figure 2.24 (excluding the cyclic pathway, in green). The strongest oxidizing agent (lowest on the vertical scale) would be photosystem II (actually the pigment component P_{680} which, after a photochemical reaction becomes a very powerful electron sink and able to oxidize water). This must be shown lower down the scale than the reaction centre of PSI.

(b) Associated with step 1 (light harvesting) are the two antenna complexes of photosystems I and II.

Associated with step 2 (photochemical reaction) are the two reaction centre complexes of PSI and PSII, containing the modified chlorophyll molecules P_{700} and P_{680}, respectively.

Associated with step 3 (electron transport) are all the carriers that transport electrons from the activated reaction centre pigments (top of the two thick grey arrows) 'downhill' to either PSI or $NADP^+$ and also from water to PSII. Proton pumping occurs during the 'downhill' transfer of electrons between PSII and PSI (more precisely, between plastoquinone and the cytochrome complex — Figure 2.23) and protons generated during the oxidation of water also contribute to the proton gradient, although no actual pumping occurs.

The ATP synthetases (coupling factors) involved in step 4 (using the PMF) are not represented directly in the Z scheme and they are not localized near to sites of proton pumping. However, the arrows showing ATP synthesis in Figure 2.24 indicate the *action* of coupling factors.

Question 12 True statements are (c) and (d).

(b) is partly true: Rubisco is the universal enzyme of carbon fixation but it catalyses the same reaction in all plants — the carboxylation of RuBP to give two molecules of PGA (C_3).

(a), (e) and (f) are false. In (a), there is a second energy-requiring reaction in the Calvin cycle — the phosphorylation of ribulose phosphate to the biphosphate (see Figure 2.26). Regarding (e), photorespiration does not release energy at any stage — it is an energy-requiring process (using ATP) which metabolizes phosphoglycolate produced by the oxygenase reaction of Rubisco (see Section 2.5.2). Statement (f) is definitely not true at low irradiances, when light is limiting, and is not necessarily true at saturating light levels, when the availability of CO_2 rather than the level or activity of Rubisco may be limiting.

Question 13 The first thing to do would be to look carefully at the form of the plant and the nature of its habitat. If it was a succulent growing in a hot, dry place, there is a high probability that it would be a CAM plant. If it was a large tree and/or if it was growing in shade or in a cold area, it would probably be a C_3 plant. As a crude check for CAM, squash some green tissue during the day and again at night and use the pH indicator paper and your sense of taste to determine if the juices are substantially more acid at night. To distinguish between C_3 and C_4 plants, hold leaves up to the light and see if there is a green ring around the veins, indicating Kranz anatomy and the C_4 pathway. Taking leaf sections and examining under the microscope for Krantz anatomy could also be carried out.

Question 14 X is a C_4 species and Y a C_3. This follows from two pieces of evidence: first, species X has a much higher rate of net photosynthesis at low CO_2 levels than species Y and attains its maximum rate when the level is still below that in the atmosphere (380–400 p.p.m.). This is because of the CO_2 concentrating effect of the C_4 pathway and the virtual elimination of photorespiration at high irradiances (which were used in the experiment shown). Second, the CO_2 compensation point (the point where the graph intercepts the horizontal axis) is close to zero for species X but around 40 p.p.m. CO_2 for species Y, and a low value is characteristic of C_4 plants (Sections 2.5.3 and 2.3.2).

Question 15 This plant is a facultative CAM species that is carrying out both CAM (as evidenced by the rise in both malic acid and net CO_2 fixation during the night) and C_3 photosynthesis (as evidenced by the CO_2 fixation during the day). The fact that daytime carbon fixation is occurring means that stomata must be open at this time and, therefore, water supply is probably adequate.

Question 16 (i), (C_4 only), applies only to (b): PEP carboxylase is dark-, not light-activated in CAM plants.

(ii), (CAM only), applies to (c) and (f): any fluctuations in C_4 plants are of malate and not malic acid and there are never *large* fluctuations of malate here. PEP is regenerated from pyruvate in C_4 plants.

(iii), (both types), applies to (a), (d), (e) and (g). Regarding (g), the CO_2 compensation point in CAM plants can be measured only at night but it is low because fixation is by PEP carboxylase.

CHAPTER 3

Question 1 (a) is false: most ions enter living cells (the symplast) before they reach the xylem in roots and are transported to shoots. Transporting elements in the xylem are, of course, dead so the statement is true for calcium, which appears to move along the cell wall (apoplast) path to root xylem. (See Sections 3.2.1 and 3.4.2.)

(b) and (c) are both true (Sections 3.2.1 and 3.2.2).

(d) is false although the second part of the statement (about anions) is true. *All* ions accumulate in roots by secondary active transport involving proton pumping (Section 3.4.1).

Question 2 The correct sequence is nitrate reductase, nitrite reductase, glutamine synthetase, glutamate synthetase.

Nitrate reductase catalyses the reduction of nitrate to nitrite and an input of reducing power is required.

Nitrite reductase catalyses the reduction of nitrite to ammonia and requires a large input of reducing power.

Glutamine synthetase catalyses the formation of the amide glutamine from ammonium ions: ATP and the substrate glutamate are required.

Glutamate synthetase catalyses the formation of two molecules of glutamate from glutamine and the organic acid α-oxoglutarate (no ATP or reducing power is required).

(All the relevant information is in Section 3.3.1.)

Question 3 (a) Boron is one of the phloem-immobile elements and so is supplied to leaves only in the transpiration stream via the xylem. Because they are usually larger and transpire more, the older leaves receive more of the scarce boron than the younger leaves — and cannot re-export it to younger leaves, as is done for the phloem-mobile elements.

(b) When supplied with nitrate-nitrogen, crops raise the pH around their roots since one OH^- ion is released for every NO_3^- ion reduced and incorporated (as an amino group, $-NH_2$) into an organic molecule. With ammonium-nitrogen, however, the reverse holds: one H^+ ion is released for every NH_4^+ ion assimilated, so pH around the roots is lowered. On an already acid soil, raising soil pH is 'better' for the soil than lowering it. (Section 3.3.1)

(c) Proton pumping could be achieved by a plasmalemma redox chain in root cells (Section 3.4.1).

Question 4 (a) The data show that initially (0–0.25 hours), both stripped and intact segments take up phosphate at similar rates (total uptake), but whereas the stripped segments maintain a similar high rate for the next 4 hours, uptake into the intact segments slows down. Furthermore, the stripped segments accumulate very little of the phosphate that is taken up — most is transported to the shoots; the intact segments accumulate the phosphate they take up and a relatively small proportion moves to shoots.

(b) The data can be explained mainly by the destruction of the endodermal cells in the stripped segments. This allows ions to diffuse along cell walls (the apoplast route) from the external solution to xylem elements, where they are carried away in the transpiration stream. Ions do not accumulate in the cells of stripped segments because they do not need to cross root cell membranes to reach the xylem. The data illustrate the key role played by the endodermis in controlling the uptake of ions.

Question 5 (a) An increase in root elongation, (iii). As explained in Section 3.5, changes in ion carriers, (i), will have little effect *in a nutrient-rich soil*. Items (ii) and (iv) will probably have much smaller effects because the zone of nutrient depletion for nitrate, a highly mobile ion, is so wide (hence extra root branches would give higher rates of uptake for a short period only, until depletion zones overlapped; and root hairs would have to be impossibly long to extend outside the zone). (v) would probably have little effect on nitrate uptake in a rich soil although, in a poor soil, it might have a substantial indirect effect (by relieving phosphorus limitation and allowing more root growth). (vi) would not influence the uptake of nitrate and neither would (vii).

(b) Because phosphate is a relatively immobile ion that binds to soil material and for which roots have a very small depletion zone, all of items (ii), (iii), (iv), (v) and (vii) would probably enhance uptake. Either (iii) or (v) could cause the greatest increase, depending on precise circumstances.

Question 6 (a) An iron-efficient plant is one in which the capacity to absorb iron from the soil, mobilize iron at the root surface and transport iron from root cortex to xylem are all increased greatly when the supply of available iron falls to a low level.

(b) The primary cause of iron-inefficiency may be *either* an inability to increase the capacity for iron uptake and mobilization when iron levels are low *or* an inability to transport iron from root cortex to xylem (Section 3.5.3).

Question 7 (a) The most helpful would probably be (ii). For plants with an abundant supply of mineral nutrients — and on a modern farm with high inputs of fertilizer they would have — this is the main way in which crop yields could be increased. The new emphasis in developed agriculture is to maintain yields but reduce inputs of fertilizer, for which (i) and (iii) would also be very helpful.

(b) For this kind of farming where there are low inputs of fertilizer and very poor soils (especially with respect to phosphorus), (iv) would probably be most helpful. (i) and (iii) would also be helpful but not (ii) (which requires high fertilizer inputs). (v) is a non-starter, even if it could be done, because depending on insects as a source of nutrients for a crop is far too risky; common insect species (such as aphids) also tend to have great variations in their numbers from year to year.

Question 8 This question relates to Section 3.7.1. (a) Growth rate would increase — inevitably since plant growth was potassium-limited. (b) Net uptake of N would increase because it is linked to growth rate. (c) $[K^+]$ and $[N]$ in leaves would probably remain unchanged, following the principle that tissue concentrations remain constant. If the plants had been severely potassium-limited, however, $[K^+]$ would have increased.

Question 9 (a) False: it is the growth rate that is insensitive to nutrient supply. Toxicity is very likely when external nutrient levels are high (Section 3.7.1).

(b) True (Section 3.8.2).

(c) False — for true halophytes, where it plays a minor role in disposing of excess NaCl (Section 3.8.3).

(d) True: succulents keep salt *concentration* in their vacuoles at a 'reasonable' level because salt accumulation is balanced by water intake — hence the succulence (Section 3.8.3).

(e) False: The link between growth rate and nutrient accumulation holds only for macronutrients. Thus irrespective of growth rate, copper toxicity may occur on copper-rich soils (Section 3.8).

(f) True (Section 3.8.3).

CHAPTER 4

Question 1 (a) False. Translocation is the transport of *solutes* (not just sugars) and it occurs in both phloem and xylem in many directions (not just from shoots to roots).

(b) True (Section 4.2.2).

(c) False. Callose is not a 'usual' major inclusion but is deposited in quantity only if the sieve tubes have been damaged.

(d) True (Section 4.2.3).

(e) False. Leaves are not always sources (when very young they are sinks) and roots may be sources if they store food reserves (as in carrots).

Question 2 The most 'appropriate' technique is (ii). The phloem was part of a living plant until the moment of fixation (freezing) and fixation was very rapid (the removal of cortical tissue helped to make it so), so a minimum time was available for callose deposition, and surging would be minimized (although not eliminated).

(i) is not at all suitable. Excision of tissue would inevitably cause disruption and injury to the phloem, as would 'puncturing'. These factors, plus the slow fixation, would tend to promote massive deposition of callose and would bring about the disorganization of cytoplasm, through surging, for example.

Question 3 (i) is a valid deduction because material (assimilate) labelled with ^{14}C was detected in stylet exudates, and stylets tap sieve tubes (Section 4.3.2). (iii) is also a valid deduction because assimilates moved to roots (a sink) from the lowest leaf (the closest source) but not from the uppermost leaf (the most distant source).

(ii) and (iv), however, are not valid deductions because this experiment does not demonstrate *simultaneous* translocation in two directions (up and down), which is the definition of bidirectional transport (Section 4.4.2) — although it suggests that it might happen. Nor were the absolute amounts of solutes measured, which would be necessary before (iv) could be concluded.

Question 4 (a) The SMT depends on two factors that might change quickly: the speed at which sap is flowing and the concentration of solutes in the sap (Section 4.3.4). Thus a rapid change in the SMT could be caused by variation in either or both of these factors.

(b) If you knew transport velocity, then all you would need to measure is sap concentration. To do this you could establish feeding aphids low down on the stem (so that translocation would be mainly to roots and not to the stem apex) and then cut off the aphid bodies to obtain exuding stylets. Collection and analysis of the exuding sap would allow you to determine its concentration and hence to estimate SMT to roots.

Question 5 (a) is false: movement is from regions of higher to those of lower *turgor* (Section 4.4.2).

(b) describes bidirectional flow in a single sieve tube, which is not predicted by the pressure flow hypothesis (Section 4.4.2).

(c) is predicted; movement between source and sink is passive according to the pressure flow hypothesis, involving no expenditure of metabolic energy and, therefore, being relatively insensitive to temperature.

(d) is predicted because it would reduce the pressure gradient between source and sink (Section 4.4.1).

Question 6 First, X may have inhibited symplastic transport of sucrose from leaf mesophyll cells to phloem. This is a rather unlikely mode of action but cannot be ruled out. Second, and more likely, X may have inhibited secondary active sucrose uptake from the apoplast into phloem by interacting with the sucrose carrier. General effects on membrane properties or on the primary active transport system (the H^+-ATPase) are ruled out by the absence of inhibition of proton extrusion or potassium uptake. There is also a third possibility which you might have suggested: X may have interfered with sucrose movement from symplast to apoplast, an essential step in the model shown in Figure 4.20 but one about which little is known.

Question 7 If removal of some sources does not affect the fruit yield, the remaining leaves must have increased their activity, and fluxes of solutes from these leaves to the fruits would have increased. Thus (ii), (iii) and (v) all help to explain the observation (Sections 4.5.1 and 4.5.3).

Question 8 (a) Girdling isolates the source–sink pair (pod–single leaf) so that the source leaf supplies only this one pod, which carries out no photosynthesis. The effect on sink activity (pod respiration) is to increase it slightly (by about 20%) and the effect on source activity (leaf photosynthesis) is to decrease it markedly (by about 80%). The large change in source activity suggests that it is responding to a decrease in sink demand and this could be explained if the leaf supplied assimilates to several pods and/or the roots before girdling. The single pod acting as a sink after girdling apparently cannot increase its activity or phloem unloading capacity (you cannot distinguish which) sufficiently to absorb all of the assimilates from one leaf. The 20% rise in sink activity would be its maximum response to a rise in assimilate input.

(b) No, the data suggest the opposite, i.e. that sink activity and/or phloem unloading control source activity. Since the relative activity of source and sink will determine the pressure gradient in the translocating phloem, sink activity also appears to control translocation rate here.

CHAPTER 5

Question 1 (a) False. Most water does not circulate but is lost by transpiration from the leaves during daylight hours (Section 5.1).

(b) False. For the greater part of the pathway from roots to leaves, water moves not by diffusion but by mass flow within xylem elements. Because xylem cells are dead, there are no protoplasts. (Section 5.2.1)

(c) True. The high surface tension of water columns in xylem, which depends on cohesion between water molecules, is essential for movement up tall plants. Lowering surface tension would, therefore, result in the columns rupturing so that water did not reach upper branches. (Section 5.1.1)

(d) False. Gradients of water potential cause water movement in *any* situation and not just in plants. The difference between animal and plant cells is that osmotic pressure is usually the only significant component of water potential for animal cells but both osmotic pressure and turgor pressure are components of water potential for plant cells.

Question 2 To answer all parts of this question you have to bear in mind that water *moves down gradients of water potential*.

(a) $\Psi_A > \Psi_B$; because water moved from glass to syringe.

(b) $\Psi_A < \Psi_B$; the increase in weight of the clay must have resulted from water movement into the lump of clay.

(c) $\Psi_A < \Psi_B$; using the same reasoning as for (b).

(d) $\Psi_A = \Psi_B$; there was no net movement of water in either direction and the system was at equilibrium.

Question 3 (a) $\Delta\Psi$ resulted from a difference in hydrostatic pressure ΔP caused by withdrawal of the syringe plunger. P was lower for the syringe barrel (negative value) than for the glass (value $= 0$).

(b) A difference in matric pressure must have been the main factor responsible for $\Delta\Psi$: there was no pressure difference and no membranes (i.e. $\sigma = 0$), so neither P nor Π could have been involved.

(c) A difference in osmotic pressure must have been the chief factor causing water movement and contributing to $\Delta\Psi$; the reflection coefficient of the differentially permeable pig's bladder must have been greater than zero and the seawater had a higher osmotic pressure than distilled water.

(d) There were no components of $\Delta\Psi$ here because no net movement of water occurred and $\Delta\Psi$ was zero.

Question 4 (a) What you need to do here is first to calculate Ψ for each solution from the given values of P nor Π (m, as we said in Section 5.2.2, can be ignored for two dilute solutions) using Equation 5.2 or, since $\sigma = 1$, Equation 5.1. The difference between Ψ_X and Ψ_Y gives $\Delta\Psi$, and water will move from the solution of higher (less negative) to that of lower (more negative) water potential.

$\Psi_X = 0 - 0.3 = -0.3\,\text{MPa}$; $\Psi_Y = 0 - 0.6 = -0.6\,\text{MPa}$

Because Ψ_Y is lower (more negative) than Ψ_X, water will move from X to Y and $\Delta\Psi$ is given by $-0.3 - (-0.6) = -0.3 + 0.6 = 0.3\,\text{MPa}$. The gradient of water potential is thus determined solely by differences in Π in this case.

(b) $\Psi_X = 0.6 - 0.6 = 0$ MPa and $\Psi_Y = 0 - 0.6 = -0.6$ MPa, as for (a). Thus Ψ_Y is still lower than Ψ_X, and water still moves from X to Y down a gradient of water potential of $0 - (-0.6 = 0.6$ MPa, determined in this case, however, by differences in hydrostatic pressure.

(c) $\Psi_X = -0.3 - 0.3 = -0.6$ MPa and $\Psi_Y = -0.6$ MPa, as for (a) and (b). Thus $\Delta\Psi = 0$ and there will be no net water movement between X and Y.

Question 5 (a) False (Section 5.3.1, Figure 5.5). Ψ is lowest (most negative) in plasmolysed cells.

(b) True. In a source tissue, water moves into sieve tubes down a gradient of water potential. Sieve tubes have lower (more negative) values of Ψ than surrounding cells because solutes are actively accumulated (Chapter 4), thus raising osmotic pressure and lowering Ψ.

(c) True (Section 5.3.1). Plasmolysed cells have zero turgor, which means that they wilt.

(d) False (Section 5.3.2). Matric pressure is higher in a dry compared with a wet soil but water potential is *lower* (more negative) because $\Psi = -m$ in soil.

(e) True (Section 5.3.3).

(f) False (Section 5.3.4). Movement of water vapour from leaf to air depends not only on the gradient of relative humidity (which would be 0 in this example) but also on the temperature gradient. If the leaf were warmer than surrounding air, water vapour could still move out, even in air of 100% r.h.

Question 6 You can use Equation 5.1 to calculate Ψ (because $\sigma = 1$ and m can be ignored within cells). Values of Ψ (in MPa) from top to bottom, are for noon, $-0.2, -0.3, -0.6, -0.7, -0.8$; and for midnight, $-0.2, -0.35, -0.26, -0.2, -0.2$. The values for Ψ_{xylem} are calculated with respect to adjacent living cells, so that solutes (Π) contribute to Ψ_{xylem}; for flow *along* xylem (i.e. using the equation $\Psi = P$ from Section 5.3.3), the values are $-0.5, -0.6$ (noon) and $-0.01, 0$ (midnight).

(a) At noon, there is a gradient of water potential for both roots and tubers to the leaves, so water will move from both these organs up to the leaves. However, Ψ_{tubers} is lower than Ψ_{root}, so water will also move into tubers from the roots, replacing some of that lost to the leaves.

(b) At midnight, the lowest (most negative) water potential occurs in the tubers, and water will move from both stems and roots into the tubers. Note that leaf cells and stem xylem 30 cm above the ground are in equilibrium ($\Delta\Psi = 0$), so, under the conditions described, leaves are not losing water to tubers.

(c) The answer to (a) suggests that more water can move out of tubers to above-ground shoots during the day than moves in from roots, which would explain tuber shrinkage. Similarly, from (b), water from both roots and stem xylem moves into tubers at night, which would explain tuber swelling.

Question 7 For (a), technique (iii) would allow direct estimation of tissue water potential (Section 5.3.1). In addition, a combination of (iv) (measuring Π) with either (ii) or (v) (measuring turgor) would allow determination of Ψ from Equation 5.1, assuming that $\sigma = 1$.

For (b), only method (i) is appropriate (Section 5.3.3).

For (c), technique (vi), which was described in Chapter 4, would provide a measure of phloem turgor when coupled to a manometer. When used to sample phloem sap and combined with technique (iv), osmotic pressure could be estimated. Hence, from Equation 5.1, water potential of the phloem could be calculated.

Question 8 (a) The first thing to appreciate is that freeze-dried seeds are dead: they have no intact cell membranes ($\sigma = 0$), and, therefore, osmosis cannot account for water movement into seeds. Dried pea seeds contain a high proportion of solids to liquids and thus behave like lumps of soil with a high matric pressure. Water enters down a gradient of water potential whose main component is matric pressure, m.

(b) Wilting is prevented because loss of water from the cuttings is minimized. This happens in the conditions described because the driving force for transpiration — the gradient of water vapour potential between leaf and air — is minimized. The polythene bag ensures high air humidity, minimizing the humidity component of the gradient; and shading prevents leaves from heating up, minimizing the temperature component of the gradient (Section 5.3.4).

(c) Rough handling increases the chance that cavitation and embolism will occur and block xylem conduits (Section 5.3.3). Root pressure that would remove such blocks does not develop in cut flowers, so embolisms accumulate during 'vase life' and wilting and death occur when xylem is totally blocked.

Question 9 (a) Membranes have a very high hydraulic resistance (Section 5.4.1). If no Casparian strip is present, radial water flow in roots can occur entirely along the apoplast (cell wall) pathway and it is not 'necessary' to cross a cell membrane. By contrast the symplast pathway involves passage across at least one cell membrane, so it will have a higher resistance to water flow. The apoplast and symplast pathways operate in parallel (see Figure 5.16) and, therefore, water will flow along the pathway of least resistance (Section 5.4.3), i.e. the apoplast.

(b) Applying Equation 5.10 or 5.11 for resistances operating in parallel, total resistance of the radial pathway at root tips is defined by:

$$\frac{1}{R_{total}} = \frac{1}{R_{apoplast}} + \frac{1}{R_{symplast}}$$

or

$$R_{total} = \frac{R_{apoplast} \times R_{symplast}}{R_{apoplast} + R_{symplast}}$$

Question 10 Factors (a), (c), (e) and (h) would all be likely to increase transpiration rates. Strong wind (a) would lower boundary layer resistance (Section 5.4.3). Dry air (c) would increase the gradient of relative humidity from leaf to air and thus the potential gradient of water vapour (the driving force — Section 5.3.4). Factor (f) would similarly increase $\Delta \Psi^{wv}$ by affecting the temperature component. And factor (h) gives minimal stomatal resistance.

Question 11 The information to answer this question is mainly in Section 3.4.2. (a) $J_v = \Delta \Psi / R$ (Equation 5.6), so with equal values of $\Delta \Psi$, the plant with the lowest resistance will have the highest flux in xylem. With equal numbers of perforated plates, conduit resistance will depend solely on area and be inversely proportional to the square of the radius—the bigger the radius, the lower the resistance. Obviously, therefore, X will have a lower resistance and higher flux than Y. Resistance in X will be proportional to $1/1^2 = 1$ and in Y to $1/0.1^2 = 100$. So water will move 100 times faster in X than in Y.

(b) Since volume flow in each conduit $= J_v \times$ area, and area is 100 times greater for X, volume flow will be $100 \times 100 = 10\,000$ times greater in X than in Y.

(c) A change in either the water potential gradient in xylem (the driving force) or in the number of xylem vessels in Y could bring about equal volume flows in the two plants. This would require an increase in $\Delta\Psi$ in Y by a factor of $10\,000$ or an increase in the number of vessels by a factor of $10\,000/100$ (since 100 vessels are pressent) = 100-fold.

Question 12 (a) Only item (ii) is *not* a common feature of guard cells (or of any other plant cells). Potassium-selective channels occur but not K^+ pumps, i.e. primary active systems linked directly to ATP use.

(b) Items (i), (v) and (vi) are essential features (Section 5.5.2). (iii) is not essential because chloride ions may replace malate as the balancing anion for K^+ and onion guard cells, for example, do not synthesize malate (Section 5.5.2). (iv) is also not essential — recall that guard cells of *Paphiopedilum* contain no chloroplasts (Section 5.5.3); ATP then derives from respiration and light-induced stomatal opening operates solely by the blue-light photosystem.

Question 13 The toxin will interfere with the proton pump in the guard cell membranes. Wilting suggests that stomatal control has been lost (plants do not close stomata in response to control signals). It would be reasonable (and correct) to suggest that the fungal toxin decouples the guard-cell proton pumps from the signals so that the pumps operate and the stomata remain open continuously. Section 5.5.3 gives the background information necessary to follow this argument.

Question 14 (a) No. If this hypothesis were correct, stomatal aperture should decline as turgor falls (i.e. during the drying period, with fresh weight giving a measure of turgor) but there should be no further decline during the period in saturated air, when turgor is constant. The data show that stomatal aperture declines only slightly during the drying period but continues to decline in saturated air. This suggests that the decline in turgor initiates a sequence of events that brings about stomatal closure but the actual control signal is not turgor *per se*. From Section 5.5.3 you know that ABA released from mesophyll cells is the most likely signal.

(b) No. The data show a very close correlation between stomatal aperture (which depends on guard cell turgor) and K^+ content. This indicates that K^+ content is probably an important factor determining guard cell turgor but does not prove that it is the only one. In fact you know from Section 5.5.2 that the content of balancing anions (malate and/or chloride) is an equally important factor. The important general point to appreciate is that a correlation cannot *prove* a causal connection.

Question 15 (a) Cold water would lower root temperature and therefore, for reasons explained in Section 5.6.2, increase hydraulic resistance and decrease the flux of water from roots to shoots. But at midday transpiration rate is high and a large absorption lag is probably still present (Figure 5.34). Root cooling would exacerbate the absorption lag causing a further decrease in leaf water potential. If osmoregulation (rise in leaf osmtotic pressure, Section 5.6.3) could not accommodate this fall in leaf water potential, leaf turgor would fall and wilting could occur.

(b) The basic reasons are the same as for (a): water supply from roots does not keep pace with transpiration losses from shoots. A large absorption lag (likely in full sun when transpiration rates are high) coupled with poor osmoregulatory ability (likely in a shade-loving species) then results in wilting.

(c) On a windy site, boundary layer resistance will be low so that, whenever stomata are significantly open, more water will be lost than on a sheltered site (see Figure 5.19). This means that the plant is more likely to wilt in drought conditions and stomata can remain open for shorter periods, hence reducing CO_2 uptake, photosynthesis and growth compared with sheltered conditions.

Question 16 (a) The two curves differ in shape after about 09.00. Up to this time they are similar and transpiration is probably limited mainly by stomatal resistance, which is decreasing as stomata open more widely. After 09.00 leaf conductance decreases quite sharply, presumably because stomatal apertures decrease. But transpiration continues to increase, which can be explained by: (i) an increase in the driving force (potential gradient of water vapour from leaf to air), due to falling air humidity and rising leaf temperature; and, possibly, (ii) a decrease in boundary layer resistance if wind speeds increase (Section 5.6.1). Only 1–2 hours after midday are these factors that tend to increase transpiration counterbalanced by the increasing leaf resistance. Other factors that could reduce transpiration include changes in leaf orientation (so that leaves are parallel to the incident light and heat up less) or shape (e.g. curling up so that stomata are enclosed see Chapters 2 and 1, respectively.

(b) Since the soil is described as moist, albeit in desert conditions, the situation is probably similar to that in the middle of the curve shown in Figure 5.37 (days 3–4). Thus leaf water potential will decline (become more negative) steadily during daylight hours. A similar but less pronounced decline will occur in the stems and roots (Figure 5.37). The development of an absorption lag will exacerbate the decline in stems where, as in leaves, shrinkage and some loss of turgor in living cells are likely (Figures 5.34 and 5.36). Some embolism and cavitation may occur in stem xylem (Section 5.3.3).

During darkness, water potential will rise in all parts of the plant as transpiration virtually stops and water enters roots. Water will be replenished in the cells that lost turgor during the day.

Question 17 (a) False. Although both types of plants can generate high osmotic pressure, which is important for survival when soil is saline (halophytes) or dry (xerophytes), only xerophytes tolerate low turgor and have well-developed mechanisms that restrict water loss (Section 5.7.2). Halophytes do not show these features.

(b) Because of the warning against simple interpretations of adaptations (Section 5.7.2), a tentative 'true' would be a reasonable answer. 'Insufficient information to answer' would be an even safer response. The rationale is that hairs trap air and therefore increase the boundary layer resistance in the same way as for sunken stomata (Section 5.7.2, point 5). However, hairs may also act as an *insulating* layer, thereby affecting the leaf's temperature and hence, in a way that is difficult to predict, the $\Delta \Psi$ component of the driving force for transpiration. In fact, empirical observations of hairless and hairy leaves do not show any clear reduction in the transpiration rate with increasing hairiness.

(c) False as a general statement; *some* desert plants (the water tappers) have such roots (Section 5.7.2) but many, including the succulents (water storers) do not.

(d) False. Xerophytes typically have features that serve to reduce transpiration and hence *delay* the point at which stomatal closure becomes necessary. When xerophytes do eventually close stomata, this occurs more completely and more rapidly than for non-xerophytes (Section 5.7.2).

Question 18 The simplest experiment would be to withhold water from the two plants: the halophyte would quite rapidly lose turgor and wilt (because it is not tolerant of dry conditions) whereas the desert succulent would not.

You could also look for evidence of CAM, which occurs in desert but not halophytic succulents. Testing sap pH during the day and at night would reveal cyclic changes in the desert plant but not in the halophyte.

Other observations which you might have suggested include looking for evidence of water-conserving features, such as a very thick, waxy cuticle (scratching and texture may reveal this) or sunken stomata (visible under a microscope). Only desert succulents possess these features.

Question 19 Similarities include (i) tolerance of low turgor and partial desiccation; (ii) features, such as rapid and complete stomatal closure, thick cuticles and sunken stomata, that reduce water loss.

Differences include (i) the mechanism whereby water loss from cells occurs: it is by controlled withdrawal into intercellular spaces in freezing-resistant plants but by transpirational loss in xerophytes; (ii) mature xerophyte leaves tolerate water shortage during any season whereas mature leaves of evergreens develop freezing resistance and tolerance of frost drought only in winter after a period of acclimation, i.e. they are in a particular biochemical state.

CHAPTER 6

Question 1 You should have marked an apical meristem at the tip of the root. There must also be a shoot apical meristem in the centre at the base of the leaves. Although it does not contribute to stem elongation (unless the plant is kept for a second year's growth), it is the source of the cells in the leaves. The third meristem, a thickening cambium between the xylem and phloem is needed to produce the typical swollen root that is eaten as a vegetable carrot.

Question 2 (a) The cells will normally divide by forming cell walls at right angles to the longitudinal axis of the pericycle, to give the extra cells needed to accommodate the increasing volume of vascular tissue.

(b) The formation of lateral roots requires the same type of division as the initiation of leaves and buds in the shoot (the cell wall forms parallel to the longitudinal axis of the pericycle as shown in Figure 6.12).

Question 3 A growing plant requires a shoot apical meristem and a root apical meristem. A shoot apex is already present as a bud in the angle between the leaf and the stem, and it must be stimulated into activity. Root apices must be initiated from cortex cells in the base of the cut shoot; as cell division is stimulated here the tissues are reorganized to become a root apex (Figure 6.6).

Question 4 This would depend on the time that had elapsed after the treatment. Lower plants have a single apical cell whose division gives rise to all the other cells. This apical cell would be tagged by colchicine treatment, and ultimately all the cells in the apex would appear tagged as the products of its division displaced 'normal' cells from the apical region.

Question 5 The outward curvature in the presence of water is due to the uptake of water by the internal cells which are under compression. The outer (epidermal) cells are stretched and under tension and cannot elongate further — recall the discussion of tissue tension in Chapter 1. A comparison of stems incubated in the presence and absence of auxin shows that the rate of expansion of the cells on the outside of the stem relative to the rate of growth of those on the inside increases the presence of auxin. There are a number of possible explanations for this. For instance, auxin might stimulate the expansion of the epidermal cells more than that of the internal cortex cells, or it might inhibit the growth of the internal but not the external cells. To distinguish between these possibilities, you would need to measure the rate of growth of each tissue before and after treatment. (The internal cells could also show a reduced response to auxin because of damage caused by the slit). In fact, it appears that auxin acts primarily by stimulating expansion of the outer epidermal cells.

Question 6 First it would be worth checking that naturally occurring cytokinins (such as zeatin) have the same effect as kinetin. You would then need to measure the cytokinin levels in the leaf discs during the course of the experiment, preferably by using a chemical assay. To establish the importance of cytokinin *in vivo*, you would need to measure cytokinin levels in leaves that yellowed naturally while still attached to the plant.

Question 7 (a) Use of artificial auxins include 2,4-D as a selective weedkiller and NAA in 'hormone' rooting promotors.

(b) Ethylene is used to induce synchronous flowering in bromeliads (such as pineapple) and increase the yield of cucumbers by increasing the ratio of female:male flowers. Often an ethylene releasing compound is sprayed onto plants. Ethylene is also used to induce fruit ripening after storage of fruits.

(c) Silver ions (an antagonist to ethylene) are used to suppress damage in bulbs.

(d) CCC suppresses gibberellin production; when applied to cereal crops it reduces stem length thereby increasing the ratio of harvestable:non-harvestable material.

Question 8 (a) False. Most of the measurements of extensibility are obtained by using isolated walls or dead cells.

(b) False. The reversible portion of expansion is due to stretching of cross-links (see Figure 6.28).

(c) This is generally true. Unless the chemical causes deplasmolysis, turgor pressure will remain zero (or at least below P_{crit}) and the growth rate $(G = E_x(P - P_{crit})$ will be zero regardless of the value of wall extensibility.

(d) False. Figure 6.32 shows that initially the microfibrils are perpendicular to the direction of growth.

(e) True. It is generally thought that cellulose microfibrils are synthesized as they are incorporated into the wall (see Section 6.4.4).

Question 9 Tissue L. Cellulose is not a flexible macromolecule, and a wall with less cellulose, present as shorter microfibrils, is likely to offer less resistance to stretching.

Question 10 The injection of sugar was intended to increase the osmotic pressure and hence the turgor pressure of the cells: in well watered plants it would indeed do so. The stimulation of growth by sugar suggests that the dwarf variety may normally grow more slowly because of lower turgor. The 41 μm difference that persists after sugar treatment could be due to differences in wall extensibility, but the experiment gives no direct evidence for this. It is equally possible that, under normal conditions, both varieties have the same turgor (P), but wall extensibility is higher in the taller plants; raising P in dwarf plants partly compensates for lower E_X.

Question 11 The graph would look like the right of Figure 6.26 ($2 \to 4$). The weight would cause both reversible and irreversible extension, and when removed, the length would decrease by the amount due to reversible extension. The estimate of E_X is probably more reliable, because you are using a single cell rather than a whole tissue. However, the unidirectional force applied still does not mimic the multidirectional turgor pressure and the cell wall *in vivo*.

Question 12 (a) At first sight, this seems to give support to the hypothesis. In the presence of such substances, H^+ ions pumped out will re-enter the cell down a concentration gradient, so there will be no accumulation of H^+ ions in the cell wall. However, such uncouplers will also inhibit the synthesis of ATP, and would therefore disrupt a wide range of activities in the cells.

(b) Acid treatment does induce growth, but if auxin acts solely by stimulating H^+ secretion, the responses to auxin and acid should have similar time courses. So (b) is not consistent with the hypothesis.

(c) This does not imply that the first action of auxin is to stimulate protein synthesis. The production of new cytoplasmic or membrane material required for cell growth will require protein synthesis, irrespective of the mechanism by which growth is triggered. So the evidence is not inconsistent with the acid growth hypothesis.

(d) A correlation between the amount of growth induced and the numbers of H^+ ions secreted is strong support for the hypothesis, particularly if it holds over a range of different treatments.

Question 13 All except (e):

(a) is a valid explanation.

(b) would explain some of the difficulties if the receptors were bound on the plasmalemma (they could equally be in the cytosol).

(c) and (d) are true. Unlike animal cells, 'target' tissues cannot be identified from their structure so it is difficult to isolate cells that would have large numbers of receptors to any particular PGR.

(e) So are many animal hormones (e.g. steroids) but receptors have been identified on membranes and within the cytosol.

Question 14 (i) This poses no problem; according to this theory, IAA has an important role in controlling the response, so it would be expected that plants insensitive to IAA would have no tropic responses.

(ii) The uneven distribution of inhibitors results in the activity of IAA being different in different zones; so again plants that are insensitive to IAA would be expected to have no tropic responses.

(iii) This type of theory suggests that the direct effect of a stimulation (such as light in phototropism) of outer cells causes cessation of growth — no growth activators/inhibitors from elsewhere in the plant need to be involved. IAA might have a controlling effect *within the cell that produces the IAA*, in which case such mutants would have no tropic responses.

Question 15 Callus will grow from the end exposed to the air. Because of the direction of auxin transport, sufficient auxin will reach a cut surface in the air only if the basal end is uppermost (i.e. the stem is inverted before placing it in the nutrient medium).

Question 16 In a word, nothing. This may appear to be a trick question, but it is not. It is an example of a common fault in experimentation: the omission of a vital treatment. Before reading further, return to the question and think again, using this clue.

The superficial reply to this question is to conclude that the inhibitor completely abolishes the stimulation to growth caused by gibberellin, because the treatment in which gibberellin and inhibitor are both included is identical with the untreated control. This is a false conclusion, made plausible by the probably fortuitous equal values. To illustrate the falseness of this conclusion, imagine a possible result for the missing vital treatment, which is 'inhibitor alone.' Being a growth inhibitor, it may, by itself, reduce growth to a low level, say to close to 0%. In this case, the presence of gibberellin has still resulted in a stimulation of growth in the presence of the inhibitor from 0 to 50%. Hence a correct conclusion would be something like 'the inhibitor reduces the growth of the plant in the presence and absence of gibberellin'. The table of data did not, of course, provide data about the response of the plant to the inhibitor alone, and so this conclusion could not be reached, nor could any other valid conclusion be reached about the interaction of the inhibitor and gibberellin.

So, as with the interpretation of data obtained from several experiments discussed in this chapter concerning PGRs and their action, we hope we have taught you to be cautious!

CHAPTER 7

Question 1 (a) False. This is only true for seeds that have a chilling requirement for breaking dormancy. Other seeds (including barley) will germinate more slowly because of the lower metabolic rate at lower temperature. (Table 7.2 shows a slight inhibition of lettuce germination with only a short chilling.)

(b) False. The site of gibberellin action is the living aleurone layer. The response of these cells is to synthesize an enzyme (α-amylase) that acts on the contents of dead endosperm cells.

(c) True. P_{fr} is the active form of phytochrome, and it is formed when P_r absorbs light — mainly in the red region (see Section 7.2.2).

(d) False. The absorption spectrum shows which light wavelengths are absorbed by a substance. Absorption of light by P_{fr} will convert it into the inactive form P_r and will therefore inhibit phytochrome-triggered responses. (Compare Figures 7.7 and 7.8.)

Question 2 In phytochrome-mediated responses, red light can mimic the effect of daylight and far-red reverses the effect of red light. 'Light' and 'red' treatments should therefore give similar results and so must correspond to dishes A and C (although it is impossible to say which is which). Far-red abolishes the effect of red light and thus corresponds to dish B. Red light, and therefore a high P_{fr} level, inhibits germination in *Phacelia*.

Question 3 Chilling is a very effective means of breaking dormancy in organs such as tubers (Section 7.3). One of the first events is the mobilization of stored reserves to provide fuel for growth of new shoots (compare this with the response of barley grains) and conversion of starch to glucose (or to the commonly transported sugar, sucrose) would give the tissue a sweet taste.

Question 4 (a) This is true for (ii) but not for (iii). Returning a plant to even illumination allows the resumption of vertical growth, so the statement is true for (i), although the kink formed by the phototropic growth remains visible.

(b) This is true only for (i). Phototropism is a response to unilateral illumination. The evidence for phytochrome-mediated responses such as those discussed in Section 7.4 comes from experiments in evenly lit environments.

(c) This is true for (iii) and to a lesser extent for (ii), but not for (i) (see Figures 7.13 and 7.15). Red light (600–700 nm) induces no phototropic curvature, and has a maximum effect on phytochrome-triggered responses and a lesser effect on long-term responses.

Question 5 (a) A decrease in the amount of light falling on the plants stimulates leaf expansion (compare treatments B and C with treatment A). The proportion of P_{fr} in the photostationary state has virtually no effect on leaf area (compare treatments B and C). The weight of the leaf blade is not significantly altered by the treatments. Simulated shade therefore causes leaves to grow larger but thinner. This response to amount of light may allow the tissue to make the best use of limited resources.

(b) Lowering light intensity has little effect on petiole weight (compare treatments A and C). However, a decreased proportion of P_{fr} causes a marked increase in petiole weight (compare treatments B and C). Reduction in P_{fr} implies that a plant is receiving less red light and more far-red light, which in nature usually indicates shading by other plants. If the increase in petiole weight stimulated by decreasing P_{fr} levels was associated with an increase in petiole length, this might allow the leaves to be carried out of the shaded zone.

Question 6 This question is intended to give you practice in interpreting data on responses that have not been described in the text. When answering questions of this type, you should always give the logic behind your answers.

(a) All four experiments support this statement. Experiment A shows that something from the seeds stimulates growth. Experiments B and C show that its effect is primarily on the adjacent tissues, and experiment D that the 'something' may be auxin.

(b) Experiment A shows that this statement is false. The continued presence of seeds is required for growth, although the change of colour associated with ripening does appear to be irreversibly triggered by 21 days (compare with the 'triggered' and 'long-term' responses mediated by phytochrome).

(c) There is no evidence for this: auxin was the only PGR tried and the results of experiment D include neither a measure of 'normality' after auxin treatment nor a comparison of applied auxin levels with those *in vivo*.

(d) Experiments B and C support this statement.

(e) Experiment D suggests that this may be true, but see the answer to (c) for the further experiments needed to confirm such a proposal.

Question 7 (a) False. Vernalization involves chilling and not a response to light (Section 7.6.1).

(b) True. If it is assumed that both environments allow reasonable growth, flowering in neutral plants is primarily controlled by age.

(c) True. Chrysanthemum is a short-day plant that requires previous vernalization.

(d) False. Although low levels of P_{fr} may be associated with high levels of P_r, all models of phytochrome action assume that the critical factor is the level of P_{fr} (see Sections 7.2.2 and 7.6.1).

(e) True. See Figure 7.26.

(f) True. It is, however, unlikely that any one of the common PGRs can be equated with florigen (see Section 7.6.2).

Question 8 The experiments confirm that photoperiod is detected by leaf tissue (Section 7.6.2). No leaf tissue on non-flowering plant (c) was exposed to inductive day lengths. Plant (a) flowers because half the leaves were exposed to inductive day lengths. Plant (b) flowers although many more leaves receive the non-inductive than the inductive treatment. This suggests that the inductive treatment cause an increase in the level of some flower-inducing stimulus. It is unlikely that plant (b) would flower if non-induced tissues contained high levels of inhibitors of flowering (see Section 7.6.2).

Question 9 (a) Not particularly. In natural environments, day length changes relatively slowly, and it is unlikely that plants will receive only a single inductive cycle.

(b) There are a number of possible answers. You could measure the number of short days needed when different numbers of leaves are left on the plant (if (i) is true, florigen levels would rise to the stimulatory level more rapidly if more leaves were present). Alternatively, you could expose different leaves on the plant to pairs of short days or expose a plant to two short days and graft on leaves from another plant given the same treatment (in both cases, plants will flower only if (i) is true). Most experimental evidence available supports (ii).

Question 10 N must be a short-day plant with a critical day length of between 12 and 14 hours (see Section 7.6.2). It is impossible to deduce anything about M. It might be a long-day plant, a short-day plant for which the value of the critical day length is greater than 14 hours or a day-neutral plant. Further increasing the day length would eventually inhibit flowering if M was a short-day plant, whereas inhibition of flowering by shorter days (e.g. 8 hours light plus 16 hours dark) would show that M was a long-day plant.

Question 11 Long-day plants flower only if the P_{fr} form of phytochrome remains above the critical level (Section 7.6.2). P_{fr} will be high in daylight or red light, will gradually fall in the dark, and will decrease immediately in far-red.

In regime (a), the P_{fr} level will be high in both 6 hour light periods and will not have time to fall to the critical level in the intervening 6 hour dark periods (a critical daylength of 11 hours is associated with a 13 hour dark period). The plants will therefore flower.

In regime (b), the P_{fr} level will be low throughout the far-red treatment, and the plants are therefore unlikely to flower.

Question 12 (a) True. Although other anatomical and biochemical events occur during ripening and abscission, both rely on the disintegration of the pectin-rich middle lamella region of the cell wall. This event is thought to be mediated by enzymes such as polygalacturonase.

(b) False. The changes that occur during senescence are the culmination of a highly organized sequence of biochemical events.

(c) True. Associated with senescence, ripening and abscission is an increase in the production of ethylene. Since inhibitors of ethylene biosynthesis or action retard these processes, it is hypothesized that ethylene plays an important role in regulating these developmental changes *in vivo*.

(d) True. Antisense genes are constructed so that they are the mirror image of a specific gene. It is only the expression of this target gene that is inhibited by the antisense gene.

Question 13 Abscission zones are restricted to precise sites in plants and only cells which comprise these zones undergo the changes resulting in cell separation. Such cells must be differentiated at these specific positions.

Abscission zone cells can be induced to undergo the process leading to cell separation by ethylene. These cells must contain ethylene receptors and have the signal transduction machinery necessary to translate the perception of ethylene into the biochemical events leading to wall breakdown. Adjacent non-abscission zone cells which are unresponsive to ethylene may lack appropriate receptors or be unable to transduce the ethylene-binding response.

FURTHER READING

GENERAL

Anderson, J.W. and Beardall, J. (1991) *Molecular Activities of Plant Cells: An Introduction to Plant Biochemistry* Blackwell Scientific Publications, hardback ISBN 0632024577, paperback ISBN 0632024585.
Salisbury, F.B. and Ross, C.W. (1985) *Plant Physiology*, 3rd edn, Wadsworth Publishing Co., California USA, ISBN 0534044824.
Taiz, L. and Zeiger, E. (1991) *Plant Physiology*, The Benjamin/Cummings Publishing Co., Inc., California USA, ISBN 0805301534.

CHAPTER 1

Cutler, D.F. (1978) *Applied Plant Anatomy*, Longman, New York, London, ISBN 0582441285.
Cutler, E.G. (1978) *Plant Anatomy*, 2nd edn, Addison Wesley, ISBN 0201012367.
Eames, A.J. and MacDaniels, L.H. (1947) *An Introduction to Plant Anatomy*, 1977 reprint of 2nd edn, Krieger, ISBN 0882755269.
Esau, K. (1988) *Plant Anatomy*, 3rd edn, Wiley, New York, ISBN 0471047651.
Rudall, P. (1987) *Anatomy of Flowering Plants*, E. Arnold, London, ISBN 0713129506.

CHAPTER 2

Gregory, R.P.F. (1989) *Biochemistry of Photosynthesis*, 3rd edn, Wiley, ISBN 0471918997.
Hall, D.O. and Rao, K.K. (1986) *Photosynthesis*, 4th edn, Studies in Biology, E. Arnold, paperback ISBN 071312945X.
Encyclopedia of Plant Physiology (New Series), Springer-Verlag, Berlin, Heidelberg, New York, Tokyo:
Vol. 6 (1979) *Photosynthesis II: Carbon metabolism and related processes*, ISBN 3540092889.
Vol 12B (1982) *Physiological Plant Ecology II: Water relations and carbon assimilation*, ISBN 0387109064.

CHAPTERS 3–5

Clarkson, D.T. (1974) *Ion Transport and Cell Structure in Plants*. McGraw Hill Book Co., New York. ISBN 070840261.

Epstein, E. (1972) *Mineral Nutrition of Plants: Principles and Perspectives*. Wiley, New York, ISBN 075165018.

Milburn, J.A. (1979) *Water Flow in Plants*, Longman, London, New York, ISBN 0582443873.

Nobel, P.S. (1970) *Plant Cell Physiology: A Physiochemical Approach*, W.H. Freeman and Co., San Francisco, Studies in Biology, E. Arnold, paperback, ISBN 0716706822.

Russell, E.W. (1988) *Soil Conditions and Plant Growth*, Longman, ISBN 0582446775.

Sutcliffe, J.F. and Baker, D.A. (1978) *Plants and Mineral Salts*, Studies in Biology, E. Arnold, paperback, ISBN 0713124520.

CHAPTERS 6 AND 7

Graham, C.F. and Wareing, P.F. (1984) *Development Control in Animals and Plants*, Blackwell Scientific Publications, paperback, ISBN 0632007583.

Roberts, J.A. and Hooley, R. (1988) *Plant Growth Regulators*, Blackie, paperback ISBN 0216924790.

A good source of review articles relevant to particular chapters in this book is the *Annual Review of Plant Physiology and Plant Molecular Biology* (called *Annual Review of Plant Physiology* prior to 1988). A selection of reviews is listed below:

Boyer, J.S. (1985) Water transport, *A. Rev. Pl. Physiol.*, **36**, 473–516.

Cosgrove, D. (1986) Biophysical control of plant cell growth, *A. Rev. Pl. Physiol.*, **37**, 377–405.

Clarkson, D.T. (1985) Factors affecting mineral nutrient acquisition by plants, *A. Rev. Pl. Physiol.*, **36**, 77–115.

Clarkson, D.T. and Hanson, J.B. (1980) The mineral nutrition of higher plants, *A. Rev. Pl. Physiol.*, **31**, 239–98.

Hedrich, R. and Schroeder, J.I. (1989) The physiology of ion channels and electrogenic pumps in higher plants, *A. Rev. Pl. Physiol. Pl. Mol. Biol.*, **40**, 539–69.

Oaks, A. and Hirel, B. (1985) Nitrogen metabolism in roots, *A. Rev. Pl. Physiol.*, **36**, 345–65.

Rennenberg, H. (1984) The fate of excess sulphur in higher plants, *A. Rev. Pl. Physiol.*, **35**, 239–98.

Spanswick, R.M. (1981) Electrogenic ion pumps, *A. Rev. Pl. Physiol.*, **32**, 267–89.

Taiz, L. (1984) Plant cell expansion: regulation of cell wall mechanical properties, *A. Rev. Pl. Physiol.*, **35**, 585–657.

Tyree, M.T. and Sperry, J.S. (1989) Vulnerability of xylem to cavitation and embolism, *A. Rev. Pl. Physiol. Pl. Mol. Biol.*, **40**, 19–38.

ACKNOWLEDGEMENTS

The series *Biology: Form and Function* (for Open University Course S203) is based on and updates the material in Course S202. The present Course Team gratefully acknowledges the work of those involved in the previous Course who are not also listed as authors in this book, in particular: Ian Calvert, Lindsay Haddon, Sean Murphy and Jeff Thomas.

Grateful acknowledgement is made to the following sources for permission to reproduce material in this book:

FIGURES

Figures 1.2b and 1.2c: from Hurry, S. (1965), *The Microstructure of Cells*, John Murray, © 1965 Stephen W.Hurry; *Figure 1.9:* from Troughton, J. and Donaldson, L. (1971), *Probing Plant Structure*, Chapman and Hall; *Figure 1.11:* courtesy of Dr B. E. Juniper, Botany School, Oxford University. *Figures 1.20a and 1.20b:* from Cutter, E. G. (1978) *Plant Anatomy*, Edward Arnold Publishers; *Figure 1.21:* from Weier, T. E., Stocking, R. C. and Barbour M. G. (1974), Botany, *An Introduction to Plant Science*, 5th edition, reproduced by permission of John Wiley and Sons Inc; *Figure 1.25:* Lowson, J.M. (1945), *Textbook of Botany*, UTP, Unwin Hyman, part of HarperCollins Publishers; *Figures 1.39, 1.40a and 1.41b:* Robbins W.W., Weier T. E. and Stocking R. (1967), *Botany, An Introduction to Plant Science*, 3rd edition, © 1971 by John Wiley and Sons. reproduced by permission of John Wiley and Sons Inc.; *Figure 1.41a:* courtesy of Heather Angel; *Figure 1.41b:* courtesy of Martin.King. *Figures 2.6 and 2.11:* Bjorkman, O. B. (1971), in Hatch, M. D., Osmond, C. B. and Slatyer, R. O. (eds), *Photosynthesis and Photorespiration*, copyright © 1971 by John Wiley and Sons Inc., reprinted by permission of John Wiley and Sons Inc., all rights reserved; *Figure 2.8:* reproduced, with permission, from the *Annual Review of Plant Physiology*, Vol 28, © 1977 by Annual Reviews Inc; *Figure 2.13 and 2.14:* reproduced, with permission, from the *Annual Review of Plant Physiology*, Vol. 31, © 1980 by Annual Reviews Inc.; *Figures 2.23 and 2.24:* from Ort, D. R. and Good, N. E. (1988), 'Textbooks ignore photosystem II-dependent ATP formation', *Trends in Biochemical Science*, No 13, 1988, Elsevier Publications. © 1988, Elsevier Publications, Cambridge; *Figure 2.33:* reproduced, with permission, from the *Annual Review of Plant Physiology*, Vol. 29, © 1978 by Annual Reviews Inc; *Figures 3.2 and 3.9:* from Clarkson D. (1974), *Ion Transport and Cell Structure in Plants*, McGraw Hill, © 1974 by David Clarkson; *Figure 3.4:* adapted from *Plant Physiology*, 3/E, by Frank B. Salisbury and Cleon W. Ross. © 1985 by Wadsworth, Inc., reprinted by permission; *Figure 3.7 and 3.12:* from Pirson, A. and Zimmermann, M. (1983), 'Physiological plant ecology III', in Lange D., Nobel P., Osmond C. and Ziegler H. (eds), *Encyclopedia of Plant Physiology*, Vol 12c, 1983, Springer-Verlag, © 1983 Springer-Verlag; *Figures 3.11 and 3.31b:* from Luttge, U. and Pitman, M. G. (1976), 'Transport in plants II, part B, tissues and organs', in Pirson, A. and Zimmermann, M. (eds), *Encyclopedia of Plant Physiology*, Vol. 2, Part B, Springer-Verlag, © Springer-Verlag; *Figure 3.13:* from Dr G. Bond, with permission from *Intermediate Botany* by L. J. F. Brimble, MacMillan; *Figure 3.14:* reproduced, with permission, from the *Annual Review of Plant Physiology*, No. 36, copyright © 1985 Annual Reviews Inc.; *Figure 3.17:* from Jackson, M. B. and Stead, A. D. (1983), *Growth Regulator in Root*

Development, Monograph No. 10, British Society for Plant Growth Regulation; *Figure 3.18:* O'Brien, T. P. and McCully, M. E. (1969), *Plant Structure and Development,* Collier Macmillan Canada Inc; *Figure 3.21:* from Romheld, V. (1979), in Harley, J. L. and Scott-Russell, R. (eds), *The Soil–Root Interface,* Academic Press, Orlando, Florida; *Figure 3.24a:* reproduced by permission of Oxford Scientific Films Ltd; *Figure 3.26:* from Stuart, F. (1982), *Journal of Ecology,* Vol. 70, No. 1, March 1982, Blackwell Scientific Publications Ltd; *Figure 3.28:* from Pitman, M. G. and Cram, W. J., *SEB Symposium XXXI,* The Company of Biologists; *Figure 4.2:* from Weier, T. E., Stocking R. C. and Barbour M. G. (1974), *Botany: An Introduction to Plant Biology,* reproduced by permission of John Wiley and Sons Inc.; *Figures 4.4a and 4.4b:* reproduced with permission, from *The Annual Review of Plant Physiology,* Vol. 23, © 1972 by Annual Reviews Inc.; *Figure 4.8:* from Rabideau and Burr (1945), *American Journal of Botany,* No. 32, Botanical Society of America; *Figure 4.9a:* from Mortimer, Dr D. C. (1965), 'Translocation of the Products of Photosynthesis in Sugar Beet Petioles', *Canadian Journal of Botany; Figure 4.9b:* reproduced by permission of the American Society of Plant Physiology; *Figure 4.10:* from Zimmermann, M. H. and Milburn, J. A. (1975), *Encyclopedia of Plant Physiology,* Vol. 1, Part 1, reproduced by permission of Springer-Verlag; *Figures 4.13 and 4.20b:* from Johnson, R. P. C. (1978), 'The microscopy of P-protein filaments in freeze-etched sieve pores', *Planta,* edition 143, 1978, reproduced by permission of Springer-Verlag,; *Figure 4.14:* from Lowson, J. M. (1945), *Textbook of Botany,* UTP, Unwin Hyman, part of HarperCollins Publishers; *Figure 4.17:* adapted with permission, from the *Annual Review of Plant Physiology,* Vol 36, © 1985 by Annual Reviews; *Figure 4.19:* from Moorey, J. M. (1977), 'Integration and regulation of translocation within the whole plant', in Jennings D.M., (ed), *Integration of Activity in the Higher Plant,* Edition No XXXI 1977, Society for Experimental Biology; *Figure 4.20c:* adapted , with permission, from the *Annual Review of Plant Physiology,* Vol. 34, © 1983 by Annual Reviews Inc.; *Figures 4.21b and 4.21c:* from Christ, R.A. (1988), 'Records of source–sink relations by means of respiration measurements', in Hall, J. L. (ed), (1989), *Journal of Experimental Biology,* edition No. 40, April 1989, by permission of Oxford University Press; *Figure 5.13:* from Cutter, E. G. (1978), *Plant Anatomy,* Edward Arnold Publishers; *Figure 5.15:* from Milburn, J. A. (1979), *Water Flow in Plants,* Longman Group UK; *Figure 5.17:* from Meidner, H. and Sherriff, D. W. (1976), *Water and Plants,* Blackie and Sons Ltd; *Figure 5.23:* from Humble and Raschke (1971), by kind permission of the authors and publishers; *Figures 5.25 and 5.26:* from *Plant Physiology,* 3/E, by Frank B. Salisbury and Cleon W. Ross, © 1985 by Wadsworth, Inc., reprinted by permission of the publisher; *Figures 5.27 and 5.39:* Schulze, Dr E. D. and Hall, Dr A. E. (1982), 'Stomatal responses, water loss and CO_2 assimilation rates of plants in contrasting environments', *Encyclopedia of Plant Physiology,* Vol. 12b, Springer-Verlag; *Figures 5.29 and 5.30:* from Rashke, K. (1979), 'Movements of stomata', *Encyclopedia of Plant Physiology,* Vol. 7, New Series, Springer-Verlag; *Figure 5.31:* from Bradford, K. J. and Hsiao, T.C. (1982), 'Physiological responses to moderate water stress', *Encyclopedia of Plant Physiology,* New Series, Ecology Ii, Springer-Verlag; *Figures 5.34 and 5.35:* © Kramer, P. J., reproduced by permission of Professor Paul J. Kramer; *Figure 5.37:* from Slatyer, R. O. (1967), *Plant-Water Relationships,* Academic Press: *Figure 5.38:* reproduced, with permission, from the *Annual Review of Plant Physiology,* Vol. 35, © 1984 by Annual Reviews Inc.; *Figure 5.42:* from Woodell, S. R. J. (1973), *Xerophytes,* © Oxford University Press 1973; *Figure 5.44:* from Lowson, F. M. (1953), *Textbook of Botany,* UTP, Unwin Hyman, part of HarperCollins Publishers; *Figure 6.16:* reproduced by permission of Dr Clive R. Spray, University of California; *Figures 6.20, 6.34, 6.37 and 6.38:* from Wareing, P. F. and Phillips, I. D. J. (1970), *The Control of Growth and Differentiation in Plants,* Pergamon Press; *Figures 6.23, 6.24 and 6.25:* reproduced, with permission, from the *Annual Review of Plant Physiology,* Vol. 36, © 1985 by Annual Reviews Inc.; *Figure 6.30:* from Zeroni, M. and Hall, M. A. (1980), 'Molecular effects of hormone treatment on tissue', in MacMillan, J. (ed), *Hormonal Regulation of Development 1,* Springer-Verlag; *Figure 6.32:* from *Plant Physiology,*

Third Edition, by Frank B. Salisbury and Cleon W. Ross © 1985 by Wadsworth, Inc., originally appearing in *Cell Ultrastructure* by Jensen, W. A. and Park, R. B. © 1967 by Wadsworth Publishing Company, Inc., reprinted by permission of the publisher; *Figures 6.41 and 6.42:* courtesy of Dr R. Firn; *Figure 6.51:* reproduced by permission of Professor P. F. Wareing, The University College of Wales, Aberystwyth; *Figure 7.5:* data, courtesy of Dr J. Roberts, Company of Biologists, on behalf of the Society for Experimental Biology; *Figure 7.17:* reproduced, with permission, from the *Annual Review of Plant Physiology,* Vol. 33, © 1982 by Annual Reviews Inc.; *Figure 7.34:* reprinted by permission from *Nature,* Vol. 334, pp. 724–6, copyright © 1988 Macmillan Magazines Ltd.

PLATES

Plates 1a, 1b, 1c and 1d: courtesy of Dr Mary Bell; *Plates 2a, 2b, 3, 4, 6a, 6b, 7, 8 and 9*: courtesy of Dr Michael Stewart; *Plates 11 and 15*: from Gunning, B. E. S. and Steer, M. W. Plant Cell Biology: an ultrastructural approach, Edward Arnold; *Plates 12, 13 and 14* courtesy of Professor B. E. S. Gunning, Research School of Biological Sciences, The Australian National University.

INDEX

Note Entries in **bold** are key terms. Indexed information on pages indicated by *italics* is carried mainly or wholly in a figure, table or plat.e

Abbreviations in sub-entries:

ABA	abscisic acid	IAA	indolyl-3-acetic acid	PAR	photosynthetically active radiation
CAM	Crassulacean acid metabolism	NAA	naphthaleneacetic acid	PGR	plant growth regulator

abscisic acid (ABA), **240**, 245
  acceleration of senescence, 326
  control of stomatal closure, **206**–7, 245
  formation in drying seed, 217
  flowering induced by, 323
  in sprouting tubers, 296, 306
abscission (leaf fall), 29, 245, **325**, 328–30, 331
  abscission zone, 226, 245, 328
absorption lag, 211–13
  acacia trees, water tappers, *218*
accessory pigments, **82**
acclimation, **73**, 77, 222
Acer platanoides, CO_2 compensation point, 75
A. pseudoplatanus (sycamore)
  control of cambial activity, 275–6
  dormancy, 295
A. saccharum (sugar maple), effect of embolism on water flow, 187
acid growth hypothesis, 258–60
'acid rain', 142
acoustic detection of cavitation, *187*
acropetal transport (auxins in roots), 277
Actinidia chinensis (kiwi fruit), accelerated ripening, 326
active secretion, 122–3
active transport, 120–22, 152, 167
  primary, **120**, **121**, 142
  secondary, *120*, **121**, 142, 169, 201
adaptation, to fluctuating light levels, 73, 88–91
adenosine triphosphate (ATP), 95
  synthesis, 84–5
ADP-glucose, 96
aerenchyma, **29**, 133
aerobic respiration, 63
after-ripening, 284
Agave spp., translocation in, 148–9
agrobacteria, 243
  use in genetic engineering, 274, *327*
alder (*Alnus* spp.), root nodules, 136
aleurone layer, 283, **284**, *286*–7
alfalfa (*Medicago sativa*), photosynthesis, 65
Aloe sp., succulent, *28*
amides, in phloem sap, 159
amino acids, in phloem sap, *159*
ammonia, formation from nitrate, *115*–16
Ammophila arenaria (marram grass), protected stomata, 220
amphibious plants, 238
α-amylase, **285**, *285*–6

amyloplast, 96
Anacyclus pyrethrum, transfer cells, *152*
Ananas comosus (pineapple)
  CAM in, 102
  ethylene-induced flowering, 243, 323
animal hormones, action, 266, *267*
anion uniport, **121**, 123
annual rhythms, flowering, 308
annual rings, *50*, **51**
annular thickening (xylem), **33**, *34*
antenna complex (energy collecting system), 82, *85*, 89
Anthriscus sylvestris (cow parsley), stem section, 38, 39, *Pl. 7*
antisense gene, **327**–8, 330
aphid stylet technique (phloem sap collection), **158**
apical dome, development, 232
apical dominance, **296**
  in potato, 305
apical meristem, 44–6, 229
  cell division in, 229–33
Apium graveolens v. dulce (celery), stalk, 31
apoplast, **111**, 192, *193*, 257, 258
apoplastic transport, 124, 142, *166*, 169–70
apple (*Malus* sp.), *Pl. 10*
  bark, *52*
aquatic environment, 12–13
Arabidopsis thaliana (thale cress), mutants, 332
Arenga sp. (palms), translocation in, 160
Aristolochia sp., (Dutchman's pipe), stem, 53, *Pl. 9*
arrack, 149
Artemisia herba-alba (sagebrush), response to humidity, 206
asexual reproduction, *see* vegetative reproduction
aspartate, formation, 99
assimilates, **148**
  partitioning of, 165–70
Aster tripolium (sea aster), proline accumulation, 215
ATP, *see* adenosine triphosphate
ATP synthetase, *84*
ATPases
  calcium-dependent, 124, 142
  proton-pumping, *120*, 167, 201, 202, 203
A. patula, *73*
  acclimation, 73
  photosynthesis, *71*, *74*, *75*

autocatalysis, in ethylene production, 243
autoradiographic studies
  of cell division, 230
  of translocation, 156, *157*
autotrophs, 8, 63
auxin receptors, 242, 246
auxins, **240**, 241–3, 332
  artificial, 242
  effect on cell wall extensibility, 256–60
  effect on ethylene action, 239, 243–4
  polar transport, **277**
  use in tissue culture, 272–3
  see also indolyl-3-acetic acid
Avena fatua (wild oat), seed dormancy, 283
A. sativa (oat), phytochrome found in coleoptiles, 291
avocado (*Persea americana*), fruit ripening, 326
axis fixation, 298

bacteria, 9
  see also agrobacteria; green sulphur bacteria; photosynthetic bacteria; nitrogen-fixing symbioses
bacteriochlorophyll *a*, 82, 83–4
bakanae (disease of rice), 244
bark, 51–52, 229
barley (*Hordeum vulgare*)
  CO_2 compensation point, *75*
  nutrient deprivation, *127*
  potassium turnover in leaves, *112*
  proline accumulation in cytoplasm, 215
  root system, *41*
    branching of, *130*
    phosphate uptake by, *126*
  seed, *283*
    cytokinins obtained from kernels, 244
    germination, 284–9
basipetal transport (auxins in stems), 277
beans (*Phaseolus* spp.), growth of leaves, 233
beech (*Fagus sylvatica*)
  bud break, 296
  changes in transpiration rate, *210*
  ectomycorrhizal root, *135*
  leaf retention by juvenile plants, 225–6
  sun and shade leaves, *26*, *27*, *210*
Bellis perennis (daisy), sun plant, 72
betaine, accumulation in cytoplasm, 215
Betula sp. (birch), rise of sap in, 189
bicollateral bundles, **53**, *54*
bidirectional flow in sieve tubes, **163**, 163

363